# Physical and mechanistic organic chemistry

*Cambridge Texts in Chemistry and Biochemistry*

# Physical and mechanistic organic chemistry

**RICHARD A. Y. JONES**
*University of East Anglia, Norwich*

*SECOND EDITION*

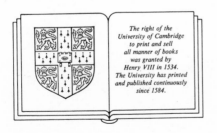

The right of the
University of Cambridge
to print and sell
all manner of books
was granted by
Henry VIII in 1534.
The University has printed
and published continuously
since 1584.

CAMBRIDGE UNIVERSITY PRESS

Cambridge

London   New York   New Rochelle

Melbourne   Sydney

Published by the Press Syndicate of the University of Cambridge
The Pitt Building, Trumpington Street, Cambridge CB2 1RP
32 East 57th Street, New York, NY 10022, USA
296 Beaconsfield Parade, Middle Park, Melbourne 3206, Australia

© Cambridge University Press 1979
© Richard A. Y. Jones 1984

First published 1979
Second edition 1984

Printed in Great Britain at the University Press, Cambridge

Library of Congress catalogue card number: 83-14317

*British Library cataloguing in publication data*
Jones, Richard A. Y.
  Physical and mechanistic organic chemistry. – 2nd ed. –
  (Cambridge texts in chemistry and biochemistry)
  1. Chemistry, Physical organic   2. Chemical reactions
  I. Title
  547.1'39   QD476

  ISBN 0 521 25863 4 hard covers
  ISBN 0 521 27886 4 paperback

(First edition
ISBN 0 521 22642 2 hard covers
ISBN 0 521 29596 3 paperback)

MP

# Contents

143x

# Introduction

It is not so long ago that organic chemistry was a subject to be learned by rote. Textbooks were largely catalogues listing, in more or less random order, the preparations and properties of each class of compound. There is no denying that still today a good memory is a valuable attribute for an organic chemist, but he does have much more insight than his predecessors into the rationale, the how and the why, of the reactions he is studying: insight, that is, into their *mechanisms*.

This insight has built up gradually over the years. Chemists have learned to recognise patterns of reactivity – similarities in the behaviour of a particular compound towards different reagents and of a series of compounds towards a particular reagent. Recognition led to rationalisation and thence to a quantitative description of the reaction process in terms of its kinetics and thermodynamics. We still have a long way to go. Ideally we should like to know the positions of the atoms, and the energies corresponding to each set of positions, at all stages from reactants to products. Such information is available via molecular orbital calculations for some very simple systems, but for most reactions we must rely on indirect evidence. Indeed, only a minority of reactions have been studied mechanistically; the knowledge we have gained from these has been extensively extrapolated to other related reactions – sometimes erroneously!

One might ask: 'Why bother? Is there any greater value, other than the satisfaction of natural curiosity, in knowing how a reaction proceeds than in merely knowing that it does so?' There are three reasons for bothering. First, as has already been indicated, because a knowledge of mechanism leads to rationalisation, to correlation, and to simplification. Instead of many thousands of individual reactions we can devote our attention to a far smaller number of reaction types. Secondly, because a knowledge of the way known reactions proceed enables us to predict the possible behaviour of unknown ones. Thirdly, because if we understand the energetics of a reaction we are more likely to be able to modify its course, say to accelerate it or to increase the yield.

For these reasons organic chemistry is nowadays taught largely in mechanistic terms. However, because the subject is so vast it is necessarily presented in an essentially didactic manner: there just is not the time in the early stages of an undergraduate course to justify each mechanism and to elaborate on the means whereby it was established.

This book is offered to advanced undergraduates and to graduate students as an attempt to remedy the omission. The first part describes the theories and techniques of physical organic chemistry; the second looks at the mechanisms of some of the major reaction types, showing how they have been established and what factors control the mechanism in particular circumstances. A book of this length cannot be comprehensive and many important areas of chemistry are omitted. I have emphasised the heterolytic and homogeneous at the expense of the homolytic and heterogeneous. In some chapters I have deliberately confined the discussion to a narrow class of reactions to keep the subject within bounds. Nevertheless, I believe that a student using the book will come to understand most of the general principles on which the study of mechanisms is based so that he can apply them in a wider sphere.

It is a pleasure to acknowledge with gratitude the help I have received from Professor K. Schofield during the preparation of the manuscript. I should also like to thank my colleagues at the University of East Anglia, especially Dr C. D. Johnson for his comments on the text and for many hours of discussion, and Professor A. R. Katritzky, a friend and mentor for more years than either of us care to remember, to whose enthusiasm I owe much of my interest in physical and mechanistic organic chemistry.

*October 1978*                                                    R. A. Y. JONES


## SECOND EDITION

In revising the book for this new edition I have rewritten many sections to try to improve their clarity and have thoroughly updated each chapter to incorporate some of the many new ideas that this ever-changing subject continues to generate: there are well over a hundred references to papers that have been published since the first edition was completed. I have also responded to suggestions that I should include more on radical reactions.

A book of problems with solutions is available to accompany this text. Enquiries should be addressed to the author at the School of Chemical Sciences, University of East Anglia, Norwich NR4 7TJ, England.

*March 1983*                                                    R. A. Y. JONES

# PART 1

## 1 Structure and mechanism

The study of organic chemistry has two aspects: static and dynamic. Static organic chemistry deals with the structures of molecules, dynamic organic chemistry with their reactions. The two are intimately linked. Mechanistic organic chemistry is largely concerned with understanding the ways in which reactivity depends on structure. Conversely, mechanistic studies depend, in the first instance, on knowledge of the structures of the species under investigation. This seems a trite observation, but the literature abounds with examples of mechanistic studies that are invalidated by an incorrect assignment of structure. Some investigations in the 1920s into the effect of strain on the ease of cyclopropane formation included work on what was thought to be the base catalysed aldol equilibrium:

$$R_2C\Big\langle {{CH_2COOH}\atop{COCOOH}} \quad\rightleftharpoons\quad R_2C\Big\langle {{CH-COOH}\atop{\underset{\underset{OH}{|}}{C}-COOH}}$$

Reinvestigation in 1959, using spectroscopic and chemical techniques not available to the early workers, showed neither compound to have the structure assigned to it! The equilibrium was in fact between two stereoisomers:

$$R_2C\Big\langle {{\overset{H}{\diagdown}\underset{}{C}\overset{COOH}{\diagup}}\atop{\underset{H}{\diagup}\underset{}{C}\underset{COOH}{\diagdown}}\Big\rangle O \qquad cis \text{ and } trans$$

1

## 1.1    The von Richter reaction

A reaction discovered in the 1870s by Victor von Richter, admittedly of no great intrinsic importance, nevertheless provides a fascinating illustration of the way in which careful structural studies can shed light on a complex reaction mechanism.

The essence of the reaction is the conversion of a nitroarene into an arenecarboxylic acid by treating it with potassium cyanide in aqueous ethanol at about 150 °C.

$$ArNO_2 \xrightarrow[\text{EtOH/H}_2\text{O}]{\text{KCN}} ArCOOH$$

Von Richter was working only a few years after Kekulé had suggested the cyclic structure for benzene, and the orientation of benzene substituents was still a controversial topic and difficult to study. Nevertheless he was able to establish that the carboxyl group did not directly replace the nitro group but entered the ring at an adjacent position:

(1.1)          (1.2)

The reaction is therefore not the simple displacement it seemed to be. Von Richter suggested it involved addition of HCN to give (1.1), followed by the loss of nitrous acid and the hydrolysis of the resulting nitrile (1.2) to the acid.

After von Richter's work the reaction was more or less ignored for the next 75 years, until J. F. Bunnett began a series of investigations. His original mechanistic proposals were similar to von Richter's own. It did seem a little surprising that it was never possible to isolate the supposed nitrile intermediate (1.2), but Bunnett was able to demonstrate that such nitriles were indeed hydrolysed to the acid under the conditions of the von Richter reaction. The mechanism requires that the hydrogen atom which replaces the nitro group should come from the solvent, and experiments using deuteriated solvents confirmed this.

Then in 1956 Bunnett found that although 2-nitronaphthalene readily undergoes the von Richter reaction to give 1-naphthoic acid, neither the 1-nitrile nor the 1-amide are hydrolysed under the same reaction conditions. This means they cannot be intermediates, and strongly suggests that the reaction with other substrates does not go through the nitrile either. Bunnett therefore proposed a new mechanism to account for this:

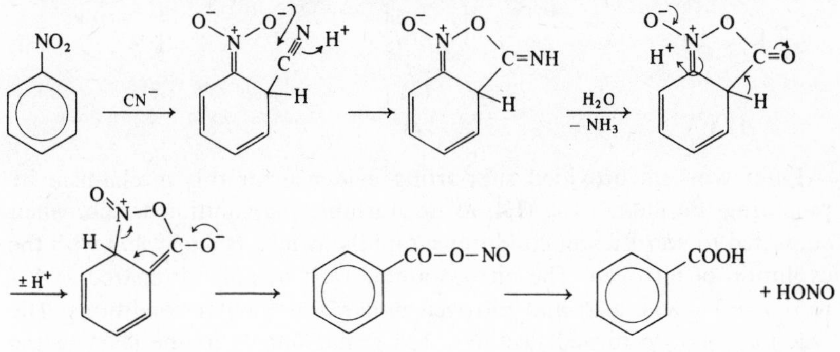

The last step of this reaction is the hydrolysis of an acyl nitrite. It was already known that in this type of reaction it is the N—O bond which is cleaved, not C—O. Consequently when it was shown that the reaction in $H_2{}^{18}O$ gave an acid in which one of the two oxygen atoms was labelled this seemed to confirm Bunnett's mechanism; one of the oxygen atoms came from the solvent during the hydrolysis of the C—N double bond, while the other came from the unlabelled nitro group.

All this time it had been tacitly assumed that the nitrogen of the nitro group was eventually expelled as nitrite ion and that the cyanide was at some stage or other hydrolysed to ammonia. But in 1960 Rosenblum for the first time paid serious attention to the by-products of the reaction and discovered that molecular nitrogen was evolved. Actually this had been noticed by Holleman as far back as 1905, but the point seems to have been ignored by later workers. Experiments using $Ar–{}^{15}NO_2$ and $C^{14}N^-$ gave nitrogen with the isotopic composition $^{15}N{\equiv}^{14}N$. It was just possible that this could arise from a reaction between ammonia and nitrite ion, but when $^{15}NH_3$ was added to the reaction mixture none of the label was incorporated into the molecular nitrogen. It follows that the nitrogen–nitrogen bond must be formed during the reaction, and Rosenblum proposed the following mechanism, which accords with all the previously established facts and with his own observations:

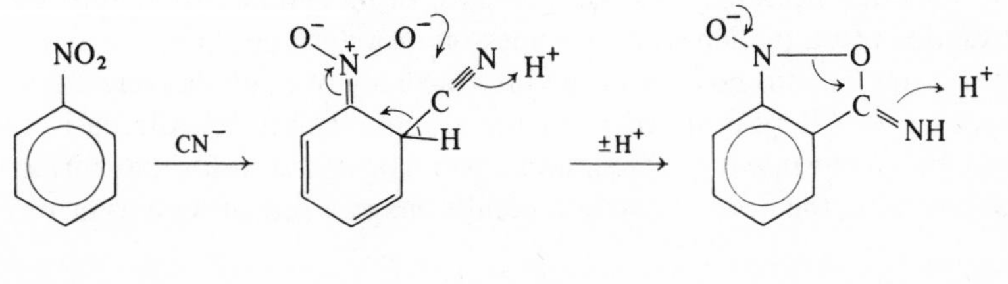

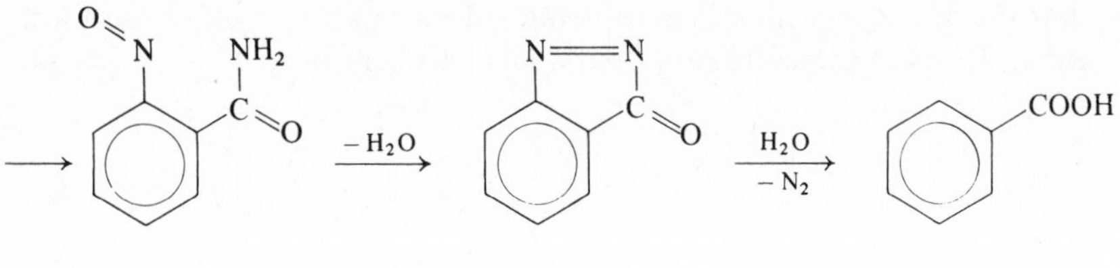

(1.3)    (1.4)

Later workers provided supporting evidence for this mechanism by preparing the indazolone (**1.4**) as an unstable red solution which, when subjected to von Richter conditions, rapidly formed benzoic acid with the evolution of nitrogen. The nitroso-amide (**1.3**) was also prepared and it too gave benzoic acid and nitrogen under von Richter conditions. The reaction mixture turned fleetingly red (indazolone?) at the start of the reaction.

Bunnett and other workers had isolated a variety of non-acidic side products from von Richter reactions, including dimeric compounds with structures such as (**1.5**) and (**1.6**). Bunnett had been unable to explain how these could be formed, but they make sense in terms of Rosenblum's mechanism as the products of oxidation-reduction reactions of the nitroso-amide (**1.3**).

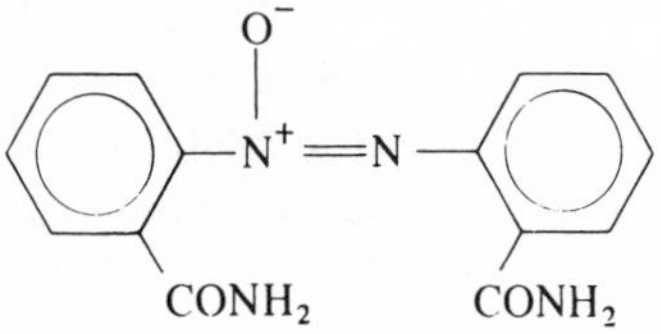

(1.5) 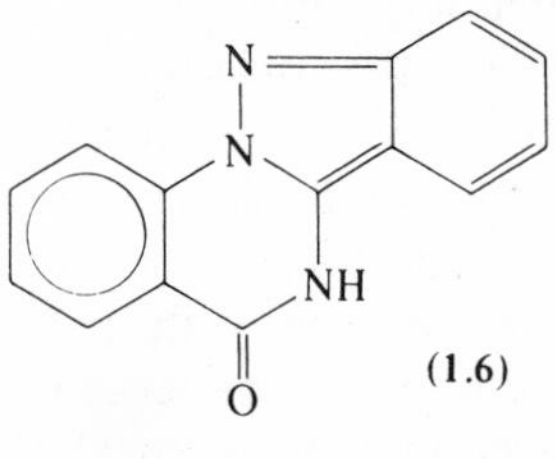    (1.6)

What can we learn about mechanistic studies from this illustration?

(*a*) It is important to characterise the products of a reaction: not just the main product which is normally isolated, but also the by-products and the products of minor side-reactions. A valid mechanism must be able to explain them all.

(*b*) It is often necessary to trace the path of individual atoms through the reaction; isotopic labelling provides one way of doing this.

(*c*) If a mechanism postulates the intermediacy of any relatively stable compound, can it be isolated from the reaction mixture? Alternatively can it be independently made and then converted into the correct products under the normal reaction conditions at a rate at least as fast as the rate of the complete reaction? Notice that this approach can easily disprove a mechanism but it can never conclusively prove it; it may be mere coincidence that some purported intermediate reacts that way – witness the behaviour of some of the nitriles in the von Richter studies.

Leading references to work on the von Richter reaction are:

V. von Richter, *Ber.*, 1871, **4**, 21, 459; 1874, **7**, 1145; 1875, **8**, 1418 (see also *J. Chem. Soc.*, 1871, **24**, 220, 686; 1872, **25**, 692; 1876, **29**, 387 for abstracts of von Richter's work in English).

J. F. Bunnett *et al.*, *J. Org. Chem.*, 1950, **15**, 481; 1956, **21**, 939, 944; *J. Amer. Chem. Soc.*, 1954, **76**, 5755.

M. Holleman, *Rec. Trav. Chim.*, 1905, **24**, 194 (*J. Chem. Soc.*, 1905, **88**, 595).

D. Samuel, *J. Chem. Soc.*, 1960, 1318.

M. Rosenblum, *J. Amer. Chem. Soc.*, 1960, **82**, 3796.

E. Cullen & P. L'Ecuyer, *Canad. J. Chem.*, 1961, **39**, 144, 155, 862.

E. F. Ullman & E. A. Barkus, *Chem. & Ind.*, 1962, 93.

K. M. Ibne-Rasa & E. Koubek, *J. Org. Chem.*, 1963, **28**, 3240.

## 1.2    Reactive intermediates

Relatively few chemical reactions are elementary, that is, proceeding from reactants to products in a single step. Much more common are multistep reactions in which intermediate species are formed on the way. Sometimes such intermediates are relatively stable molecules in their own right and it may be possible to isolate them from the reaction mixture either by stopping the reaction short or by using milder conditions than usual.

Other intermediates may be too reactive to be isolated but their presence in the reaction mixture may be detected by spectroscopic or other physical methods. If reactants and products both absorb in the ultraviolet or visible, and if their spectra are sufficiently different, then there will be wavelengths where the two spectra cross (figure 1.1(*a*)). During a reaction in which no other species are present in significant concentration the absorbance at these wavelengths will remain constant,

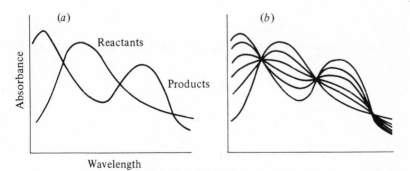

Figure 1.1. Isosbestic points. (*a*) Reactant and product spectra. (*b*) Appearance of chart after repeated scanning during reaction.

the decay of reactants being matched by the increase in the products. Repeated scanning as the reaction proceeds reveals a very characteristic pattern (figure 1.1(*b*)); the crossovers are called *isosbestic points*. The absence of isosbestic points must mean that some additional species is present during the reaction in sufficient concentration for its own spectrum to become apparent, superimposed on those of reactants and products. However, the presence of isosbestic points does not prove that the reaction is elementary; there could be highly reactive intermediates present in concentrations too low for their spectra to be significant.

Infrared spectra are usually more complex than ultraviolet and repeated scanning of a full spectrum during a reaction tends to give a chartful of scribble. The principle, however, remains valid, and information can be gleaned if there are strong bands which change markedly during a reaction. During the formation of oximes from ketones with hydroxylamine the rate of decay of the carbonyl stretching band is often much faster than the appearance of C=N absorbance, showing that some other species (actually the adduct $R_2C(OH)NHOH$) must be building up in the early stages of the reaction and then dying away with the formation of the ultimate products.

These and other spectroscopic and physical methods of observation are limited by the sensitivity of the particular technique. Below a certain concentration an intermediate may be present but undetectable. Thus n.m.r. spectroscopy, so valuable in structural studies of stable molecules, is less important in detecting reactive intermediates (though Fourier transform techniques are beginning to change this).

A further difficulty is that spectra measured during a reaction are often so complicated, with absorbances from reactants and products present as well as those from intermediates, that they seldom provide much help in determining the actual structures of the intermediates. The conditions needed to characterise a suspected intermediate may differ so much from the normal reaction conditions that the comparison is invalid. The formation of a cyclic bromonium ion (**1.7**) from $CH_3CHFCH_2Br$ with $SbF_5$ in liquid sulphur dioxide can be demonstrated by n.m.r. spectroscopy and it confirms that such ions can exist; but it does not prove the

$$\underset{Me}{\overset{Br^+}{\underset{H\text{----}C\text{------}C\text{----}H}{\triangle}}}\overset{H}{} \qquad SbF_6^- \qquad (1.7)$$

hypothesis that they are intermediates in the reaction of bromine with alkenes in aqueous or alcoholic solution (chapter 9).

In this context e.s.r. spectroscopy is rather special, affording a valuable means of studying the nature of radical intermediates. It is a particularly sensitive technique; radicals can be detected in concentrations as low as $10^{-9}$M. And since non-radical species do not generate e.s.r. spectra there is much less chance of complication from the superimposition of spectra of several different species.

The most fleetingly unstable reaction intermediates may not be detectable by available physical methods but it is often possible to recognise their presence by diverting them from the normal reaction path with some suitable trapping agent. One of the pieces of evidence that an electrophilic species is an intermediate in the addition of bromine to alkenes is the formation of alternative products in the presence of added anions:

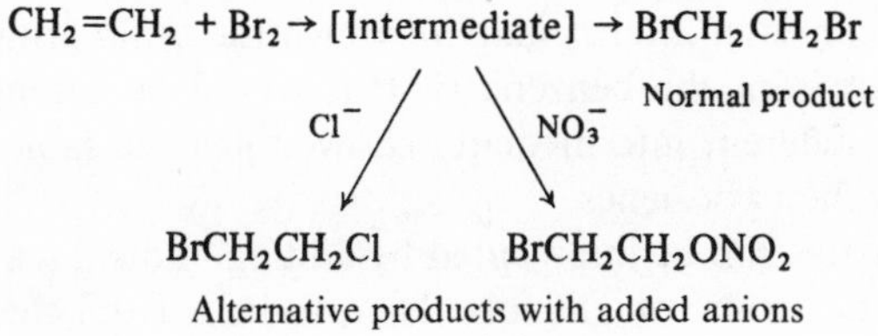

$$CH_2{=}CH_2 + Br_2 \rightarrow [\text{Intermediate}] \rightarrow BrCH_2CH_2Br$$

Normal product

$$Cl^- \diagup \qquad \diagdown NO_3^-$$

$$BrCH_2CH_2Cl \qquad BrCH_2CH_2ONO_2$$

Alternative products with added anions

In general such a technique cannot, by itself, give detailed information about the structure of an intermediate. In the present example one could hypothesise various structures such as:

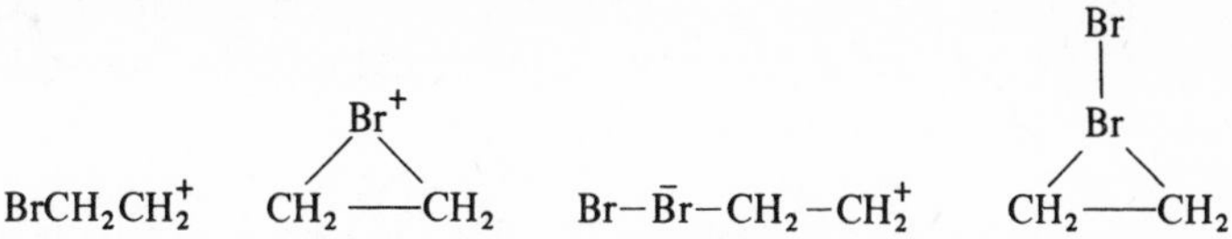

$$BrCH_2CH_2^+ \qquad \overset{\overset{+}{Br}}{\underset{CH_2 - CH_2}{\diagup \diagdown}} \qquad Br{-}\bar{Br}{-}CH_2{-}CH_2^+ \qquad \overset{\overset{\displaystyle Br}{\underset{\displaystyle Br}{\mid}}}{\underset{CH_2{-\!\!-}CH_2}{\diagup \diagdown}}$$

In some cases stereochemical studies may help (see below). In others one can learn more from more detailed trapping experiments. For example, the occurrence of benzyne as an intermediate has been probed by its Diels–Alder reaction with an added diene, such as furan:

However, the isolation of such an adduct is not conclusive proof that the species which has been trapped is actually benzyne. Suppose the reaction being studied involved the diazotisation of sodium anthranilate. One could certainly postulate that this would lead to benzyne:

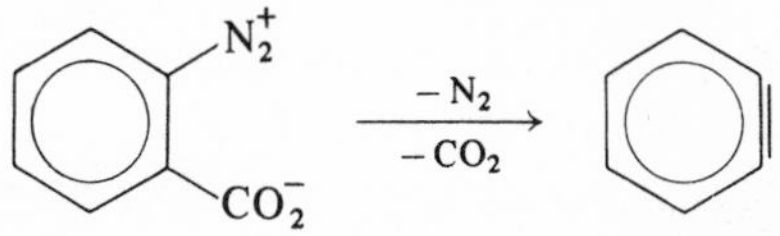

But it could be that the furan adduct was formed by the attack of some other reactive species:

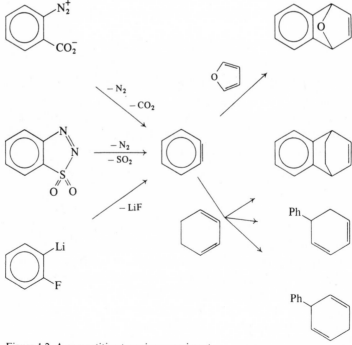

That this is unlikely was shown by a series of competitive trapping experiments, using different methods of generating benzyne and intercepting it with a mixture of furan and cyclohexadiene (figure 1.2) [63TL1017]. The different trapped products were produced in virtually identical proportions from each precursor. The implication must be that all three reactions lead to a common intermediate in which the *ortho* groups have been completely lost from the benzene ring; it would be a remarkable coincidence if three different intermediates showed just the same relative reactivities towards the two dienes.

Radical intermediates can be intercepted by adding 'radical scavengers' such as oxygen, nitric oxide, or phenols. The products from these traps are not always easy to isolate, but one can at least observe that trapping

Figure 1.2. A competitive trapping experiment.

has occurred because the direct reaction is inhibited. Alternatively one can use the technique of spin-trapping – converting a highly reactive radical into a less unstable one that is sufficiently long lived for its e.s.r. spectrum to be measured [80APO1]. 'Nitroso-t-butane' is commonly used:

$$R^{\cdot} + t\text{-Bu}-N{=}O \longrightarrow \begin{array}{c} R \\ \diagdown \\ N{-}O^{\cdot} \\ \diagup \\ t\text{-Bu} \end{array}$$

The hyperfine structure of the spectrum can give information about the nature of the group R.

One very important word of caution must be expressed here. Even if a species can be detected in a reaction mixture or isolated from it, this is not conclusive proof that it is a true intermediate, lying on the path from reactants to products. It could well arise from a side-equilibrium, unrelated to the main reaction:

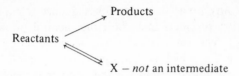

For example, during the ethoxide catalysed Claisen condensation of ethyl acetate one might be able to detect the presence of an ethoxide adduct, but this is irrelevant to the actual condensation:

$$CH_3COOEt \xrightarrow[-H^+]{EtO^-} (CH_2COOEt)^- \xrightarrow[-EtO^-]{CH_3COOEt} CH_3COCH_2COOEt$$

$$\Big\updownarrow EtO^-$$

$$\begin{array}{c} O^- \\ | \\ CH_3{-}C{-}OEt \\ | \\ OEt \end{array}$$

## 1.3    Stereochemical correlations

The stereochemistry of many reactions provides important insights into their mechanisms. We have seen that trapping experiments point towards an electrophilic intermediate in the addition of bromine to alkenes (§1.2). In general the reaction with geometrically isomeric alkenes leads to diastereoisomeric adducts, formed by the addition of bromine atoms to opposite sides of the carbon–carbon double bond (chapter 9).

This indicates that there is a stereochemical cohesion between the two steps of the reaction, and any proposed mechanism must account for this. If the intermediate had an open chain structure, such as Br-CHMe–CHMe$^+$, it would seem likely that rapid rotation about the central carbon–carbon bond would destroy this cohesion, so the stereochemical evidence points towards a bridged structure (e.g. **(1.8)**) with a bromine atom tying the two reaction sites together and transmitting information about the direction of attack in the first step through to the second:

$$(1.8)$$

If chirality is lost at any point in a reaction sequence it cannot be recovered, so the reaction products must be either achiral or racemic. Thus in the classical $S_N1$ reaction a chiral substrate is converted into a planar, achiral carbenium ion which is then open to nucleophilic attack from either side giving racemised products. The important mechanistic implications of *partial* racemisation, which accompanies many $S_N1$-type reactions, is discussed in §7.2.

Stereochemically different arrangements of the same atoms may have widely different energies. The eclipsed conformer **(1.9b)** of ethane is some 12 kJ mol$^{-1}$ less stable than the staggered conformer **(1.9a)**. The difference is largely *stereoelectronic*; that is, it relates to the disposition in space of the bonding orbitals [83Acc207]. In butane the fully eclipsed *syn*-periplanar conformer **(1.10)** is at least 25 kJ mol$^{-1}$ less stable than the *anti*-periplanar conformer. Here a large part of the difference is *steric*; the inner hydrogen atoms of the methyl groups approach each other considerably closer than their van der Waals radii so they are subject to repulsive interactions between their electron clouds.

In a similar manner the transition states for reactions can be subject to

(1.9a)          (1.9b)          (1.10)

steric and stereoelectronic influences, and reaction mechanisms must take these into account.

Steric hindrance is a term used widely but also often rather vaguely. Its meaning, however, is precise. If steric interactions in the transition state are greater than in the ground state of the reactants then the system will have to supply extra energy to overcome the difference and the reaction will be retarded. Less common is steric acceleration, where steric strain is relieved in the transition state. When an acetoxy group is attached axially to a cyclohexane ring it is hydrolysed more slowly than when it is equatorial because of increased steric interactions in the transition state **(1.11)**. However, the rates of hydrolysis of tosylates are the other way round; the reaction has considerable $S_N1$ character leading to reduced steric strain in the axial transition state **(1.12)**.

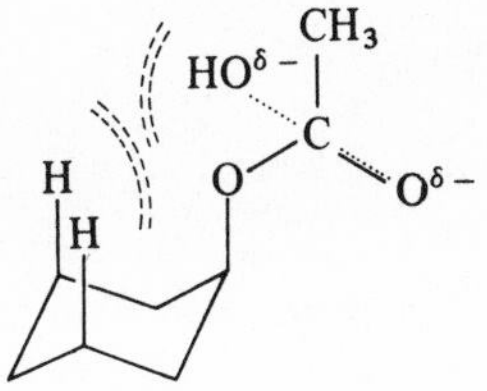

(1.11) 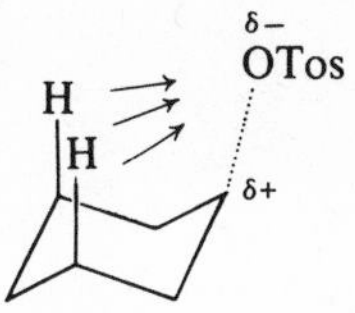          (1.12)

Steric interactions may cause molecules to adopt geometries that are stereoelectronically favourable for one mode of reaction and unfavourable for another, as with the totally different reactions of the three isomeric bromohydrins **(1.13–1.15)** with $Ag_2O$. These are governed by the stereoelectronic requirement that the group displacing the bromide should approach it from the rear, and the steric interactions that lead the phenyl group preferentially to take up an equatorial conformation [60JA2357].

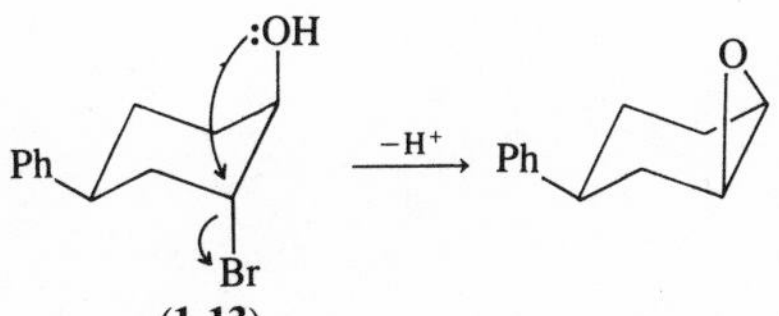

(1.13)

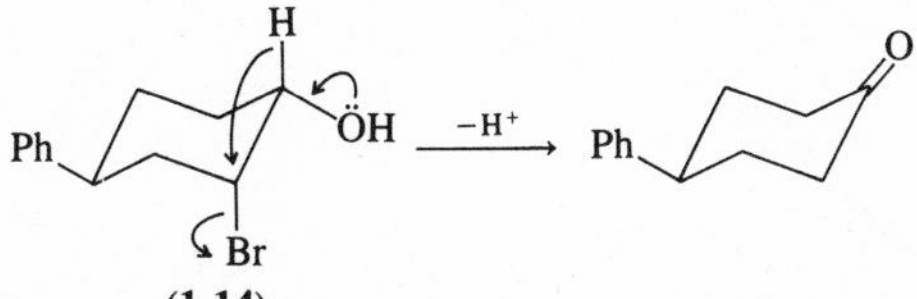

(1.14)

**(1.15)**

## 1.4 Further reading

E. S. Lewis & C. E. Boozer (Products), M. L. Bender (Intermediates), W. H. Saunders (Use of isotopes), S. L. Friess (Stereochemistry), in *Investigation of Rates and Mechanisms of Reactions*, part 2, ed. S. L. Friess, E. S. Lewis & A. Weissberger, 2nd edn., Wiley-Interscience, New York, 1963 (vol. 8 of *Technique of Organic Chemistry*).

S. P. McManus, *Organic Reactive Intermediates*, Academic, New York, 1973.

E. L. Eliel, *Sterochemistry of Carbon Compounds*, McGraw-Hill, New York, 1962.

E. L. Eliel, N. L. Allinger, S. J. Angyal & G. A. Morrison, *Conformational Analysis*, Wiley-Interscience, New York, 1965.

D. Whittaker, *Stereochemistry and Mechanism*, Oxford University Press, Oxford, 1973.

## 2.1    Rate equations

### 2.1.1    *Kinetic order*

The analysis of reaction rates and the factors which influence them is one of the most important tools in the study of reaction mechanisms. For many reactions there is a simple correlation between the rate at any point of time and the concentration of reactant species:

$$\text{Rate} = [A]^x[B]^y[C]^z \qquad [2.1]$$

where $x$, $y$ and $z$ are small integers whose sum, the *order* of the reaction, seldom exceeds 4.

The principle on which all kinetic analysis is based is that the rate of an elementary chemical process (§1.2) is proportional to the concentrations of the species actually participating in it, is independent of other species, and is unaffected by other chemical processes that may be occurring at the same time.

The only common elementary reactions are unimolecular or bimolecular, and in accord with this principle they clearly follow the behaviour of equation [2.1]:*

A       $\rightarrow$ products: Rate $= k_1[\text{A}]$

A + A $\rightarrow$ products: Rate $= k_2[\text{A}]^2$

A + B $\rightarrow$ products: Rate $= k_2[\text{A}][\text{B}]$

Elementary termolecular processes, requiring the synchronous coming together of three molecules, are very rare. Thus whereas first or second order kinetics may arise from simple one-step reactions, the observation of third or higher orders is a sure sign that the reaction is not elementary.

On the other hand even if a reaction does follow low order kinetics that is no guarantee of a simple reaction mechanism. Examples in later sections of this chapter illustrate multistep reactions with simple first or

---

* The symbol $k_1$ conventionally is taken to mean *either* the rate coefficient for a first order reaction *or* the rate coefficient for the first step in a reaction sequence. The context usually makes clear which is intended.

second order kinetics. Moreover, in many cases the conditions under which the reaction is carried out disguise the kinetic dependence on one or more reactant. For example, acid catalysed esterification follows third order kinetics, typically:

$$\text{Rate} = k_3[\text{AcOH}][\text{EtOH}][\text{H}^+]$$

However, the reaction may well be carried out in ethanol as the solvent so that $[\text{EtOH}]$ is effectively constant, and the acid, being a catalyst which is regenerated as fast as it is consumed, also remains at constant concentration; consequently the observed behaviour is that of a first order reaction (pseudo-first order kinetics):

$$\text{Rate} = k_1'[\text{AcOH}]$$

### 2.1.2 Integrated rate equations

The study of reaction kinetics entails measuring the concentration of some species, spectroscopically, titrimetrically or in any other way, as it changes during the course of the reaction. Normally one observes either the declining concentration of a reactant or the growing concentration of a product. In any case the experimental data are in terms of concentration and time, and the rate equation to which they must be fitted, commonly resembling equation [2.1], is a differential equation in these quantities:

$$\text{Rate} = \pm d[\text{X}]/dt = f[\text{X}]$$

To obtain the rate coefficients and kinetic order from this equation it is usually necessary to integrate it, preferably manipulating it in such a way as to give a linear correlation of some concentration-dependent term with time. The method of manipulation depends, in turn, on the kinetic order and also on whether one is observing reactant or product concentrations and other experimental details. To some extent, therefore, the analysis of kinetic data is a matter of trial and error.

(a) *First order* equations in which the reactant is monitored integrate as follows:

$$\text{Rate} = -d[\text{A}]/dt = k_1[\text{A}]$$

Therefore

$$\ln[\text{A}] - \ln[\text{A}]_0 = -k_1 t$$

where $[\text{A}_0]$ is the initial concentration of A at time $t = 0$. Thus a plot of $\ln[\text{A}]$ against time is linear with slope $-k_1$ and intercept $\ln[\text{A}_0]$.

If in a reaction $\text{A} \rightarrow \text{P}$ it is the product, P, that is monitored then stoicheiometry gives the relation:

$$[\text{P}] = [\text{A}_0 - \text{A}]$$

whence

$$\text{Rate} = d[P]/dt = k_1[A] = k_1[A_0 - P]$$

Therefore

$$\ln[A_0] - \ln[A_0 - P] = k_1 t$$

which again gives a linear plot.

Because of the simplicity of first order analysis it is often desirable to carry out kinetic experiments under pseudo-first order conditions. For example, a second order process, $A + B \rightarrow$ products, can be investigated first by monitoring the reaction of A in a large excess of B, and then that of B in a large excess of A.

(b) *Second order* kinetics can, however, be followed directly. The simplest case is when the rate depends on the square of the concentration of a single reagent (and this, of course, is not amenable to pseudo-first order simplification):

$$\text{Rate} = -d[A]/dt = k_2[A]^2$$

Therefore

$$1/[A] - 1/[A_0] = k_2 t$$

If a product is being monitored $[P] = [A_0 - A]$, and the integrated expression becomes:

$$1/[A_0 - P] - 1/[A_0] = k_2 t$$

For a second order reaction of the type $A + B \rightarrow P$:

$$\text{Rate} = -d[A]/dt = -d[B]/dt = k_2[A][B]$$

In its simplest form, where $[A_0] = [B_0]$, this reduces to the case of $2\,A \rightarrow P$, since stoicheiometry requires that the two species remain in equal concentrations. Otherwise the difference, $\Delta$, in the initial concentrations $(\Delta = [B_0 - A_0])$ will remain constant throughout the reaction so that

$$-d[A]/dt = k_2[A][A + \Delta].$$

Therefore

$$\ln\frac{[A + \Delta]}{[A]} - \ln\frac{[A_0 + \Delta]}{[A_0]} = \Delta k_2 t \qquad [2.2]$$

If it is the product that is being monitored:

$$\text{Rate} = d[P]/dt = k_2[A_0 - P][B_0 - P]$$

Therefore

$$\ln\frac{[A_0][B_0 - P]}{[B_0][A_0 - P]} = \Delta k_2 t$$

(c) *More complex rate equations.* Clearly this type of treatment can be extended to the integration of more complex differential equations; details may be found in standard texts on kinetics. The difficulty is often not in deriving the necessary linear equation, but in using it with adequate precision. For example, it may prove experimentally difficult to carry out a second order reaction between two different substrates either under pseudo-first order conditions or with exactly equal initial concentrations. Consequently equation [2.2] must be used, but it is obvious that this is very imprecise at small values of Δ. There are mathematical ways of easing the difficulties which again are discussed in more specialised books. But it is important to be aware that the kinetic analysis of complex reactions can be very uncertain, relying as it does on the statistical fitting to a hypothetical straight line of results that are inevitably subject to some experimental error. To take a relatively simple example, the benzidine rearrangement follows mixed second and third order kinetics (§2.2.2; §14.1.2),

$$\text{Rate} = k_2[\text{benzidine}][\text{H}^+] + k_3[\text{benzidine}][\text{H}^+]^2$$

but the results frequently correlate satisfactorily with fractional order expressions such as:

$$\text{Rate} = k[\text{benzidine}][\text{H}^+]^{1.6}$$

## 2.2 The kinetics of composite reactions

### 2.2.1 Sequential reactions

Most reactions of organic chemistry are not elementary. Instead they involve two or more discrete steps with the formation of intermediate species *en route* from reactants to products. Consider the following two-step sequence:

$$\text{Reactant(s)} \underset{k_{-1}}{\overset{k_1}{\rightleftharpoons}} \text{Intermediate} \overset{k_2}{\rightarrow} \text{Products}$$

The rate of formation of products is $k_2$ [Intermediate], but the concentration of the intermediate is unknown. Provided it is a relatively high-energy species, its concentration at all stages during the reaction will be small, and we may make the reasonable assumption that it is constant. This approximation is sometimes named after its proponent, Bodenstein, but is more commonly called simply the 'steady-state approximation'. It follows that the rate of formation of the intermediate is the same as its rate of destruction:

$$k_1[\text{Reactants}] = k_{-1}[\text{Intermediate}] + k_2[\text{Intermediate}]$$

$$\text{Rate} = k_2[\text{Intermediate}] = \frac{k_1 k_2[\text{Reactants}]}{k_{-1} + k_2} \qquad [2.3]$$

The detailed kinetics of the reaction depend primarily on the *partition factor*, $k_{-1}/k_2$, which measures the extent to which the intermediate reverts to reactants rather than continuing to products.

If $k_2 \gg k_{-1}$ then $k_{-1}$ can be ignored in the denominator of equation [2.3] which simplifies to:

$$\text{Rate} = k_1[\text{Reactants}]$$

In other words, every molecule of intermediate is whisked through to products and the *rate-limiting* step of the reaction is the formation of the intermediate. In such a reaction sequence nothing that happens after the first step can have any kinetic effect. The concentrations of any species that are not involved in the first step do not appear in the rate equation, nor is it possible to say from kinetic analysis how many steps there are in the sequence from intermediate to products.

Kinetically, therefore, such a sequence is indistinguishable from a simple elementary reaction and one must use non-kinetic methods to demonstrate that it is indeed composite. Frequently this is obvious; in a reaction of the type:

$$\left\{ \begin{array}{ll} \text{Step 1} & \text{A} + \text{B} \xrightarrow{\text{Slow}} \text{C} \\ \\ \text{Step 2} & \text{C} + \text{D} \xrightarrow{\text{Fast}} \text{Products} \end{array} \right.$$

the species D appears in the products but not in the rate equation. For example, some nitrations of arenes fall into this category (chapter 10): the rate is controlled by the slow conversion of nitric acid into the nitronium ion, $NO_2^+$, and is unaffected by the concentration of the arene itself.

Moreover, in such a reaction sequence even the *nature* of the species D cannot affect the overall rate of reaction, and it is therefore not unusual to find that a particular compound undergoes a series of different reactions all at the same rate. The base-promoted halogenations of acetone in aqueous solution, whether with chlorine, bromine or iodine, all occur at the same rate and all follow the rate equation:

$$\text{Rate} = k[\text{acetone}][\text{HO}^-]$$

This accords with the mechanism:

$$CH_3COCH_3 + HO^- \xrightarrow[\text{Slow}]{k} (CH_3COCH_2)^- \xrightarrow[\text{Fast}]{Hal_2} CH_3COCH_2Hal$$

Consider now a reaction sequence for which $k_{-1} \gg k_2$; equation [2.3] reduces to:

$$\text{Rate} = (k_1 k_2/k_{-1})[\text{Reactants}] = k_2 \cdot K_1[\text{Reactants}]$$

That is, there is a rapidly established *pre-equilibrium* between reactants and intermediate (equilibrium constant $K_1$) with a rate-limiting decomposition of the intermediate to products. In this case a reagent that does not enter the reaction until the second step *does* appear in the rate equation. This is a situation which commonly gives rise to third order kinetics:

$$\begin{cases} \text{Step 1} \quad A + B \underset{\text{Fast}}{\overset{K_1}{\rightleftharpoons}} C \\ \text{Step 2} \quad C + D \underset{\text{Slow}}{\overset{k_2}{\longrightarrow}} \text{Products} \end{cases}$$

$$\text{Rate} = K_1 k_2 [A][B][D]$$

It is essentially this behaviour that causes the third order kinetics of acid catalysed esterification (§2,1.1). The acid is protonated in a rapid pre-equilibrium and the alcohol attacks the protonated intermediate in a subsequent rate-limiting step.

When the intermediate is partitioned between reactants and products in comparable proportions equation [2.3] must be used without further simplification, and this can lead to complex kinetic behaviour:

$$\begin{cases} \text{Step 1} \quad A + B \underset{k_{-1}}{\overset{k_1}{\rightleftharpoons}} C \\ \text{Step 2} \quad C + D \overset{k_2}{\rightarrow} \text{Products} \end{cases}$$

$$\text{Rate} = \frac{k_1 k_2 [A][B][D]}{k_{-1} + k_2 [D]}$$

Such a reaction is at all times first order in both A and B, but its dependence on [D] is variable. If D is present in low enough concentration then $k_{-1} \gg k_2[D]$ and the reaction follows overall third order kinetics with step 2 rate-limiting. At higher concentrations of D it may be possible that $k_2[D] \gg k_{-1}$, control switches to step 1, and the term in [D] drops out of the rate equation. Between the two extremes the rate shows no simple dependence on [D].

If a reaction sequence passes through a relatively stable intermediate then the steady-state approximation may be inapplicable and one may encounter behaviour such as that illustrated in figure 2.1. Detailed kinetic analysis of such a system is not difficult, but for our present purposes the most important feature to note is the build up and subsequent decay of the intermediate during the reaction, behaviour which can often be detected spectroscopically (§1.2).

### 2.2.2 Competitive reactions

It often happens that a reactant undergoes two or more reactions simultaneously. This can be an inconvenience (especially to the synthetic chemist), but analysis of such behaviour may provide valuable mechanistic information. Competitive kinetics are particularly important for

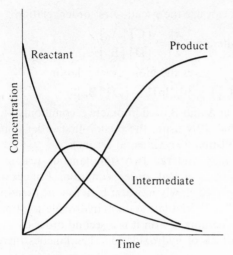

Figure 2.1. Build up of intermediate during consecutive reactions.

studying extremely fast reactions, for which direct measurements are often difficult and unreliable: aromatic nitration (chapter 10) has been much studied this way. The technique also affords a good way of detecting small differences in reactivity, and is therefore often used for investigating kinetic isotope effects (§2.4).

If a single reactant decomposes by two different first order paths then, provided they are not in equilibrium with each other, the two different products are formed in direct proportion to the two rate coefficients:

$$A \overset{k_1}{\rightarrow} B$$
$$A \overset{k_2}{\rightarrow} C \qquad k_1/k_2 = [B]/[C]$$

This type of situation is not very common, but competitive pseudo-first order reactions are important. Two different substrates, A and B, both in excess, are treated with a common reagent, X. Thus a mixture of two different arenes may be nitrated together using only a small quantity of nitric acid.

$$X + A \overset{k_1}{\rightarrow} C$$
$$X + B \overset{k_2}{\rightarrow} D$$

If A and B are not kept in excess, then the one that reacts the faster will be completely consumed before the reaction is complete. The other substrate could then continue to react with the balance of X which remained, so the product ratio would be unrelated to their relative rates of formation. If the excess is so large that the concentrations of A and B are effectively constant we can use the pseudo-first order relation:

$$\frac{k_1[A_0]}{k_2[B_0]} = \frac{[C]}{[D]} \quad \text{Therefore} \quad \frac{k_1}{k_2} = \frac{[C]/[A_0]}{[D]/[B_0]}$$

(A true second order analysis gives the more exact relation:

$$k_1/k_2 = \{\ln(1 - [C]/[A_0])\}/\{\ln(1 - [D]/[B_0])\}$$

but with the sort of excess of A and B used in practice, commonly at least five-fold, the errors introduced by using the pseudo-first order approximation are no more than normal experimental errors.)

Sometimes a substrate may undergo two simultaneous reactions of different order, resulting in a mixed order equation. For example, isopropyl bromide is hydrolysed slowly in water by a first or pseudo-first order mechanism (see chapter 7). In dilute sodium hydroxide solution this solvolysis continues, but superimposed on it is a second order hydrolysis involving the direct $S_N2$ attack of hydroxide ion. The kinetics therefore follow the equation:

$$\text{Rate} = k_1[\text{i-PrBr}] + k_2[\text{i-PrBr}][\text{HO}^-]$$

This behaviour can be detected by plotting Rate/[i-PrBr] against $[\text{HO}^-]$ to find the straight line correlation:

$$\text{Rate}/[\text{i-PrBr}] = k_1 + k_2[\text{HO}^-]$$

A special case of competitive kinetics occurs when two different reactants are in rapid equilibrium with each other:

$$A_1 + X \xrightarrow{k_1} B_1$$

$$K \, \Updownarrow$$

$$A_2 + X \xrightarrow{k_2} B_2$$

$A_1$ and $A_2$ may, for example, be conformational or tautomeric isomers. If their rate of interconversion is much faster than their rates of reaction to give products $B_1$ and $B_2$, then their relative concentrations are at all times in the proportion defined by the equilibrium constant: $[A_2]/[A_1] = K$. Therefore the relative rates of reaction are given by:

$$\frac{d[B_2]}{dt} \bigg/ \frac{d[B_1]}{dt} = \frac{k_2[A_2][X]}{k_1[A_1][X]} = K \cdot k_2/k_1$$

The two reactions necessarily continue for the same length of time (because one of the reactants, A, cannot be consumed before the other), so that this constant ratio of rates must also be the ratio in which the products are formed. It is not possible to measure $k_1$ and $k_2$ because $A_1$ and $A_2$ are not separately accessible. Consequently we cannot determine $K$, the equilibrium constant for the reactants, from the observed product ratio. (It may, however, be possible to estimate the values of $k_1$ and $k_2$ from model compounds.) This conclusion is known as the *Curtin–Hammett principle* [54RCP111; 83CR83].

If the rate of interconversion of the reactants is very much slower than the rates of reaction, then of course the system behaves as two quite independent reactions. For intermediate rates of interconversion the kinetic analysis is more complicated [80T1173].

A variation on this behaviour, often called *Winstein–Holness* kinetics [55JA5562] and frequently found in conformational studies, occurs when products $B_1$ and $B_2$ are in equilibrium with one another. (For example, cyclohexanol, which exists as an equilibrating mixture of equatorial and axial isomers, may be esterified to give *eq* and *ax* cyclohexyl acetates, also in equilibrium with each other.) In this case the experimental observations give only an overall rate of reaction, $d[B_1 + B_2]/dt$:

$$\text{Rate} = k_1[A_1][X] + k_2[A_2][X]$$
$$= k_{obs}[A][X]$$

where $[A]$, the measured concentration of cyclohexanol (say), is $[A_1] + [A_2]$. Therefore

$$k_{obs} = \frac{k_1[A_1] + k_2[A_2]}{[A_1] + [A_2]} = \frac{k_1 + k_2 K}{1 + K}$$

Or

$$K = (k_1 - k_{obs})/(k_{obs} - k_2)$$

Again the implication is that we need some independent estimate of the rate coefficients $k_1$ and $k_2$, which are not directly measurable, in order to investigate the equilibrium between $A_1$ and $A_2$.

For example, the *cis* and *trans* isomers of 4-t-butylcyclohexanol, held rigid by the bulky *eq* t-butyl group, can be used as models for *ax* and *eq* cyclohexanol, respectively.

### 2.2.3    Chain reactions

Chain reactions are frequently encountered in radical processes, because once an unpaired electron has been generated it can only be paired again by reaction with another radical. Since most radicals are of high energy their concentrations are low and their rates of recombination are therefore very small.

Take as a typical example the bromination of an alkane, following thermal or photolytic generation of bromine atoms:

$$Br_2 \xrightarrow{k_1} 2\ Br^{\bullet}$$

The chain sequence is then:

$$\begin{cases} Br^{\bullet} + R-H \xrightarrow{k_2} HBr + R^{\bullet} \\ R^{\bullet} + Br_2 \xrightarrow{k_3} RBr + Br^{\bullet} \end{cases}$$

Kinetic analysis is simplified if one step of the chain is considerably faster than the other. In this example we can calculate values of $\Delta H^{\ominus}$ for the two steps from tables of bond energies as about $+50$ and $-100 \text{ kJ mol}^{-1}$, respectively; it seems reasonable to suppose that step 3 is very much faster than step 2 and that therefore bromine atoms are present in relatively much higher concentration than are alkyl radicals. Consequently the chain reaction is terminated largely by the recombination of bromine atoms:

$$2\,Br^{\bullet} \xrightarrow{k_4} Br_2$$

In the two steps of the chain there is an internal balance between the formation and destruction of bromine atoms, so the steady-state approximation requires that initiation and termination should also balance:

$$k_1[Br_2] = k_4[Br^{\bullet}]^2$$

Therefore

$$\text{Rate} = k_2[RH][Br^{\bullet}] = k_2 \cdot (k_1/k_4)^{\frac{1}{2}} \cdot [RH][Br_2]^{\frac{1}{2}}$$

Since $k_1/k_4$ is the measurable equilibrium constant for the dissociation of molecular bromine, it is possible to determine $k_2$, the rate coefficient for the rate-limiting step of the chain.

In this particular example there is no ambiguity about the rate-limiting step, but in fact the information can be found from the form of the rate equation. If step 3 were rate-limiting then termination would occur largely by the dimerisation of $R^{\bullet}$ (rate coefficient $k_5$). We would now assume that the rapidity of step 2 would whisk away each bromine atom as fast as it was formed, so the initiation step would also control the rate of formation of $R^{\bullet}$. Analysis as above then leads to:

$$\text{Rate} = k_3 \cdot (k_1/k_5)^{\frac{1}{2}} \cdot [Br_2]^{\frac{1}{2}}$$

These fractional kinetic orders are typical of radical reactions. For further discussion of their kinetics see §15.3.

### 2.2.4 *Data from the analysis of reverse reactions*

Some reactions are difficult to study because their equilibria are biased strongly towards the reactants. In such a case one may be able to obtain kinetic information from a study of the reverse reaction, using the relation between equilibrium constant and rate coefficients:

$$K = k_{forward}/k_{reverse} \qquad\qquad [2.4]$$

This relation is strictly applicable to elementary processes, but it is also valid for composite reactions provided the equilibrium between products and reactants is rapidly established and there is no significant build up of any reaction intermediates.

Consider for example the base catalysed aldol reaction of ketones: acetone dimerises to diacetone alcohol, but equilibrium is reached after only a few per cent conversion:

$$2\,CH_3COCH_3 \underset{k_r}{\overset{k_f}{\rightleftharpoons}} (CH_3)_2C(OH)CH_2COCH_3$$

‘A’                          ‘D’

The rate equation determined experimentally for the reverse reaction is:

$$Rate_r = k_r[D][HO^-] \qquad\qquad [2.5]$$

At equilibrium the rates of the forward and reverse reactions are the same; combining this with equations [2.4] and [2.5] gives:

$$Rate_f = Rate_r = (k_r/K)\cdot[D][HO^-] = (k_f\cdot[A]^2/[D])\cdot[D][HO^-]$$

Whence the rate equation for the forward reaction is:

$$Rate_f = k_f[A]^2[HO^-]$$

### 2.2.5    Autocatalysis

Autocatalysis is the catalysis of a reaction by one of its own products; for example, C may be a catalyst for the reaction:

$$A \rightarrow B + C \qquad\qquad \text{Therefore Rate} = k_2[A][C]$$

Clearly there must be a trace of C present initially or else there must be a concurrent first order uncatalysed reaction; otherwise the reaction would never start. Ignoring this, however, we can set $[A] = [A_0 - C]$, whence:

$$Rate = d[C]/dt = k_2[A_0][C] - k_2[C]^2$$

At the beginning of the reaction when $[C] = 0$ the rate is zero; again at the end, when all the A is converted into products and $[C] = [A_0]$, the rate falls back to zero. The resulting sigmoid curve (figure 2.2) is characteristic of autocatalysed reactions.

## 2.3    Absolute reaction rates

### 2.3.1    The Arrhenius equation

Kinetic analysis of a reaction gives the form of the rate equation, the manner in which the rate is influenced by the concentrations of the reacting species. The actual rate also depends on the rate coefficient. Most of our understanding of the numerical significance of rate coefficients derives from analysis of their variation with temperature according

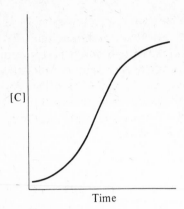

Figure 2.2. Changes in catalyst concentration during autocatalysis.

to the Arrhenius equation:

$$k = A \cdot \exp(-E_a/RT) \qquad\qquad [2.6]$$

Arrhenius was not actually the discoverer of this relation; it was first put forward empirically by Hood in 1878. Later both van't Hoff and Arrhenius interpreted it in thermodynamic terms. Arrhenius [1889ZP(4)226] recognised the connection between the exponential dependence of the rate coefficient on temperature and the similar form of the Maxwell–Boltzmann distribution, and he suggested that before molecules could react they had to take up an 'activated' form whose energy was higher than that of the reactants by the amount $E_a$. The rate of reaction would be proportional to the number of these activated molecules, which would in turn be related to the concentration of reactants by the term $\exp(-E_a/RT)$. Arrhenius himself offered no suggestions about the physical significance of the 'pre-exponential factor' $A$, nor did he attempt to predict what might govern the magnitude of $E_a$.

### 2.3.2 Collision theory

Collision theory, based on the kinetic theory of gases, leads to an expression resembling the Arrhenius equation:

$$k = Z \cdot \exp(-E_a/RT)$$

where $Z$, which is the collision number, the frequency of collisions between reactant molecules, can be calculated from molecular mass and diameter, and $\exp(-E_a/RT)$ is the fraction of those collisions which occur with a kinetic energy of $E_a$ or more, that is, which can provide enough energy to lead to the formation of an *activated complex*. The only significant difference in form from the Arrhenius equation is that $Z$, unlike $A$, is temperature-dependent, being proportional to $\sqrt{T}$. However, in practice it is difficult to study reactions over a very wide temperature range and the variation in $Z$ is unimportant in comparison with the exponential term. Values of $Z/\text{mol}\,l^{-1}\,s^{-1}$ are typically of the order of $10^{10} \times \sqrt{T}$. For some simple gas phase reactions the observed $A$-values do approach this, but for other reactions they fall short by many powers of ten. The reason is not hard to find, but its quantitative resolution is virtually impossible. The problem is that collisions between reactant molecules must not only have sufficient energy, they must also be correctly oriented. The simple gas phase reaction

$$CH_3^{\bullet} + H_2 \rightarrow CH_4 + H^{\bullet}$$

occurs a thousand times slower than its $Z$-value predicts. Obviously most of the collisions of the methyl radical occur at its peripheral hydrogen atoms and only a few strike at the carbon atom and lead on to methane, but there seems no way of calculating this extra probability factor.

### 2.3.3 Transition state theory

Transition state theory treats an elementary reaction as if it were a two-stage process: first the formation of the activated complex or *transition state** at the point of maximum energy on the reaction path, and then the collapse of the transition state to products:

$$\text{Reactants} \underset{k_{-1}}{\overset{k_1}{\rightleftharpoons}} [TS]^{\ddagger} \overset{k_2}{\rightarrow} \text{Products}$$

(The symbol $\ddagger$ is customarily used to denote a transition state or properties related to it.) Applying the steady-state approximation to the transition state gives:

$$\text{Rate} = \frac{k_1 k_2}{k_{-1} + k_2}[\text{Reactants}]$$

The empirical rate equation is simply Rate $= k[\text{Reactants}]$, whence:

$$k = k_1 k_2/(k_{-1} + k_2) \qquad\qquad [2.7]$$

The activated complex is precariously balanced on its energy summit. On either side of it are structures of lower energy and the smallest molecular motion in either direction can tip it down the slope. In other words, the activated complex can be envisaged as differing from a normal molecule in that one of its vibrational modes has been converted into translational motion: for example, during an $S_N2$ attack on $CH_3Br$ the stretching vibration of the progressively weakening C–Br bond reaches a point beyond which the separating nuclei no longer reverse their motion but continue to move apart. As this limit is approached the vibrational

---

* The terms 'activated complex' and 'transition state' are usually regarded as synonymous; the distinction is sometimes made that 'activated complex' refers to the actual molecular entity whereas 'transition state' describes the set of states or energy levels.

energy levels lie very close together and all of them are accessible (whereas for normal vibrations only the lowest levels are significantly populated). The total vibrational energy approaches the classical value calculated from kinetic theory: $E = k_B T$, where $k_B$ is the Boltzmann constant, kinetic and potential energy terms each contributing $\frac{1}{2}k_B T$. Applying Planck's equation, $E = h\nu$, for the energy of a quantised oscillator, gives the mean limiting frequency of the dissociation of the activated complex as $\nu^\ddagger = k_B T/h$. This frequency is the same for the collapse of any transition state in any reaction; therefore $k_{-1} = k_2 = \nu^\ddagger$. Thus equation [2.7] may be written:

$$k = \tfrac{1}{2}\nu^\ddagger \cdot (k_1/k_{-1})$$

The term $k_1/k_{-1}$ is equivalent to an equilibrium constant, $K^\ddagger$, for the 'equilibrium' between reactants and transition state, and is governed by the normal thermodynamic relation: $K^\ddagger = \exp(-\Delta G^\ddagger/RT)$. Therefore:

$$k = \tfrac{1}{2}\nu^\ddagger K^\ddagger = \tfrac{1}{2}(k_B T/h) \cdot \exp(-\Delta G^\ddagger/RT) \qquad [2.8]$$

In fact this is not a true equilibrium: as the atoms move towards the structure of the transition state their momentum may easily carry them over the energy barrier and down on the side of the products. The extent to which this happens cannot be evaluated thermodynamically: it depends on the detailed geometry of the moving atoms. In the extreme all the molecules that reached the transition state would be carried across and none would revert to reactants. Thus the number $\frac{1}{2}$ in equation [2.8] should be replaced by a variable parameter, $\kappa$, called the *transmission coefficient*, which measures the proportion of molecules carried over the energy barrier, and which may be expected to have a value between $\frac{1}{2}$ and 1; it is common practice to assume a value of 1. The resulting equation (with or without the inclusion of $\kappa$) is known as the Eyring equation:

$$\begin{aligned} k &= (k_B T/h) \cdot \exp(-\Delta G^\ddagger/RT) \\ &= (k_B T/h) \cdot \exp(\Delta S^\ddagger/R) \cdot \exp(-\Delta H^\ddagger/RT) \end{aligned} \qquad [2.9]$$

Again this accords with the form of the Arrhenius equation, save for the relatively insignificant temperature term in the pre-exponential factor. The appearance of the entropy of activation, $\Delta S^\ddagger$, accords with the ideas of collision theory on the probability of a suitable encounter. The exponential term is now expressed as an enthalpy rather than an energy of activation. The two thermodynamic functions are related by $\Delta H = \Delta E + P\Delta V$. For reactions in solution $\Delta V$ is negligible and therefore so is the effect on it of changing temperature; thus we may compare the Arrhenius and Eyring equations directly. The Arrhenius equation [2.6] may be written logarithmically as:

$$\ln k = \ln A - E_a/RT$$

from which it can be seen that $E_a$ is obtained empirically from the slope of a plot of $\ln k$ against $1/T$:

$$\frac{d(\ln k)}{d(1/T)} = -E_a/R$$

Treating the Eyring equation similarly gives:

$$\ln k = \ln(k_B/h) + \Delta S^{\ddagger}/R + \ln T - \Delta H^{\ddagger}/RT$$

$$\frac{d(\ln k)}{d(1/T)} = -T - \Delta H^{\ddagger}/R$$

A comparison of these two differential equations shows that $\Delta H^{\ddagger}$ (and hence $\Delta S^{\ddagger}$) can be obtained at any temperature from the relation: $\Delta H^{\ddagger} = E_a - RT$.

For gas phase reactions one cannot ignore the effect of changing temperature on the term $P\Delta V$. If $\Delta n^{\ddagger}$ is the change in the number of molecules in going from reactants to transition state, then $P\Delta V^{\ddagger} = \Delta n^{\ddagger} \cdot RT$, and the relation between $\Delta H^{\ddagger}$ and $E_a$ becomes:

$$\Delta H^{\ddagger} = E_a + (\Delta n^{\ddagger} - 1)RT$$

The transition state is always a single species, so $\Delta n^{\ddagger}$ is 0 for unimolecular reactions and 1 for bimolecular ones.

For some simple gas phase reactions it is possible to calculate values for $\Delta S^{\ddagger}$ and $\Delta H^{\ddagger}$ so as to compare them with the values determined experimentally. For example, the chlorination of methane is controlled by the reaction step:

$$CH_4 + Cl^{\bullet} \rightarrow CH_3^{\bullet} + HCl$$

and the rate of this reaction can be calculated by making various assumptions and approximations. In essence the procedure is as follows.

(a) The problem is simplified to that of a three body system by assuming that the methyl group remains unchanged during the course of the reaction.

(b) The potential energy of any geometrical arrangement of the three body system [Me, H, Cl] is calculated by first taking their interactions two at a time: [Me, H], [H, Cl], [Me, Cl]. The potential energies of these systems (figure 2.3) are known from the vibrational spectra and dissociation energies of the corresponding molecules, MeH, HCl and MeCl, and these experimental data can be used to fit adjustable parameters in any appropriate VB or MO quantum mechanical calculation. These parameters can then be inserted into the calculations for the complete three body system.

In all such calculations the energy of a linear arrangement of the three bodies is lower than that of an angular one, so we can simplify the procedure further by considering only two geometrical variables: the Me—H and H—Cl distances in the linear [Me—H—Cl] entity.

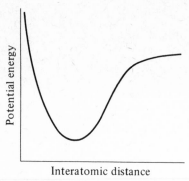

Figure 2.3. Potential energy function for a two body system.

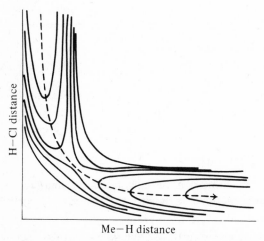

Figure 2.4. Potential energy surface for linear three body system [Me–H–Cl].

These calculations result in a three-dimensional graph of energy against the two bond distances, usually represented in contour form as in figure 2.4. In geographical terms this corresponds to two valleys joined by a pass or col, and the reaction can be represented as a journey from one valley to the other – the dotted line in figure 2.4. A vertical cross-section along this line gives the familiar diagram showing how the potential energy changes during the reaction (figure 2.5).

(c) One further stage of calculation is needed for a quantitative estimation of the reaction rate. The potential energy surface of figure 2.4 and the portion of it shown in profile in figure 2.5 represent the classical energy minimum for each geometrical arrangement of the [Me–H–Cl] system. The energy of the real system will be greater than this because of

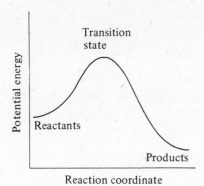

Figure 2.5. Reaction profile for MeH + Cl$^\bullet$ → Me$^\bullet$ + HCl.

its zero-point (the ground state vibrational energy which all molecules retain even at absolute zero) and because not all the molecules will be in the lowest translational or rotational energy levels. Statistical mechanical calculations can take this into account; they require values for the moments of inertia and vibrational force constants of the transition state, and in the present example these can be estimated from its geometry and from the shape of the potential energy surface, as calculated in the previous stage.

The results of such a calculation can be remarkably accurate: for the chlorination of methane calculated values of $\Delta H^\ddagger$ and $\Delta S^\ddagger$ are 2.8 kJ mol$^{-1}$ and $-21.2$ J K$^{-1}$ mol$^{-1}$, respectively; the experimentally observed values are 2.7 and $-17.8$. However, a quantitative interpretation of transition state theory in complicated reactions is beyond our present capabilities.

Nevertheless, the qualitative ideas of the theory have wide applicability (see, for example, the discussion of solvation effects on $\Delta S^\ddagger$, §5.1.4) and in limited areas, especially in the study of kinetic isotope effects, they can be used semi-quantitatively even in fairly complex systems.

## 2.4    Kinetic isotope effects

### 2.4.1    *Primary kinetic isotope effects*

The reaction of $CH_2D_2$ with chlorine gives a mixture of products: $CH_2DCl + DCl$ and $CHD_2Cl + HCl$. At room temperature about nine times as much hydrogen chloride as deuterium chloride is produced. Thus the transfer of deuterium from methane to a chlorine atom is some nine times slower than the transfer of protium. This is an example of a kinetic isotope effect and it arises in the following way.

The potential energy curves of figures 2.3–2.5 are the same for both protiated and deuteriated systems, but the greater mass of the deuterium atom manifests itself in a different distribution of the vibrational energy

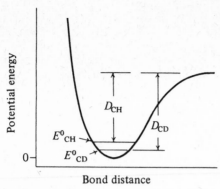

Figure 2.6. Differences in zero-point vibrational energies and bond dissociation energies of C–H and C–D bonds.

levels. In particular, the zero-point energy of a bond to deuterium is significantly lower, and the bond dissociation energy is correspondingly higher (figure 2.6).

The zero-point energy of a vibration is given by $E^0 = \frac{1}{2}h\nu$, where $\nu$ is the vibrational frequency. For carbon–hydrogen bonds the stretching frequency (which is the vibrational mode that can lead to dissociation) lies at around 3000 cm$^{-1}$, so $E^0_{CH} \simeq 18$ kJ mol$^{-1}$. The frequency is related to the properties of the bond by Hooke's law: $\nu \propto \sqrt{(k/\mu)}$, where $k$, the force constant, is not altered by changing protium to deuterium; but $\mu$, the reduced mass, is increased by a factor of about 2. (The reduced mass, $\mu_{XY}$, for a bond between atoms X and Y is $m_X m_Y/(m_X + m_Y)$. If one of the masses, say $m_X$, is much smaller than the other, as with nearly all bonds to hydrogen or deuterium, then $\mu_{XY} \simeq m_X$.) Thus the stretching frequency for carbon–deuterium bonds falls to around 2200 cm$^{-1}$, and $E^0_{CD}$ is about 13 kJ mol$^{-1}$. The bond dissociation energy for C–D is consequently about 5 kJ mol$^{-1}$ more than for C–H, and using the Arrhenius equation we can predict the relative rates of dissociation:

$$\frac{k_H}{k_D} = \frac{A \cdot \exp(-D_{CH}/RT)}{A \cdot \exp(-D_{CD}/RT)} = \exp(5000/RT) \simeq 7.5 \text{ at } 25\,^\circ\text{C}$$

This calculation also illustrates the dramatic effect of temperature on isotope effects. Doubling the temperature (in this example from 300 K to 600 K) reduces the isotope effect to its square root (2.7).

In fact this calculation is rather crude and underestimates the differences in reactivities. In particular, for the dissociation of a polyatomic molecule like methane we should include contributions not only from the stretching vibration but also from the bending modes, and we ought too to consider the redistribution of vibrational energies in the methyl fragment consequent on removal of the hydrogen atom. Nevertheless, the value calculated does agree fairly well with the magnitude of the H/D kinetic isotope effect commonly observed.

Simple dissociation is not a particularly common reaction. More usual, as exemplified by the methane–chlorine example, is a reaction in which a hydrogen atom (or in a heterolytic process a proton or hydride ion) is in the process of being transferred between two different sites at the transition state. We now need to consider not only the zero-point energies of the reactants, but also those of the transition state itself. In the present example we can consider three different types of vibration:

$$\overset{\uparrow}{Me}\cdots\overset{\uparrow}{\underset{\downarrow}{H}}\cdots Cl \qquad \overset{\leftarrow}{Me}\cdots\overset{\rightarrow}{H}\cdots\overset{\leftarrow}{Cl} \qquad \overset{\leftarrow}{Me}\cdots H\cdots\overset{\rightarrow}{Cl}$$

$$\text{A} \qquad\qquad \text{B} \qquad\qquad \text{C}$$

Vibrations of type A are similar to the bending modes of the ground state methane. In a detailed treatment they ought to be considered, but to a first approximation we shall assume that the ground state and transition state frequencies are comparable so that any isotope effect cancels out.

The type B vibration, antisymmetric stretching, is the mode that in the transition state is converted into translational motion along the reaction coordinate; the hydrogen atom is being transferred from Me to Cl (broken line in figure 2.7). This is therefore no longer an oscillatory motion and does not have any associated zero-point energy.

The symmetrical stretching vibration, type C, takes place at right angles to the reaction coordinate, with the Me–H and H–Cl distances increasing and decreasing in phase (double arrow in figure 2.7). The extent to which the hydrogen atom participates in this vibration depends

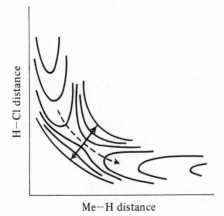

Figure 2.7. Stretching modes in [Me–H–Cl] transition state.

on the detailed structure of the transition state. For the generalised system X···H···Y the frequency of the type C vibration is given by:

$$\nu = \frac{1}{2\pi}\left\{\frac{k_X}{m_X} + \frac{k_X + k_Y - 2\sqrt{k_X k_Y}}{m_H} + \frac{k_Y}{m_Y}\right\}^{\frac{1}{2}} \qquad [2.10]$$

where $k_X$ and $k_Y$ are the force constants for the X–H and Y–H partial bonds. It follows that when $k_X = k_Y$ (in casual phraseology, if the hydrogen is half transferred in the transition state) then the term in $m_H$ disappears from the equation and there is no contribution from the hydrogen/deuterium mass difference to the zero-point energy of the transition state. The situation (figure 2.8) is equivalent to the simpler one of figure 2.6 and the kinetic isotope effect approaches its maximum value.

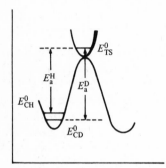

Figure 2.8. Kinetic isotope effect for a reaction in which a hydrogen atom is about half transferred at the transition state. The reaction coordinate is shown with the ground state vibrational energy levels superimposed. The type C vibration of the transition state is represented by the upper potential energy curve. The mass effect contributes nothing to this (equation [2.10] with $k_X = k_Y$), and $E^0_{TS}$ is the same for protium and deuterium. The difference in zero-point energies of the isotopically distinct ground states is therefore reflected fully in the corresponding activation energies.

If, on the other hand, the hydrogen atom is still strongly bonded to its reactant site in the transition state ($k_X \gg k_Y$), or if it is almost totally transferred to the product site ($k_X \ll k_Y$), then equation [2.10] simplifies to:

$$\nu = (1/2\pi)\sqrt{(k_X/\mu_{XH})} \qquad \text{or} \qquad \nu = (1/2\pi)\sqrt{(k_Y/\mu_{HY})}$$

The full difference in zero-point energies now appears in the transition state as well as the ground state, so that there is no net difference in activation energies for the two isotopic systems. In the intermediate situation, with the bond less or more than half broken in the transition state, a partial kinetic isotope effect is seen (figure 2.9).

This analysis suggests that in a series of related reactions involving

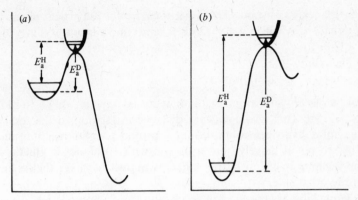

Figure 2.9. Partial kinetic isotope effects are found in less symmetrical reactions where the mass effect differentiates the zero-point energies of the isotopically distinct transition states but to a smaller extent than in the ground states. This is illustrated for (a) a reactant-like transition state with little transfer of hydrogen, and (b) a product-like transition state in which hydrogen transfer is almost complete.

hydrogen transfer one might observe a variable H/D isotope effect, passing through a maximum when the transition state is symmetrical. For example, in the transfer of a proton to a basic substrate from a range of different acids the effect should be a maximum for an acid whose $pK_a$ is the same as the $pK_{BH^+}$ of the base.

Such behaviour is indeed found [e.g. 78JA5954; 5956], and it has been widely used to rationalise reaction mechanisms in terms of the structure of the transition state. However, in some cases the maximum is much less well defined than simple theory predicts. This may arise if there is significant interaction between groups X and Y in the transition state X⋯H⋯Y. Such interaction introduces an additional term, not containing $m_H$, into the expression [2.10] for the frequency of the type C vibration; the isotopic distinction in the transition state is thereby reduced and a large isotope effect may be found even for an unsymmetrical transition state [78T1619].

A kinetic isotope effect may also arise from the phenomenon of quantum mechanical tunnelling [80MI]. In classical mechanics a potential energy barrier is precisely defined: a system whose kinetic energy is less than the barrier height is unable to cross. In quantum mechanics, however, there is a finite possibility that the barrier can be penetrated. This can be visualised in terms of the Heisenberg uncertainty principle expressed in the form:

$$\Delta x \cdot \Delta p \simeq h/2\pi$$

where $\Delta x$ is the uncertainty in a particle's position (considered for simplicity along a single axis), $\Delta p$ is the uncertainty in its momentum, and

$h$ is Planck's constant. The kinetic energy of a particle of mass $m$ is $E_k = p^2/2m$, whence $p = \pm(2mE_k)^{\frac{1}{2}}$, and the uncertainty in $p$ is thus $2 \times (2mE_k)^{\frac{1}{2}}$.

Therefore

$$\Delta x \simeq h/4\pi(2mE_k)^{\frac{1}{2}} \qquad [2.11]$$

For a hydrogen atom with kinetic energy of, say, $10\,\text{kJ mol}^{-1}$, $\Delta x \simeq 7$ pm. This is a significant distance in relation to the total movement of a hydrogen atom during a proton transfer reaction and indicates, to put it naively, that a proton with $10\,\text{kJ mol}^{-1}$ kinetic energy, approaching an energy barrier higher than this, may nevertheless find itself on the other side.

Tunnelling has the effect of accelerating a reaction relative to a classical mechanical model. The acceleration is usually expressed as a *tunnel correction coefficient*, $Q_t$, and depends markedly on the height of the energy barrier, the exact shape of the potential energy surface at the transition state, and the geometry and masses of the nuclei. It also increases rapidly with a decrease in temperature. For many reactions at room temperature, values of $Q_t$ lie between 1 (no significant tunnelling) and about 4.

In the context of kinetic isotope effects two particular aspects of tunnelling are noteworthy. First, $\Delta x$, and therefore $Q_t$, depends on mass (equation [2.11]): $Q_t^H$ is invariably greater than $Q_t^D$ so that tunnelling can enhance the zero-point kinetic isotope effect. Secondly, the values of $Q_t$ (and hence of the ratio $Q_t^H/Q_t^D$) for hydrogen transfer reactions are thought to pass through maxima at symmetrical transition states where the hydrogen atom is half-way between its reactant and product sites [71TF1995]. This mimics the behaviour predicted by the zero-point model and there has recently been some debate as to whether this invalidates some of the conclusions derived from that model.

It is not easy to dissect a kinetic isotope effect into its two components. The clearest indication of significant tunnelling is to find differences in Arrhenius activation energies, $E_a^D - E_a^H$, that are greater than the isotopic differences in zero-point energies of the reactants. Recent studies of base-induced elimination reactions showed that the contribution of tunnelling to the total kinetic isotope effect can be significant but that $Q_t^H/Q_t^D$ varied but little with changes in the extent of hydrogen transfer at the transition state [79JA7594; 81JO4247].

Of course, the observation of kinetic isotope effects is not limited to the comparison of protium and deuterium. Tritium is sometimes used further to magnify the mass difference. On the other hand one can also observe kinetic isotope effects with heavy elements: $^{12}C/^{13}C/^{14}C$, $^{16}O/^{18}O$,

$^{32}S/^{34}S$ and $^{35}Cl/^{37}Cl$ differences, in particular, have been widely studied. However, with these elements the proportional change of mass is very small. H/D isotope effects commonly lie in the range 2–8 and H/T effects are often around 20; S and Cl effects can be significant at as low a level as 1.001, but observations of such minute differences demand meticulous technique.

### 2.4.2 Secondary kinetic isotope effects

Primary kinetic isotope effects arise when the bond that is actually being broken in the reaction is isotopically modified. One can also observe changes in reactivity when isotopic changes are made elsewhere in the molecule. These can arise from a source we deliberately neglected in discussing primary effects – changes in the vibrational behaviour of bonds not involved in the reaction. Very pronounced secondary effects can be observed in reactions that involve a change in carbon hybridisation. The out-of-plane bending vibration of trigonal C–H bonds occurs at around $800\ cm^{-1}$; for tetrahedral carbon the corresponding mode is about $1400\ cm^{-1}$:

This corresponds to a difference of some $3.5\ kJ\ mol^{-1}$ in zero-point energy, so replacement of protium by deuterium could lead to a change of about $3.5(1 - 1/\sqrt{2}) \simeq 1\ kJ\ mol^{-1}$. In a reaction which involves conversion of an $sp^2$ carbon to $sp^3$ or vice versa the transition state will presumably involve only partial rehybridisation, but even a $\frac{1}{2}\ kJ\ mol^{-1}$ difference in activation energy leads to an isotope effect of 1.2 at room temperature.

This type of effect can operate in either direction. Primary effects invariably involve bond weakening (lowering of frequency) in moving to the transition state. The secondary effect can entail raising or lowering the frequency, so deuterium can retard or accelerate the reaction. This is neatly illustrated by the reversible Diels–Alder cyclo-addition of tetracyanoethylene to anthracene, in which the 9,10-carbons of the anthracene are rehybridised [65T1993]:

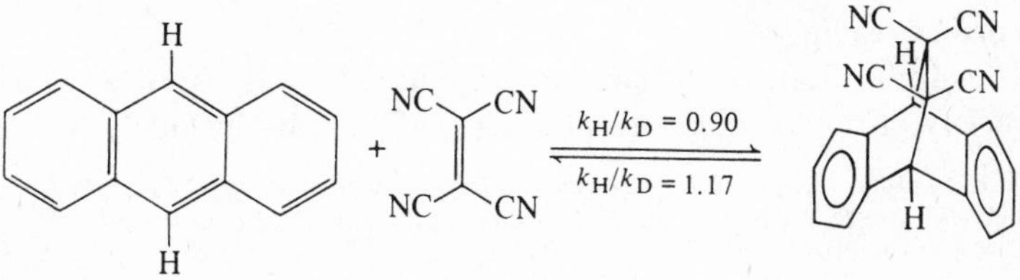

Similar effects are observed in $S_N2$ reactions, where again the C—H bonding in the transition state differs significantly from that in the ground state:

$$
\text{Nu:} \quad \overset{H}{\underset{|}{\diagdown}} C{-}Z \longrightarrow \left[ \text{Nu} \cdots \overset{H}{\underset{|}{C}} \cdots Z \right]^{\ddagger}
$$

In these reactions the C—H bending frequencies are susceptible to both steric and electronic effects, which can be finely balanced. The aqueous solvolyses of methyl halides exhibit an inverse H/D isotope effect $(k_H/k_D \simeq 0.96)$, whereas for isopropyl halides the ratio is around 1.06 (§7.31).

Surprisingly, some solvolyses exhibit a pronounced $\beta$-deuterium effect – that is, an effect arising from isotopic replacement of hydrogen one bond away from the reaction centre. For the solvolyses of isopropyl bromides in water at 60 °C the relative rates are: $(CH_3)_2CHBr$ 1.00; $(CH_3)_2CDBr$ 0.94; $(CD_3)_2CHBr$ 0.76. There remains some doubt about the origin of these $\beta$-effects [79JA5532]. They are probably due to differential hyperconjugation between the C—H or C—D bonds and the incipient carbenium ion centre of an ion-pair intermediate (chapter 7), an explanation which is certainly in accord with their relative magnitudes: $3° > 2° > 1°$. This in turn may lead to differential solvation with consequent changes in $\Delta S^{\ddagger}$ as well as $\Delta H^{\ddagger}$ (compare §5.1.4). It seems that some $\beta$-effects are less strongly influenced by temperature than is usual, and this points to a significant entropy contribution to $\Delta G^{\ddagger}$ (compare equation [2.9]).

In any case, because of the exponential relation between temperature and the magnitude of normal kinetic isotope effects, a large effect is proportionally more susceptible to temperature changes than is a small one. Consequently kinetic isotope effects in high-temperature reactions need to be assessed with caution: the normal maximum for a primary H/D effect at 500K is only about 2.2, and an unsuspected secondary effect could be contributing significantly to an observed difference of this order of magnitude.

## 2.5 Further reading

The appropriate bookshelf in any library will carry a wide variety of books on chemical kinetics. Examples include:

G. B. Skinner, *Introduction to Chemical Kinetics*, Academic, New York, 1974.

J. W. Moore & R. G. Pearson, *Kinetics and Mechanism*, 3rd edn, Wiley, New York, 1981.

For more detailed treatments:

S. W. Benson, *Foundations of Chemical Kinetics*, McGraw-Hill, New York, 1960.

S. L. Friess, E. S. Lewis & A. Weissberger (eds), *Investigation of Rates and Mechanisms of Reactions*, 2nd edn, Wiley-Interscience, New York; part 1, 1961; part 2, 1963 (vol. 8 of *Technique of Organic Chemistry*).

C. H. Bamford & C. F. H. Tipper (eds), *Comprehensive Chemical Kinetics*, section 1 (vols. 1, 2, 3); *The Theory and Practice of Kinetics*, Elsevier, Amsterdam, 1969.

Historical development:

H. Eyring, *Chem. Rev.*, 1935, **17**, 65.

Isotope effects:

L. Melander, *Isotope Effects on Reaction Rates*, Ronald Press, New York, 1960.

L. Melander & W. H. Saunders, *Reaction Rates of Isotopic Molecules*, Wiley, New York, 1980.

E. A. Halevi, *Progr. Phys. Org. Chem.*, 1963, **1**, 109 (secondary isotope effects).

C. J. Collins & N. S. Bowman (eds), *Isotope Effects in Chemical Reactions*, Van Nostrand Reinhold, New York, 1970.

W. H. Saunders, *Survey Progr. Chem.*, 1967, **3**, 109.

R. A. More O'Ferrall, in *Proton-Transfer Reactions*, ed. E. F. Caldin & V. Gold, Chapman & Hall, London, 1975.

H. Kwart, *Acc. Chem. Res.*, 1982, **15**, 401.

# 3    Linear Gibbs energy relations

## 3.1    The Hammett equation

### 3.1.1    *The equation*

We are accustomed to the idea of a general relation between structure and reactivity: toluene is more reactive than benzene towards nitration and this implies that it is also more reactive towards halogenation, sulphonation, Friedel–Crafts reactions, and all other electrophilic substitutions. Phenol is more reactive still, and nitrobenzene is decidedly less reactive. These generalisations, though, are all qualitative and although experimental results can be quantified in terms of partial rate factors these cannot be used predictively; given the partial rate factors for the nitration of toluene we cannot deduce those for chlorination.

In the 1920s and 1930s there were many attempts to devise quantitative expressions of these relations. The most important and far-reaching is due to L. P. Hammett. He dealt with systems of the type:

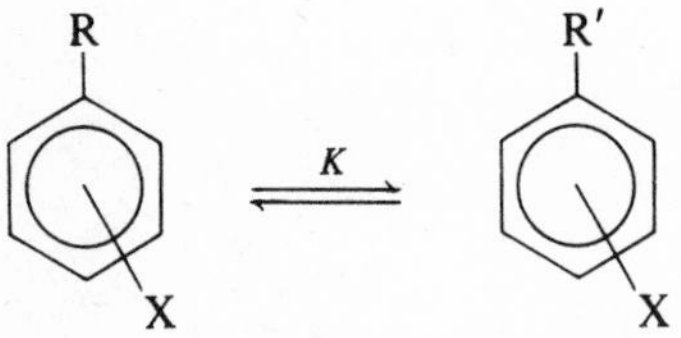

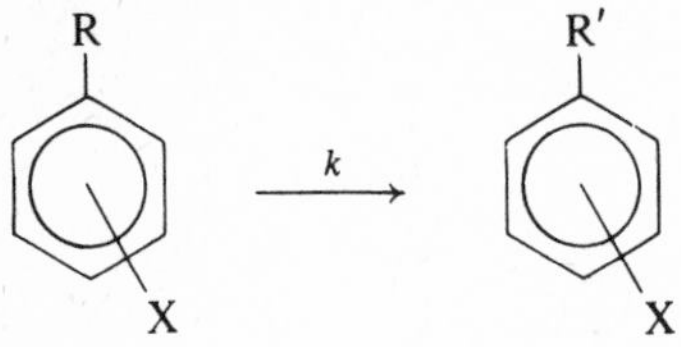

R is a reaction site in the side-chain attached to a benzene ring and X is a *meta* or *para* substituent. Hammett excluded *ortho* substituents on the grounds that there would be specific steric interaction between reaction site and substituent which would not be amenable to a regular quantitative treatment.

He found that the thermodynamic or kinetic behaviour of many such reactions could be correlated by a relation now known as the *Hammett equation* [35CR(17)125; 37JA96]:

$$\log(\mathbf{k}/\mathbf{k}_0) = \rho\sigma \qquad [3.1]$$

The symbol **k** can represent either the equilibrium constant $K$ or the rate coefficient $k$; $\mathbf{k}_0$ is the corresponding value for the unsubstituted system (i.e. for X = H). The *reaction constant*, $\rho$, is constant for a particular reaction and for the conditions under which it is carried out (temperature, solvent); it is independent of the nature of X. The *substituent constant*, $\sigma$, depends only on X and on whether it is *meta* or *para*; it is independent of the nature of R.

The relation may be exemplified by the rates of saponification of substituted ethyl benzoates (table 3.1 and figure 3.1).

Table 3.1. *Relative rates of saponification of substituted ethyl benzoates at 30 °C in 80% aqueous ethanol* [26LA(450)1]

| Substituent X | Relative rate $(k/k_0)$ | $\sigma$ | Substituent X | Relative rate $(k/k_0)$ | $\sigma$ |
|---|---|---|---|---|---|
| H | 1 | 0 | $m$-NO$_2$ | 63 | 0.71 |
| $m$-CH$_3$ | 0.71 | −0.07 | $p$-NO$_2$ | 104 | 0.78 |
| $p$-CH$_3$ | 0.47 | −0.17 | $m$-Br | 8.1 | 0.39 |
| $m$-NH$_2$ | 0.41 | −0.16 | $p$-Br | 4.9 | 0.23 |
| $p$-NH$_2$ | 0.023 | −0.66 | $m$-Cl | 7.4 | 0.37 |
| $p$-OCH$_3$ | 0.21 | −0.27 | $p$-Cl | 4.3 | 0.23 |

The source of the $\sigma$ values in the table is described later. A plot of log $(k/k_0)$ against $\sigma$ gives a good straight line of slope $+2.50$, which from the Hammett equation must be the $\rho$ value for this reaction.

The Hammett equation is an example of a *linear Gibbs energy relation* (commonly called a linear free energy relation, or LFER). Equilibrium constants and rate coefficients are related to Gibbs energy differences by the equations:

$$\log K = -\Delta G^{\ominus}/2.3\,RT$$
$$\log k = \log(k_B T/h) - \Delta G^{\ddagger}/2.3\,RT$$

(compare eqn 2.9). Therefore the Hammett equation can be written:

$$\log(\mathbf{k}_X/\mathbf{k}_0) = \log\mathbf{k}_X - \log\mathbf{k}_0 = \Delta\Delta\mathbf{G}_X/2.3\,RT = \rho\sigma_X$$

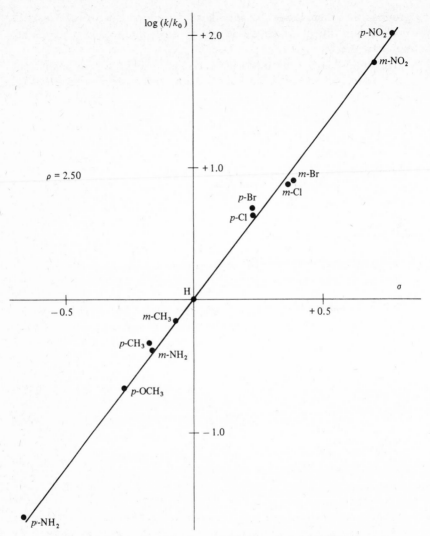

Figure 3.1. Hammett plot for the saponification of ethyl benzoates.

where the suffix X refers to the substituted system and 0 refers to the unsubstituted one. The expression $\Delta\Delta G_X$ is a second difference: the difference between $\Delta G$ values for the substituted and unsubstituted reactions.

If we compare the effects of two different substituents, A and B, on the same reaction at the same temperature we have:

$$\frac{\Delta\Delta G_A/2.3\,RT}{\Delta\Delta G_B/2.3\,RT} = \frac{\rho\sigma_A}{\rho\sigma_B}$$

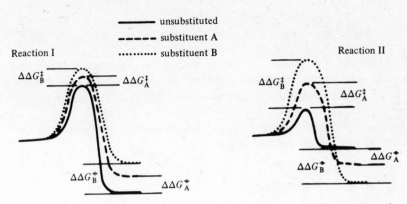

Figure 3.2. Hypothetical reaction profiles to illustrate the Hammett equation.

Therefore

$$\Delta\Delta G_A = (\sigma_A/\sigma_B) \cdot \Delta\Delta G_B \qquad [3.2]$$

which is, as the expression says, a linear relation between Gibbs energy terms.

Figure 3.2 represents schematically the significance of the Hammett equation. Here are represented two reactions which are affected very differently by substituents. Reaction I is slowed somewhat by substituent A, rather more by B; its equilibrium constant is also reduced by both substituents. The substituents have a much greater influence on the rates of reaction II, and they increase the equilibrium constant. The Hammett equation asserts that, however different the reactions we are comparing, the ratio $\Delta\Delta G_A/\Delta\Delta G_B$ remains constant. In the diagrams the ratio is shown as about 0.5, so $\sigma_B$ is about 2 $\sigma_A$, and this ratio applies to kinetic and thermodynamic processes, and it applies whether the process is favoured or hindered by the substituents.

Equation [3.2] shows that we can determine a ratio between two substituent constants but not absolute values. To do this we need to define a further arbitrary point of reference. (The first is the choice of hydrogen as a standard substituent.) Hammett chose the ionisation of substituted benzoic acids in aqueous solution at 25 °C; for these equilibria he defined $\rho$ as unity:

$$\log\left\{\frac{K_a(X)}{K_a(H)}\right\} = \sigma = \Delta pK_a \qquad [3.3]$$

The Hammett $\sigma$ constant is therefore simply the difference between the $pK_a$ values of benzoic acid itself and the substituted benzoic acid. Values of $\sigma$ for various substituents are given in table 3.2 at the end of this chapter.

For any other reaction the appropriate $\rho$ value is obtained from the Hammett plot, as exemplified in figure 3.1.

### 3.1.2 Sigma and rho values

Benzoic acid ionisation is encouraged by electron-withdrawing substituents and hindered by electron donation. Since a strong acid has a small $pK_a$ value this means that an electron-withdrawing substituent has a positive $\sigma$ and an electron-donating one has a negative $\sigma$. (This is opposite to the qualitative sign convention used in describing M and I substituent effects.) The magnitude of $\sigma$ is a measure of the strength of the electron-withdrawing or donating ability of the substituent.

If the signs of $\rho$ and $\sigma$ are the same then $\log(k/k_0)$ is positive and $\mathbf{k}$ is greater than $\mathbf{k_0}$. Therefore a positive $\rho$ value indicates that the reaction is accelerated or the equilibrium favoured by electron withdrawal; a negative $\rho$ shows the opposite. The magnitude of $\rho$ shows the sensitivity of the reaction towards substituent changes. In the hypothetical reactions represented in figure 3.2 the magnitude of $\rho$ for the rate of reaction II is greater than for reaction I, and the sign of $\rho$ for the equilibrium in reaction II is opposite to that for the other three processes.

The value of $\rho$ can be used to measure the efficiency with which various groups transmit electronic effects. For the ionisation of some side-chain carboxylic acids the $\rho$ values are:

| | | |
|---|---|---|
| $X-C_6H_4-COOH$ | 1.00 | $X-C_6H_4-CH_2CH_2-COOH$ 0.21 |
| $X-C_6H_4-CH_2-COOH$ 0.49 | | $X-C_6H_4-CH=CH-COOH$ 0.47 |

These values may be used as a measure of the attenuation of electronic effects along the side-chains: a single $CH_2$ group is similar in its effect to the two-carbon $CH=CH$ fragment for example. Including various other reaction sequences and taking average values we can assess the attenuation factors of various groups as: $CH_2$, 0.43; $CH=CH$, 0.48; $C\equiv C$, 0.39; $p\text{-}C_6H_4$, 0.24.

The magnitude of $\rho$ is markedly dependent on solvent. For the ionisation of benzoic acids in ethanol the value is 1.96. The difference from the value in water can be ascribed to the difference in solvation. Ethanol is less able to stabilise the benzoate anion by solvation (§5.1.2), so the substituents have a greater role to play. If they are electron-withdrawing they can help to satisfy the demand of the ion for charge-dispersal; if they are electron-donating the carboxylate group suffers accordingly because it is less able to seek help from the solvent.

The value of $\rho$ can frequently be of help in mechanistic studies. Consider the saponification of ethyl benzoates (figure 3.1):

$$\underset{HO^-}{Ar-C\overset{O}{\underset{OEt}{\diagup}}} \rightleftharpoons Ar-\overset{O^-}{\underset{OH}{\underset{|}{C}}}-OEt \longrightarrow Ar-C\overset{O}{\underset{OH}{\diagup}}$$

The large positive value of $\rho$ shows that the reaction is markedly accelerated by electron-withdrawing substituents. This implies that the first step is rate-limiting: electron withdrawal facilitates nucleophilic attack but it would retard the expulsion of ethoxide.

### 3.1.3 Non-linear Hammett plots

For some reactions a plot of $\log(k/k_0)$ against $\sigma$ fails to give a straight line. This implies a breakdown in one or other of Hammett's fundamental postulates; either $\rho$ is being influenced by the substituent or $\sigma$ by the reaction. It is not always easy to distinguish the two either conceptually or empirically, but two cases are fairly clear-cut.

(a) *Mechanistic change: variation in $\rho$.* In a two-step consecutive sequence the overall reaction rate is limited by the slower of the two steps:

$$\text{Reactants} \rightarrow \text{Intermediate} \rightarrow \text{Products}$$

Conversely, if there are two alternative routes the dominant one, that on which the overall rate largely depends, is the faster:

With either of these mechanistic patterns it is possible for substituent changes to invert the relative rates of the two steps, leading to characteristic non-linear Hammett plots.

The formation of benzaldehyde semicarbazones is a two-stage reaction:

$$\underset{(1)}{ArCHO} \rightleftharpoons \underset{(2)}{\overset{H^+}{ArCHOH-NHX}} \rightarrow ArCH=NX$$

$$(X = NHCONH_2)$$

The second step is acid-catalysed dehydration; in strongly acidic media this is fast and step 1 is rate-limiting. This involves nucleophilic attack on the carbonyl group and, like the ethyl benzoate saponification, it is accelerated by electron withdrawal; $\rho$ is about $+1$. In neutral solution step 2 is slower and becomes rate-limiting. The overall rate is now proportional to $K_1 \cdot k_2$ and these two terms have about equal and opposite values of $\rho$ so the rate is virtually independent of substituents and the observed $\rho$ value is zero.

At an intermediate acidity, around pH 4, one can observe the transition from one mechanism to the other. Electron-withdrawing groups (positive $\sigma$) accelerate step 1 and leave step 2 rate-limiting with $\rho = 0$; with electron-donating substituents step 1 is retarded and becomes rate-limiting so that $\rho = 1$. The resulting Hammett plot is convex upwards (figure 3.3) [60JA1773].

log $(k/k_0)$

Figure 3.3. Hammett plot for benzaldehyde semicarbazone formation at pH 4.

The other type of non-linearity is exemplified by the hydrolysis of ethyl benzoates in 99.9% sulphuric acid [60P84] (§12.1.3). The reaction proceeds via protonated intermediates that can decompose by two alternative routes.

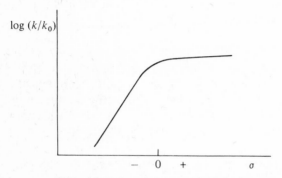

In general, acylium ions, $ArCO^+$, are less unstable than the highly energetic ethyl cation and the upper route is preferred. However, very

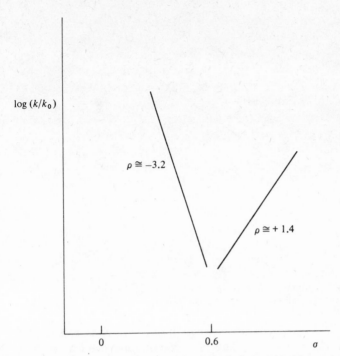

Figure 3.4. Hammett plot for the hydrolysis of ethyl benzoates in 99.9 per cent $H_2SO_4$.

strongly electron-withdrawing substituents destabilise the acylium ion to the extent that the lower route takes over. This process is actually accelerated by electron withdrawal in the Ar group, because this facilitates the expulsion of the ethyl cation. The resulting Hammett plot (figure 3.4) is convex downwards.

(*b*) *Through-conjugation: variation in σ*. Hammett himself pointed out a discrepancy in some reactions of anilines and phenols [37JA96]. For example, a plot for the ionisation of anilines ($pK_{BH^+}$) shows a generally good correlation but with some points markedly off the line (figure 3.5). Thus the σ value for the nitro group is 0.78, but to fit it to the straight line in figure 3.5 would require a value of 1.27.

This is not really surprising. In *p*-nitroaniline there is a direct mesomeric interaction between the $NH_2$ and $NO_2$ groups (**3.1**). This is not possible in the anilinium ion; on the contrary, there is a mutual repulsion between the two substituents [81PPO1], so *p*-nitroaniline is a much weaker base than aniline. Moreover, this differential mesomerism is absent from the reference reaction, the ionisation of benzoic acids. That

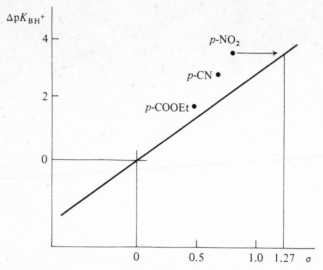

Figure 3.5. Ionisation of anilines in water at 25 °C. The straight line ($\rho = 2.77$) is determined by all *meta* substituents and *para* electron donors.

is, there is a substituent effect in one reaction series which is absent from another, so the Hammett postulate of a reaction-independent $\sigma$ must break down.

Notice the difference between this enhanced *through-conjugation* from substituent to reaction site and the normal resonance interaction between a nitro group and the benzene ring which is an intrinsic property of the group and is accounted for in its normal $\sigma$ value (**3.2**).

$$
\begin{array}{cc}
(3.1) & (3.2)
\end{array}
$$

To accommodate such through-conjugation a series of special $\sigma$ values called $\sigma^-$ ($\sigma^*$ in early literature) was drawn up. In general $\sigma^-$ differs from $\sigma$ only for *para* substituents that can mesomerically withdraw electrons; some typical values of $\sigma^-$ are listed in table 3.2.

By definition $\sigma^-$ was to be used for the ionisation of anilines and phenols, but it was found that many other reactions involving through-conjugation correlate better with $\sigma^-$ than with $\sigma$; nucleophilic aromatic substitution is an example (chapter 13):

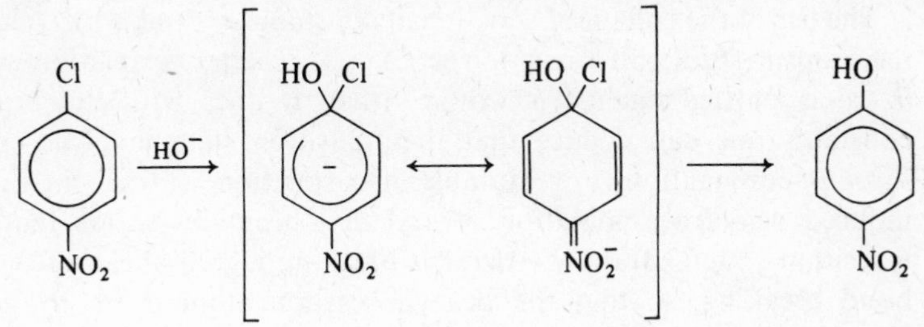

Analogous to $\sigma^-$ is $\sigma^+$ which is used to correlate reactions during which there is a change in through-conjugation between an electron-demanding reation site and a para resonance-donor such as MeO or $NH_2$. The reaction used to define $\sigma^+$ is the $S_N1$ solvolysis of 2-arylprop-2-yl chlorides in 90 per cent acetone at 25 °C [58JA4979]:

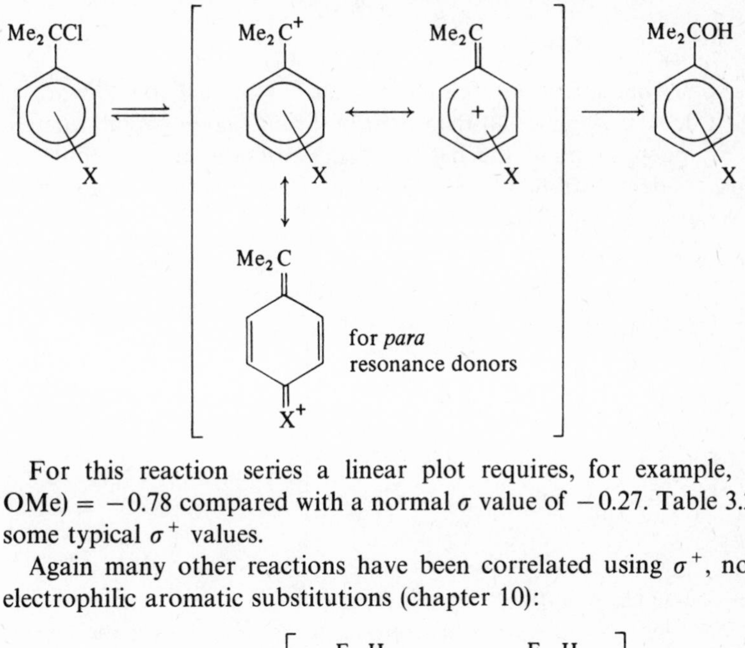

For this reaction series a linear plot requires, for example, $\sigma^+(p\text{-OMe}) = -0.78$ compared with a normal $\sigma$ value of $-0.27$. Table 3.2 lists some typical $\sigma^+$ values.

Again many other reactions have been correlated using $\sigma^+$, notably electrophilic aromatic substitutions (chapter 10):

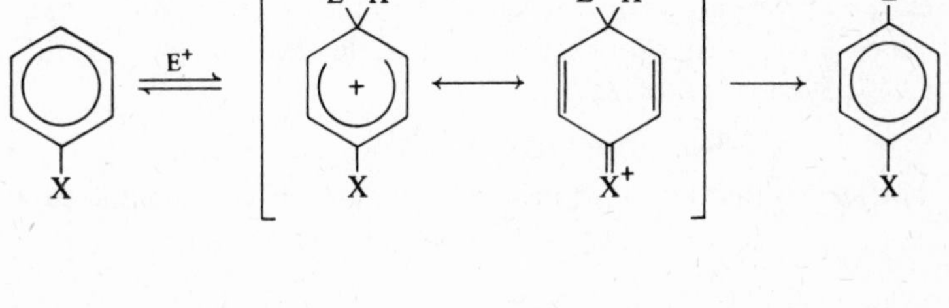

In these reactions $(k_X/k_0)$ is the partial rate factor $f$, so that

$$\log f_m = \rho\sigma_m \qquad \log f_p = \rho\sigma_p^+$$

The use of the enhanced substituent constants $\sigma^-$ and $\sigma^+$ is valuable in mechanistic studies. If a reaction series gives a better correlation when log $(k/k_0)$ is plotted against $\sigma^-$ or $\sigma^+$ than it does with the original $\sigma$ constants one can deduce that the transition state involves extensive through-conjugation. For example, a correlation with $\sigma^-$ in the base-induced dehydrobromination of arylethyl bromides shows that in the transition state (**3.3**) the C—H bond breaking is well ahead of the C—Br bond breaking, so that the benzylic carbon atom bears considerable negative charge, which is susceptible to through-conjugation with $-M$ groups in the *para* position [57JA3712]:

(**3.3**)

It is somewhat surprising that the $\sigma, \sigma^-, \sigma^+$ approach should work as well as it does; it implies that the extent of through-conjugation, where it exists, is independent of the nature of the reaction. It takes but one example to demonstrate that this is false. The standard reaction itself involves some through-conjugation:

But of course by definition this follows $\sigma$, not $\sigma^+$.

In fact one would expect a continuum of $\sigma$ values between the extremes, varying with the demands of the reaction. A number of sliding scales have been proposed, for example the Yukawa–Tsuno equation [3.4] [59BJ971; 66BJ2274]:

$$\log(k/k_0) = \rho\{\sigma + r(\sigma^+ - \sigma)\} \qquad [3.4]$$

The idea is that $(\sigma^+ - \sigma)$ measures the extent to which the substituent is capable of resonance donation and $r$, which varies with the reaction as does $\rho$, measures the demand placed on this resonance by the reaction. If $r$ is zero the correlation is simply with $\sigma$; if it is unity the correlation is with $\sigma^+$.

One might expect $r$ to be more or less proportional to $\rho$ since both relate to the demand placed by the reaction site on the substituent. This is found for a variety of $S_E$Ar reactions of biphenyl [63APO35]:

$\Rightarrow S_E\,Ar$

The biphenyl rings are normally twisted. The more demand a reaction places on electron donation from the biphenyl $\pi$ system, the more the rings respond by moving towards coplanarity. The result is that a plot of $\log f$ against $\rho$ gives a parabolic curve, as predicted by eqn 3.4 with $r$ and $\rho$ proportional to each other. This idea is confirmed by a comparison with the behaviour of fluorene in the same reactions; now that the rings are forced to be coplanar all the time $\log f$ correlates linearly with $\rho$.

In other systems, however, analyses of data by the Yukawa–Tsuno equation (or the corresponding $\sigma^-$ version [66J(B)842]) give values of $r$ which appear to bear no relation to $\rho$ and one is tempted to doubt their validity; the introduction of an extra independently adjustable parameter is bound to improve the correlation so in some cases $r$ may be little more than a statistical artefact.

Several attempts have been made to define new $\sigma$ values which exclude any through-conjugation contribution. Taft devised the $\sigma^0$ scale using 'insulated' reaction series, namely the ionisation of $ArCH_2COOH$ and $ArCH_2CH_2COOH$ and the saponification of $ArCH_2COOEt$ and $ArCH_2OCOEt$ [60JPC1805]. His procedure was:

(a) To plot $\Delta pK_a$ or $\log (k/k_0)$ against $\sigma$ using only *meta* substituents to exclude any possibility of through-conjugation.

(b) To derive $\rho$ from these plots.

(c) To determine statistically the best values of $\sigma^0$ for other substituents so as to give the best fit to the four straight lines.

The resulting $\sigma^0$ values should be more 'fundamental' than the original $\sigma$ but they are less easy to measure and the low $\rho$ values which are consequent on the introduction of the insulating group in the side-chain lead to reduced precision. Some values of $\sigma^0$ are given in table 3.2. It can be seen that the substituents for which $\sigma^0$ differs markedly from $\sigma$ are the strong resonance-donors ($NMe_2$, OMe, etc.) when they are in the *para* position; in other words, the groups for which through-conjugation in the benzoic acid is significant.

### 3.2    Polar and mesomeric effects

Even when through-conjugation is separated out, as in the Taft procedure, there still remains a variety of mechanisms whereby a substituent can exert its influence on a reaction [71CEd427]. We can divide these into two main groups:

(*a*) *Mesomeric effects* (usually symbolised by *R* for 'resonance'), which involve a direct transfer of charge between ring and substituent via the $\pi$ system (compare (**3.2**)); hyperconjugative interactions are included.

(*b*) *Polar effects*, in which the dipolarity of the C–X bond modifies the electrical behaviour of the rest of the molecule. Polar effects in turn can be considered in two main categories. The *inductive effect*, *I*, operates through the $\sigma$ bonded framework. It depends on the electronegativity of the atom X which generates a partial charge on the adjacent carbon atom; this in turn polarises the next carbon–carbon bond, and so on:

$$\overset{\delta-}{X}\text{——}\overset{\delta+}{C}\text{——}\overset{\delta\delta+}{C}\text{——}\overset{\delta\delta\delta+}{C}-$$

For a long time it was widely assumed that this through-bond effect was dominant; indeed, the total polar effect was usually referred to as the inductive effect. It is now clear that the main contributor is the *field effect*, *F*, a direct through-space polarisation of the reaction site by the substituent dipole [68JA336; 72CEd400; 72JA3080; 77JO534; 80J(P2)985; 81JA39; 82CC273]. The clearest evidence for this assertion comes from comparisons between compounds in which substituents are separated from the reaction site by the same number of bonds (implying similar inductive effects) but which differ in their orientation in space. The field effect is proportional to the cosine of the angle between the reaction site and the substituent dipole [38JCP513]:

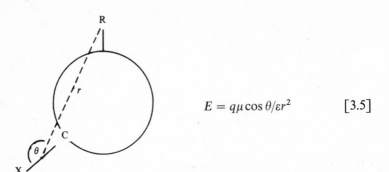

$$E = q\mu\cos\theta/\varepsilon r^2 \qquad\qquad [3.5]$$

This expression gives the energy of interaction between a reaction site bearing a charge $q$ (as in the removal of a proton from an acid) and a substituent with electric dipole moment $\mu$: $\varepsilon$ is the effective dielectric constant (relative permittivity) of the molecular cavity between R and C−X, a quantity which has been, and to some extent remains, a matter of contention [82JA4793; see also 82JO2318].

The p$K_a$ values of the brominated cyclophane acids (**3.4**) may be compared with the value of 7.59 for the unsubstituted acid. Through-bond inductive interactions should be identical for the pseudo-*ortho* and

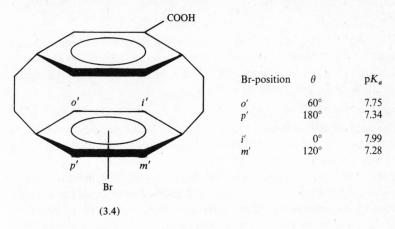

| Br-position | $\theta$ | p$K_a$ |
|---|---|---|
| $o'$ | 60° | 7.75 |
| $p'$ | 180° | 7.34 |
| $i'$ | 0° | 7.99 |
| $m'$ | 120° | 7.28 |

(3.4)

pseudo-*para* isomers, and for the pseudo-*meta* and pseudo-*ipso*, and in each case the electron-withdrawing bromine atom should increase the acid strength relative to the parent compound. The experimental values of p$K_a$ do not agree with this model. They show a strong angular dependence, and in those isomers for which $\theta < 90°$ the bromine sub-stituent actually reduces the acid strength [77CC608].

It is of considerable interest to make quantitative estimates of the separate polar and mesomeric effects of substituents; not only does such an analysis give us further insight into the nature of the substituent groups, but it can also help unravel our understanding of complicated reaction mechanisms. (See §13.2.1 for a particularly good example.) The methods of carrying out this quantitative separation, however, are controversial.

### 3.2.1 The extended Hammett approach

The polar effect of a substituent can be measured by investigat-ing its behaviour in resonance-free systems. Particularly important in this context are studies of 4-substituted bicyclo [2,2,2] octanecarboxylic acids

[53JA2167; 64JA5188; 67JA5677; 68JA336] in which the geometry is closely similar to that of *para* substituted benzoic acids but where there is no $\pi$ system for mesomeric interactions (**3.5**).

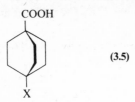

(**3.5**)

The $pK_a$ values of these acids measured in 50 w/w per cent aqueous ethanol at 25 °C have been used to define a set of substituent constants originally called $\sigma'$ but now usually called $\sigma_I$ ($I$ for inductive, but actually encompassing all polar effects):

$$\Delta pK_a = 1.56\,\sigma_I \qquad [3.6]$$

The coefficient 1.56 is used to put $\sigma_I$ on a scale as closely comparable to Hammett's $\sigma$ as possible [81PPO119]. (Slightly different values were used in earlier work.)

Values of $pK_a$ are available for only a few bicyclooctane acids. However, it has been shown that the ionisation of substituted acetic acids, $XCH_2COOH$, in water at 25 °C correlates excellently with $\sigma_I$ and they have therefore been used to extend the range of $\sigma_I$ values [81PPO119]:

$$\sigma_I = 0.246\,\Delta pK_a \qquad [3.7]$$

Values of $\sigma_I$ are given in table 3.3 at the end of this chapter. Notice that the table does not give separate values of $\sigma_I$ for *meta* and *para* substituents. There have been several attempts, both theoretical and empirical, to assess the relative magnitudes of *meta* and *para* polar effects; all conclude that in general they are not very different because the contributions of the $\cos\theta$ and $1/r^2$ terms in equation [3.5] largely cancel each other, but quantitative estimates do not even agree which is the larger.

One cannot devise a reaction series complementary to that of the bicyclooctane acids, with only mesomeric interactions and no polar ones, so the mesomeric contribution to substituent effects must be obtained indirectly. The observed effect of a substituent is the sum of the polar and mesomeric contributions; thus one may write:

$$\sigma_p = \sigma_I + \sigma_R$$
$$\sigma_m = \sigma_I + \alpha\sigma_R \qquad [3.8]$$

This assumes that the polar effect has the same force at the *meta* and *para* positions. The mesomeric effect, on the other hand, is felt fully at the *para* position, but attenuated ($\alpha < 1$) in transmission to the *meta* position (**3.6**).

$$(3.6)$$

*meta* position experiences secondary effect, induced by charges at neighbouring atoms

Direct mesomeric effect at *para* position

With $\sigma_m$, $\sigma_p$ and $\sigma_I$ previously determined, the two equations [3.8] are sufficient to obtain values of $\sigma_R$ and $\alpha$.

One can use this method to separate out the polar contributions from any type of combined substituent constant: $\sigma$, $\sigma^-$, $\sigma^+$, or $\sigma^0$. Values of $\sigma_R^0$ (table 3.3) may be taken as a measure of the mesomeric interaction between the substituent and the ring, uncontaminated by contributions from through-conjugation.

This approach to separating polar and mesomeric effects is rooted in Hammett's original equation [3.1]. A substituent influences reactions in a characteristic way that can be summarised in a single parameter, $\sigma$ (treating *meta* and *para* substituents as separate entities). This $\sigma$ value is applicable to all reactions except those involving through-conjugation, and its separation into polar and mesomeric components according to equations [3.8] does not detract from the essential simplicity of the analysis of experimental results as illustrated in figure 3.1. For this reason the approach is often referred to as the single substituent parameter method.

### 3.2.2 Multiple substituent parameter equations

An alternative treatment suggests that different reactions may respond differently to polar effects on the one hand and mesomeric effects on the other. The examples of through-conjugation could be taken as exemplifying this. Such a line of thought leads to an equation of the type:

$$\log(k/k_0) = \rho_I\sigma_I + \rho_R\sigma_R \qquad [3.9]$$

in which $\rho_I$ and $\rho_R$ are independent reaction constants, measuring the sensitivity of the reaction to the different classes of substituent effect. There are no separate $\sigma$ values for *meta* and *para* substituents: the differences in reactivity of *meta* and *para* isomers are taken to be due to differences in the way $I$ and $R$ effects are transmitted to the reaction site and are accommodated by different $\rho$ values. In effect, the *meta* and *para* reactions are treated as belonging to different reaction series.

The interpretation of experimental results by equation [3.9] is no longer a simple graphical procedure; it requires multiple regression analysis to determine the best values of $\rho_I$ and $\rho_R$. In the most widely used of these dual parameter equations, developed by Taft and his coworkers [73PPO1], the values of $\sigma_I$ are derived as before from direct measurements on resonance-free systems, but the $\sigma_R$ values cannot be obtained simply by subtraction, as in equations [3.9], because that would imply a fixed ratio of $\rho_I$ to $\rho_R$. Two different approaches have been used instead.

At first Taft analysed a number of reaction series statistically to obtain $\sigma_R$ values that gave the best overall fit. The disadvantage of this is that as new results are collected the whole set of $\sigma_R$ values changes. The alternative is to select a particular reference system (as Hammett chose the ionisation of benzoic acids) and use that to define $\sigma_R$. Taft has recently used $^{13}$C n.m.r. chemical shifts to derive $\sigma_R^0$ values (that is, $\sigma_R$ values applicable to systems in which there is no through-conjugation) from the equation: $\Delta\delta_p^C = 4.0\,\sigma_I + 19.8\,\sigma_R^0$, where $\Delta\delta_p^C$ is the difference between the chemical shift of the *para* carbon of a monosubstituted benzene and that of benzene itself [79JO4766].

A second derivation of $\sigma_R^0$, due to Katritzky and Topsom [77CR639], is based on the intensity, $A$, of the infrared band that appears at about $1600\,\mathrm{cm^{-1}}$ in benzene derivatives. This relates to $\sigma_R^0$ by the equation:

$$A = 17\,600\,(\sigma_R^0)^2 + 100$$

In this vibration the ring moves through stretched and squashed extremes. If there is resonance interaction between substituent and ring this will be accentuated in the squashed extreme:

The change in dipole moment during this vibration is thus exaggerated and the intensity enhanced. The correlation is with the square of $\sigma_R^0$ because the intensity is enhanced by both electron donors and attractors.

In addition to $\sigma_R^0$ values, Taft has found it necessary to use other $\sigma_R$ values in equation [3.9], depending on the nature and extent of through-conjugation: $\sigma_R^-$, $\sigma_R^+$, and, for reactions in which through-conjugation resembles that in benzoic acid, $\sigma_{R(BA)}$. Experimental results are then analysed using whichever of the $\sigma_R$ sets gives the best correlation.

The essential difference between the single and dual parameter approaches can be seen by rewriting equation [3.9] as:

$$\log(k/k_0) = \rho(\sigma_I + \lambda\sigma_R) \qquad [3.10]$$

where $\lambda \equiv \rho_I/\rho_R$. The single parameter equations [3.8] imply that for *para* substituents the polar and resonance effects contribute equally to all reactions ($\lambda = 1$) and that the transmission of the secondary resonance effect to the *meta* position is the same in all reactions as it is in the ionisation of benzoic acids ($\lambda = \alpha$). The only flexibility is the use of enhanced resonance parameters to allow for through-conjugation. In the dual parameter approach $\lambda$ can vary freely from one reaction to another, and within a particular reaction series there can be different $\rho$ and $\lambda$ values for *meta* and *para* interactions.

The extended Hammett approach and Taft's dual substituent parameter equation are both based partly on experiment and partly on intuition. A fairly simple model is developed to describe the way substituent effects might operate, and substituent constants are then derived from systems which, there is reason to think should reflect some part of that model. The bicyclooctane acids, for example, are assumed to respond to polar effects, and the infrared intensities are assumed to measure $(\sigma_R^0)^2$ because they give values comparable to those from other sources that fit the same model.

A quite different approach, following a suggestion by Jaffé [53CR(53)191], is to abandon a preconceived model and to subject a large number of experimentally reliable sets of data to statistical analysis to discover those substituent parameters that best fit the whole range. (Taft's statistical derivation of $\sigma_R$ was a small step in this direction.) A survey of 570 different values of $\log(k/k_0)$ embracing 76 reaction series and 17 substituents concluded that three independent parameters are needed fully to characterise the effect of each substituent [79J(P2)537]. The experimental results could all be described by the triple substituent parameter equation:

$$\log(k/k_0) = \rho_I\sigma_I + \rho_R\sigma_R + \rho_E\sigma_E \qquad [3.11]$$

In this expression $\sigma_I$ has roughly the same significance as before, and $\sigma_R$ is comparable to Taft's $\sigma_R^0$, although the numerical values do not exactly coincide. The third parameter, $\sigma_E$, does not appear to have any simple physical significance but it is presumably related to interactions between the substituent and the reaction site because $\rho_E$ turns out to be negative for reactions that correlate with $\sigma^-$, zero for $\sigma^0$, and positive for $\sigma^+$. This type of analysis discards the untidy proliferation of $\sigma_R$ values that Taft had to use, but it requires an enormous amount of experimental information to produce sets of parameters for only a small number of substituents.

### 3.2.3 Interpreting polar and mesomeric parameters

The relative merits of the different approaches to separating polar and mesomeric parameters are hotly disputed. Taft and his co-workers argue that their analysis is most in accord with the actual mode of action of the substituents and therefore allows realistic interpretation of chemical behaviour; a good illustration of this is the decomposition of diazonium salts (§13.2.1) where apparently anomalous substituent behaviour is shown to accord with a mechanism in which there is an unusual blend of polar and mesomeric effects working in opposite directions.

On the other side are those who fight to retain the essential simplicity of Hammett's original idea. They accept that a single parameter equation cannot correlate all reactions, but they draw attention to the wide range of reactions that can be correlated without invoking dual substituent parameters (see, for example, [81J(P2)409]) and maintain that over-sophistication is unjustified in the light of the theoretical approximations and experimental inaccuracies that inevitably accompany structure–reactivity correlations; if enough extra parameters are introduced correlations are improved, but the resulting numerical values, as with $\sigma_E$, may have little apparent chemical significance.

Both schools of thought, however, make use of separate polar and resonance parameters in one form or another to interpret the qualitative behaviour of substituents. Thus substituents like $Me_2N$, $MeO$ and $F$ show polar electron withdrawal (positive $\sigma_I$) and mesomeric donation (negative $\sigma_R$). Their net effect depends on the relative magnitudes of the two effects and on the value of $\alpha$ (equation [3.8]) or $\lambda$ (equation [3.10]); the methoxyl group, for example, is usually an electron donor to the *para* position and an acceptor from the *meta* position. The nitro group is strongly electron-withdrawing ($\sigma_I$ and $\sigma_R$ both positive) but its predominant influence is polar, because of the large positive charge on nitrogen, and the mesomeric interaction, though significant, is less important.

There have been prolonged controversies about the polar and mesomeric (hyperconjugative) effects of alkyl groups. It was believed for many years that hyperconjugative electron donation decreased along the series methyl, ethyl, isopropyl, t-butyl; in other words, a C–H bond was thought to be a better $\pi$ donor than C–C. This is now known to be false. There is little difference in the true $\pi$-releasing powers of different alkyl groups. The apparent effect was a consequence of steric hindrance to solvation [59T(5)194].

Hyperconjugation is the major contributor to the electron-releasing properties of alkyl groups, but the argument about the residual polar effect is unresolved. Taft, on the basis of $\sigma_I$ values derived from ester hydrolyses (§3.3), asserts that there is a small but regular increase in electron donation in the series methyl, ethyl, isopropyl, t-butyl (that is, in the opposite sense to the purported hyperconjugative effect). Other workers believe that Taft's values do not accurately measure polar effects but include contributions from steric interactions (especially in the way they perturb solvation), and that the polar effect of all alkyl groups is approximately zero [64PPO323; 77JA5687; 79JO903; 80JA7988]. Taft has shown that even in the gas phase there is a regular increase in the basicities of aliphatic amines, $RNH_2$, and alcohols, $ROH$, as R is changed from methyl to t-butyl [77JO916; 78JA7765]. Part of this effect, however, arises from the increasing *polarisability* of the larger groups (the ease with which their electron clouds can be distorted). This can function in either direction, facilitating both electron donation and electron withdrawal depending on the demands of the reaction site, as is shown by the fact that the gas phase *acidity* of the alcohols also increases along the series from methanol to t-butanol [78JA7765].

### 3.2.4 The Dewar–Grisdale equation

The Hammett equation and its extensions are all empirical. There have been several attempts to produce a non-empirical analysis of substituent effects, notably by Dewar and Grisdale [62JA3539, 3548]. They also separate $\sigma$ into field and mesomeric effects ($F$ and $M$) but they endeavour to analyse their relative contributions theoretically.

$F$, being predominantly a through-space effect, was taken to vary as $1/r$, where $r$ is the distance between the points of attachment of the substituent and reactive side-chain measured in units of benzene $C–C$ bond lengths and assuming regular hexagonal geometry. (Contrast the $1/r^2$ term of equation [3.5].)

$M$ varies as $q$, the charge delocalisability of the system. This is the magnitude of the charge produced at the point of attachment of the side-chain by replacing the substituent by $CH_2^+$ (or $CH_2^-$); it is calculated from NBMO coefficients (§6.2.2). Thus:

$$\sigma = F/r + Mq$$

Whence for simple benzene derivatives:

$$\sigma_m = F/\sqrt{3}$$
$$\sigma_p = F/2 + M/7 \qquad [3.12]$$

From these equations the values of $F$ and $M$ can be determined for individual substituents. The significance of the method is that it can readily be extended to polycyclic systems. For example, taking the methoxyl group:

$$\sigma_m = 0.12 \quad \text{therefore} \quad F = 0.21$$
$$\sigma_p = -0.27 \quad \text{therefore} \quad M = -2.62$$

For reactions at position 1 of X-substituted naphthalenes (ignoring positions 2 and 8 in which steric interactions interfere) the values of $r$ and $q$ and the $\sigma$ values calculated from them are:

| $X =$ | 3 | 4 | 5 | 6 | 7 |
|---|---|---|---|---|---|
| $r$ | $\sqrt{3}$ | 2 | $\sqrt{7}$ | 3 | $\sqrt{7}$ |
| $q$ | 0 | 1/5 | 1/20 | 0 | 1/17 |
| $\sigma$(OMe) | 0.12 | $-0.42$ | $-0.05$ | 0.07 | $-0.07$ |

Correlations using these values are remarkably good, considering the simplicity of the approach and the approximations involved. By the same method $\sigma^+$ and $\sigma^-$ values may be resolved into $F$ and $M$ components and have been used to investigate $S_E$Ar and $S_N$Ar reactions in polycyclic systems. Typical values of $F$ and $M$ are given in table 3.3.

### 3.2.5 Theoretical studies of substituent effects

Dewar and Grisdale used a theoretical approach to analyse experimentally determined substituent parameters. Since that time the tecniques of theoretical chemistry, aided by powerful computer programs, have advanced dramatically and it is now possible to make fairly accurate calculations of the energies of interaction between substituents and reaction sites, though often without an accompanying conceptual model to give us a mental picture of the nature of the interaction. Such calculations, for example, suggest that the stabilising interaction between the groups in $p$-nitroaniline amounts to 9.2 kJ mol$^{-1}$ whereas the destabilising interaction in the $p$-nitroanilinium cation is 52.7 kJ mol$^{-1}$; it is the latter effect, therefore, that is largely responsible for the base-weakening influence of the nitro substituent [81PPO1]. The two interactions together (61.9 kJ mol$^{-1}$) correspond to a difference of 10.9 between the $pK_{BH^+}$ values of aniline and $p$-nitroaniline, but the calculations relate to isolated molecules, uninfluenced by solvation: the $pK_{BH^+}$ difference in water (in which $\sigma^-$ values were determined) is only 3.6. The basicity of $p$-nitroaniline has not yet been measured in the gas phase, but where comparisons are available between experimental gas phase measurements of substituent effects and theoretical calculations there is generally good agreement [80JO818, 1056; 81PPO1]. As our understanding of the processes of solvation improves (compare §5.1.4) the importance of such theoretical calculations in interpreting substituent effects in normal solution reactions will undoubtedly increase.

### 3.3 Steric effects: the Taft equations

In aliphatic systems substituents are usually closer to the reaction site than in *meta* and *para* substituted benzenes and steric effects cannot be ignored. Taft set out to quantify these [52JA2729, 3210; 53JA4231; 56MI]. Following a much earlier suggestion of Ingold

[30J1032] he compared the acid and base catalysed hydrolyses of aliphatic carboxylic esters. The respective transition states resemble:

For acid catalysis

$$\left[ \begin{array}{c} OH \\ | \\ R-C-OR' \\ | \\ OH_2 \end{array} \right]^+$$

For base catalysis

$$\left[ \begin{array}{c} O \\ | \\ R-C-OR' \\ | \\ OH \end{array} \right]^-$$

Taft postulated first that steric interactions in the two reactions are similar since the transition states differ by only two protons. (This is debatable; solvation effects could be markedly different.) He also supposed resonance effects to be negligible. If these two assertions are accepted it follows that any difference between substituent effects in the acid and base catalysed reactions must be attributed to their polar effects, which clearly influence the oppositely charged transition states very differently.

Thus he wrote:

$$\log (k/k_0)_B - \log (k/k_0)_A = \rho^* \sigma^*$$

where $k$ is the hydrolytic rate coefficient for a particular ester RCOOR', $k_0$ the value for the corresponding acetate (R = $CH_3$), B and A refer to base and acid catalysis, and $\sigma^*$ is a substituent constant relating to the polar effect of R relative to the standard $CH_3$. To define an absolute scale of $\sigma^*$ values Taft set $\rho^* = 2.48$ which gave values of $\sigma^*$ comparable to Hammett $\sigma$ values.

Later he found that $\sigma^*$ for R = $XCH_2$ correlated well with $\sigma_I$ for X as measured by the $pK_a$ values of bicyclooctanecarboxylic acids and used the expression $\sigma_I(X) = 0.45 \sigma^*(CH_2X)$ to measure new $\sigma_I$ values.

The next step in Taft's argument was to note that substituent effects in the acid catalysed hydrolyses of *meta* and *para* substitued benzoate esters are very small; for example, the $\rho$ value in 60 per cent aqueous ethanol at 100 °C is only 0.14. The rate-limiting step is the nucleophilic attack of a water molecule on the conjugate acid of the ester. This is accelerated by electron withdrawal but the same effect reduces the concentration of the conjugate acid; presumably the two effects are finely balanced as in semicarbazone formation (§3.1.3(*a*)).

It seems likely that electronic effects will likewise be insignificant in acid catalysed ester hydrolyses in aliphatic systems, so that any substituent effect that is observed must be steric in origin. Taft therefore proposed his second equation:

$$\log (k/k_0)_A = E_s \qquad [3.13]$$

where $E_s$ is a steric substituent constant: a positive value of $E_s$ implies that the group R is smaller than the reference $CH_3$ group, a negative value that it is larger. Qualitatively the values of $E_s$ do accord with our intuitive ideas of size, and for a number of groups they correlate well with van der Waals radii [69JA615]. A revised set of $E_s$ values, adjusted to make hydrogen rather than methyl the reference group, is given in table 3.4.

A number of aliphatic reactions have been correlated by the relation: $\log(k/k_0) = \rho^*\sigma^*$, which implies that the substituents are exerting their influence through polar effects alone, steric effects being minimal. Other reactions correlate with:

$$\log(k/k_0) = \rho^*\sigma^* + \rho_s E_s$$

where $\rho_s$ is a reaction constant measuring the susceptibility of the reaction to steric effects (see, for example, §11.1).

Taft's analysis is the basis of most current work on the separation of steric and electrical substituent effects, though it has been subject to criticism and modification. His $\sigma^*$ values are now little used; $\sigma_I$ is generally used in preference. The $E_s$ parameter, however, remains the most widely used measure of steric effects, and attempts to derive alternatives appear to offer no advantage [80JO5166].

Taft has also used the same $\sigma^*/E_s$ treatment for *ortho* substituted aromatic systems, assuming that it can overcome the steric problems that forced Hammett to confine his attention to *meta* and *para* systems. His standard reaction was the hydrolysis (acid and base catalysed) of *o*-toluate esters. Using the same equations as for aliphatic systems he derived the corresponding substituent constants. However, whereas most commentators agree with Taft that, despite the rather sweeping assumptions he makes, his analysis of aliphatic systems is broadly speaking justified by the results, few would extend this to *ortho* substituted aromatics. Several other attempts to quantify *ortho* substituent effects have been made, but there is only poor agreement amongst the different methods. It is almost certainly unreasonable to expect to find a simple widely applicable $\sigma_o$ scale. The balance between field, inductive, mesomeric, steric and solvent effects is complex. Some excellent correlations have been made with equations containing three or more independently adjustable parameters, but the line between physical significance and statistical coincidence becomes a little difficult to discern in such cases [71PPO235; 72MIc; 76PPO49; 80J(P2)1350].

### 3.4     Linear Gibbs energy relations and transition state studies

For a kinetic correlation the reaction constant $\rho$ is related to the structure of the transition state (or more strictly to the difference between ground and transition states), reflecting the extent to which charge builds up at the reaction site and demands a response from a substituent. The fact that linear Gibbs energy relations *are* linear implies that $\rho$ is truly a constant, characteristic of the reaction in question, and that therefore the transition state has an invariant structure, independent of substituent changes.

This conclusion is diametrically opposed to a long-established principle of physical organic chemistry, commonly referred to as the *Hammond postulate*. This asserts, in its simplest form, that in an endothermic reaction step the transition state tends to resemble the products, and in an exothermic one it tends to resemble the reactants. The implication is that in a series of related reactions the structure of the transition state depends on $\Delta H^{\ddagger}$ (figure 3.6), and therefore to a reasonable approximation on $\Delta G^{\ddagger}$.

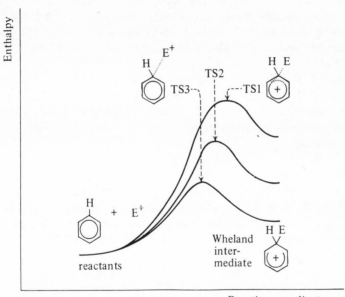

Figure 3.6. The Hammond postulate. For the three related reaction steps (differing, for example, in the nature of the electrophile $E^{+}$) the transition states TS1, TS2, TS3 become progressively more reactant-like as the enthalpy differences $\Delta H_1^{\ominus}$, $\Delta H_2^{\ominus}$, $\Delta H_3^{\ominus}$ (and therefore $\Delta H_1^{\ddagger}$, $\Delta H_2^{\ddagger}$, $\Delta H_3^{\ddagger}$) decrease.

Using electrophilic aromatic substitution (chapter 10) as an example, we could envisage a reaction which has a high energy transition state (TS1), closely similar to the Wheland intermediate, with extensive charge delocalisation, very sensitive to the electrical effects of substituents, so that $|\rho|$ is large. Another reaction may have a lower energy, more reactant-like transition state (TS3); much less charge is delocalised into the ring so little demand is placed on substituents and $|\rho|$ is small. There is indeed such a mechanistic spectrum: for bromination in acetic acid the $\rho$ value is $-12$; for Friedel–Crafts ethylation with $EtBr/GaBr_3$ it is $-2.4$.

This behaviour is sometimes called the *reactivity–selectivity principle*. The reactive ethylating agent is relatively unselective towards different substrates as shown by the low $|\rho|$ value; toluene, for example, is attacked only some two and a half times faster than benzene. The much less reactive bromine is much more selective, attacking toluene some 600 times faster than benzene. Applying the argument in reverse, we deduce that a fast reaction has a more reactant-like transition state than has a slow one.

Hammond's original discussion [55JA334] dealt with the gross structure of a transition state. It has been widely interpreted, however, as meaning that a small change in reactants which increases their reactivity will consequently decrease their selectivity and vice versa. In this sense it is sometimes called the Hammond–Leffler postulate [53SC(117)340; 77APO(14)69].

Herein lies the dilemma. Hammett analyses are used to investigate the nature of the transition state (as measured by $\rho$) by varying the substituents and hence by varying the reactivity. Hammond asserts that the inevitable consequence of changing the rate is to change the nature of the transition state.

The debate is also linked to a number of controversies about the changeover between two different reaction mechanisms. For example (§7.3), is the borderline between $S_N1$ and $S_N2$ reactions best represented as a single mechanism with a gradually changing transition state or as a region in which two distinct mechanisms, each characterised by a particular transition state structure, are running in parallel?

The problem remains unresolved. The evidence is ambiguous. Hammett plots certainly *are* linear; the saponification of ethyl benzoate (figure 3.1) covers a reactivity range of 4000:1 and ranges of $10^6:1$ are not uncommon. Deviations from linearity are often explicable in terms other than a changing transition state, and where there is an obvious change in the transition state it is usually because there is a fundamental mechanistic change. There do seem to be reactions in which the reactivity–selectivity principle breaks down: bromine in trifluoroacetic acid is about a million times more reactive an electrophile than it is in acetic acid, yet in electrophilic aromatic substitutions the selectivities are virtually identical. If anything it is the more reactive system which has the higher $\rho$ value.

On the other hand there are many reaction series that are difficult to rationalise without invoking the concept of a variable transition state whose structure responds to any change in the system; compare for example the discussion of the E2 transition state in §8.1.3. It is hard to see how a transition state that is demonstrably altered by relatively small changes in the reagent (as revealed by changes in $\rho$) can fail to be influenced by similar changes in the substrate.

Part of the difficulty lies in the need most of us have for conceptual pictures of phenomena that we cannot directly observe. The Hammett equation and the Hammond–Leffler postulate are statements relating two sets of experimental observations: for the former the relation is between $\log(k/k_0)$ for the system under investigation and $\log(K_a/K_{a0})$ for benzoic acid ionisations; for the latter it is between reactivity, $\log(k/k_0)$, and selectivity, $d\log(k/k_0)/d\log(K/K_0)$. When we seek to delve beneath the surface of these macroscopic observations and to rationalise them in molecular terms we need to construct mental models; we may then forget that they are only models and treat them as if they, rather than the results they were designed to illuminate, are the reality. (The analysis of electrical substituent effects (§3.2.3) presents the same problem. The customary models envisage two major types of effect, polar and mesomeric: when an analysis of the experimental observations calls for three (equation [3.11]) the models cannot cope.)

The implication is not that we should refrain from using models, but that we should not become wedded irrevocably to one in particular: we should be prepared to use different models for different situations.

It is also helpful to develop models that emphasise the experimentally observable quantities rather than their supposed molecular basis. A step in this direction was taken by Lewis and More O'Ferrall [81J(P2)1084], who have pointed out that when pushed to extremes the correlation between rates and equilibria cannot be linear because $\Delta G^{\ddagger}$ can be neither negative nor less than $\Delta G^{\ominus}$ (figure 3.7). Their analysis is closely related to the Marcus theory relating rates of proton transfer to acid dissociation

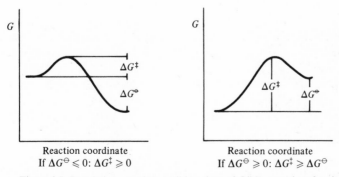

Figure 3.7. Restrictions on the possible values of Gibbs energies of activation.

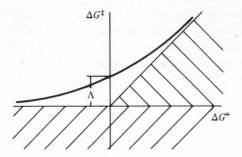

Figure 3.8. The hatched areas cover the inaccessible values of $\Delta G^{\ddagger}$ as shown by figure 3.7. No linear correlation between $\Delta G^{\ddagger}$ and $\Delta G^{\ominus}$ can avoid these areas. The curve shows a possible hyperbolic correlation. The quantity $\Lambda$ is the 'intrinsic' Gibbs energy of activation for the reaction, that is, the value of $\Delta G^{\ddagger}$ when $\Delta G^{\ominus} = 0$ (compare figure 4.7).

equilibria (§4.2.3). Figure 3.8 shows that a possible consequence of this in a one-step process is a *hyperbolic* rather than linear Gibbs energy relation. The slope of the curve is a measure of selectivity. On the left, $\Delta G^{\ddagger}$ is small (high reactivity) but it changes only slowly with $\Delta G^{\ominus}$ (low selectivity). On the right reactivity is low but selectivity is high because $\Delta G^{\ddagger}$ responds fully to changes in $\Delta G^{\ominus}$. In practice many reactions lie well to the right of the curve, the rate-limiting step having a large positive $\Delta G^{\ominus}$, and they also commonly involve large values of $\Lambda$. Under these conditions the observed relation between $\Delta G^{\ddagger}$ and $\Delta G^{\ominus}$ approaches linearity (and it is not possible to distinguish a linear plot from a nearly linear one on the basis of experimental results that are bound to contain some random fluctuations), and there is little change in selectivity over a wide range of reactivity.

## 3.5 Tables of substituent constants

Table 3.2. *Hammett substituent constants, $\sigma_m$ and $\sigma_p$, and modified values,* $\sigma_p^-$, $\sigma_p^+$ *and* $\sigma_p^0$

| Substituent | $\sigma_m$ | $\sigma_p$ | $\sigma_p^-$ | $\sigma_p^+$ | $\sigma_p^0$ |
|---|---|---|---|---|---|
| Me | −0.07 | −0.17 | | −0.31 | −0.12 |
| Et | −0.07 | −0.15 | | −0.30 | −0.13 |
| i-Pr | −0.07 | −0.15 | | −0.28 | −0.16 |
| t-Bu | −0.10 | −0.20 | | −0.26 | −0.17 |
| $CH_2Ph$ | −0.08 | −0.09 | | | |
| $CH{=}CH_2$ | 0.05 | −0.02 | | | |
| $C{\equiv}CH$ | 0.21 | 0.23 | | | |
| Ph | 0.06 | −0.01 | | −0.18 | 0.04 |
| CHO | 0.35 | 0.42 | 1.13 | | |
| COMe | 0.38 | 0.50 | 0.87 | | 0.46 |
| COPh | 0.34 | 0.43 | | | |
| $CO_2^-$ | −0.10 | 0.00 | | | |
| COOH | 0.37 | 0.45 | 0.73 | | |
| COOEt | 0.37 | 0.45 | 0.64 | | 0.46 |
| CN | 0.56 | 0.66 | 1.00 | | 0.69 |
| $CF_3$ | 0.43 | 0.54 | (0.74) | | (0.53) |
| $CCl_3$ | 0.32 | 0.33 | | | |
| $NH_2$ | −0.16 | −0.66 | | −1.3 | −0.38 |
| $NMe_2$ | −0.15 | −0.83 | | −1.7 | −0.44 |
| NHCOMe | 0.21 | 0.00 | | −0.6 | (0.03) |
| $NO_2$ | 0.71 | 0.78 | 1.27 | | 0.82 |
| $N_2^+$ | 1.76 | 1.91 | 3.2 | | |
| $NH_3^+$ | 0.86 | 0.60 | | | |
| $NMe_3^+$ | 0.88 | 0.82 | | | (0.80) |
| $O^-$ | −0.47 | −0.81 | | | |
| OH | 0.12 | −0.37 | | −0.92 | −0.16 |
| OMe | 0.12 | −0.27 | | −0.78 | −0.10 |
| OPh | 0.25 | −0.03 | | −0.5 | |
| OCOMe | 0.39 | 0.31 | | | |
| $OSO_2Ph$ | 0.36 | 0.33 | | | |
| SH | 0.25 | 0.15 | | | |
| SMe | 0.15 | 0.00 | | −0.60 | 0.08 |
| SOMe | 0.52 | 0.49 | 0.73 | | |
| $SO_2Me$ | 0.60 | 0.72 | (1.05) | | (0.69) |
| $SMe_2^+$ | 1.00 | 0.90 | (1.16) | | |
| F | 0.34 | 0.06 | | −0.07 | 0.17 |
| Cl | 0.37 | 0.23 | | 0.11 | 0.27 |
| Br | 0.39 | 0.23 | | 0.15 | 0.26 |
| I | 0.35 | 0.18 | | 0.14 | 0.27 |

[$\sigma_m$ and $\sigma_p$ values are taken from 73Med1207; $\sigma_p^-$ mainly from 53CR(53)191 but values in parentheses are from 51JA2181, 56JA87, 57JA717; $\sigma_p^+$ from 58JA4979; $\sigma_p^0$ mainly from 60JPC1805, 66BJ2274, but values in parentheses are from 59JA5352, 59R815.]

Table 3.3. *Polar and mesomeric substituent constants*

| Substituent | $\sigma_I$ | | $\sigma_R^0$ | $F$ | $M$ |
|---|---|---|---|---|---|
| Me | −0.01 | (−0.04) | −0.10 | −0.12 | −0.77 |
| Et | −0.01 | (−0.05) | −0.10 | −0.12 | −0.77 |
| i-Pr | 0.01 | (−0.06) | −0.12 | −0.12 | −0.77 |
| t-Bu | −0.01 | (−0.07) | −0.13 | −0.17 | −0.79 |
| $CH_2Ph$ | 0.03 | | −0.10* | −0.14 | −0.08 |
| $CH=CH_2$ | 0.11 | | −0.05* | 0.09 | −0.44 |
| $C\equiv CH$ | 0.29 | | −0.09 | 0.36 | 0.34 |
| Ph | 0.12 | | −0.10 | 0.10 | −0.43 |
| CHO | | | 0.24 | 0.61 | 0.82 |
| COMe | 0.30 | | 0.22 | 0.66 | 1.20 |
| COPh | | | 0.19 | 0.59 | 0.95 |
| $CO_2^-$ | −0.19 | | | −0.17 | 0.61 |
| COOH | 0.30 | | 0.29 | 0.64 | 0.91 |
| COOEt | 0.30 | | 0.18 | 0.64 | 0.91 |
| CN | 0.57 | | 0.09 | 0.97 | 1.23 |
| $CF_3$ | 0.40 | | 0.10 | 0.74 | 1.17 |
| $CCl_3$ | 0.36 | | 0.07 | 0.55 | 0.37 |
| $NH_2$ | 0.17 | | −0.47 | −0.28 | −3.65 |
| $NMe_2$ | 0.17 | | −0.53 | −0.26 | −4.90 |
| NHCOMe | 0.28 | | −0.42 | 0.36 | −1.27 |
| $NO_2$ | 0.67 | | 0.17 | 1.23 | 1.16 |
| $NH_3^+$ | | | | 1.49 | −1.01 |
| $NMe_3^+$ | 1.07 | | −0.14 | 1.52 | 0.41 |
| $O^-$ | | | −0.59 | −0.81 | −2.82 |
| OH | 0.24 | | −0.40 | 0.21 | −3.32 |
| OMe | 0.30 | | −0.43 | 0.21 | −2.62 |
| OCOMe | 0.38 | | −0.24 | 0.68 | −0.19 |
| SH | 0.27 | | −0.19 | 0.43 | −0.47 |
| SMe | 0.30 | | −0.25 | 0.26 | −0.91 |
| SOMe | | | −0.07 | 0.90 | 0.28 |
| $SO_2Me$ | 0.59 | | 0.06 | 1.04 | 1.40 |
| $SMe_2^+$ | 0.90 | | −0.09 | 1.73 | 0.24 |
| F | 0.54 | | −0.34 | 0.59 | −1.64 |
| Cl | 0.47 | | −0.22 | 0.64 | −0.63 |
| Br | 0.47 | | −0.23 | 0.68 | −0.75 |
| I | 0.40 | | −0.22 | 0.61 | −0.86 |

[$\sigma_I$ values are taken from 81PPO119 except that the additional values in parentheses, illustrating Taft's interpretation of the polar effects of alkyl groups (§3.2.3), are from 78JA7765. $\sigma_R^0$ values, obtained by infrared intensity method (§3.2.2), are from 77CR639; in this method the sign of $\sigma_R^0$ for the values marked with an asterisk is ambiguous and the signs shown here are those derived from other sources. Dewar–Grisdale $F$ and $M$ values are calculated from the data of table 3.2 using equations 3.12.]

Table 3.4. *Steric substituent parameters*

| Substituent | $-E_S$ | Substituent | $-E_S$ |
|---|---|---|---|
| $CH_3$ | 1.24 | Ph (depth) | 1.01 |
| | | Ph (width) | 3.82 |
| $CH_2Me$ | 1.31 | CN | 0.51 |
| $CH_2CMe_3$ | 2.98 | | |
| $CH_2Ph$ | 1.6 | $NH_2$ | 0.61 |
| $CH_2F$ | 1.48 | $NO_2$ (depth) | 1.01 |
| $CH_2Cl$ | 1.48 | $NO_2$ (width) | 2.52 |
| $CH_2Br$ | 1.51 | $NMe_3^+$ | 2.84 |
| $CH_2I$ | 1.61 | | |
| | | OH/OMe | 0.55 |
| $CHMe_2$ | 1.71 | SH/SMe | 1.07 |
| $CHPh_2$ | 2.7 | F | 0.46 |
| $CHF_2$ | 1.91 | Cl | 0.97 |
| $CHCl_2$ | 2.78 | Br | 1.16 |
| $CHBr_2$ | 3.1 | I | 1.4 |

$E_S$ values are taken from [76PPO91]. In this compilation the values have been adjusted from Taft's original scale so as to set $E_S$ for hydrogen at zero. For a modified list of $E_S$ parameters see [78T3553].

| | |
|---|---|
| $CMe_3$ | 2.78 |
| $CEt_3$ | 5.04 |
| $CPh_3$ | 5.92 |
| $CF_3$ | 2.4 |
| $CCl_3$ | 3.3 |
| $CBr_3$ | 3.67 |

## 3.6 Further reading

M. Charton, *Progr. Phys. Org. Chem.*, 1982, **13**, 119.

J. Hine, *Structural Effects on Equilibria in Organic Chemistry*, Wiley, New York, 1975.

C. D. Johnson, *The Hammett Equation*, Cambridge University Press, London, 1973.

N. B. Chapman & J. Shorter (eds), *Advances in Linear Free Energy Relationships*, Plenum, London, 1972; *Correlation Analysis in Chemistry*, Plenum, New York, 1978.

R. D. Topsom, *Acc. Chem. Res.*, 1983, **16**, 292.

S. Ehrenson, *Progr. Phys. Org. Chem.*, 1964, **2**, 195.

C. D. Ritchie & W. F. Sager, *Progr. Phys. Org. Chem.*, 1964, **2**, 323.

D. H. McDaniel & H. C. Brown, *J. Org. Chem.*, 1958, **23**, 420.

H. H. Jaffé, *Chem. Rev.*, 1953, **53**, 191.

Polar and resonance effects:

J. Shorter, *Correlation Analysis of Organic Reactivity*, Wiley, Chichester, 1982.

W. F. Reynolds, *Progr. Phys. Org. Chem.*, 1983, **14**, 165.

R. D. Topsom, *Progr. Phys. Org. Chem.*, 1976, **12**, 1.

S. Ehrenson, R. T. C. Brownlee & R. W. Taft, *Progr. Org. Chem.*, 1973, **10**, 1.

L. M. Stock, *J. Chem. Educ.*, 1972, **49**, 400.

A. R. Katritzky & R. D. Topsom, *Angew. Chem. Internat. Ed.*, 1970, **9**, 87.

## Steric effects and *ortho* substituted aromatics:

R. J. Gallo, *Progr. Phys. Org. Chem.*, 1983, **14**, 115.
H. Förster & F. Vögtle, *Angew. Chem. Internat. Ed.*, 1977, **16**, 429.
T. Fujita & T. Nishioka, *Progr. Phys. Org. Chem.*, 1976, **12**, 49.
S. H. Unger & C. Hansch, *Progr. Phys. Org. Chem.*, 1976, **12**, 91.
J. Shorter, *Quart. Rev.*, 1970, **24**, 433.
R. W. Taft, in *Steric Effects in Organic Chemistry*, ed. M. S. Newman, p. 556, Wiley, New York, 1956.

## Polar effects of alkyl groups:

L. S. Levitt & H. F. Widing, *Progr. Phys. Org. Chem.*, 1976, **12**, 119.

## Reactivity and selectivity. The Hammett equation and the Hammond postulate:

C. D. Johnson, *Chem. Rev.*, 1975, **75**, 755; *Tetrahedron*, 1980, **36**, 3461.
A. Pross, *Adv. Phys. Org. Chem.*, 1977, **14**, 69.

# 4 Acids and bases

## 4.1 Dilute solutions

### 4.1.1 Acid and base strengths

According to the definitions of Brønsted [23R718] and Lowry [23CI43] an acid is a proton donor and a base a proton acceptor.* Thus an acid–base reaction is simply a proton transfer:

$$HA + B \rightleftharpoons A^- + HB^+$$

Acid   Base

The products of this reaction form a second acid–base pair; $A^-$ is called the conjugate base of HA, and $HB^+$ is the conjugate acid of B. The terminology is used even if the reaction is effectively irreversible; one seldom thinks of molecular hydrogen as an acid, but $H^-$ is certainly a base and so $H_2$ is its conjugate acid. The charges shown in the equation are purely conventional. An acid or base can be neutral or bear positive or negative charges, and many species can behave as an acid in one reaction and a base in another. Thus the hydrogen sulphate anion, $HOSO_3^-$, can accept a proton to form sulphuric acid or donate one to form the sulphate dianion. Likewise the hydrazinium ion, $NH_2NH_3^+$, can generate $N_2H_4$ or $N_2H_6^{2+}$.

The position of an acid–base equilibrium defines the relative strengths of the two acids and two bases present. In the reaction between hydrogen chloride and sodium acetate the equilibrium lies well to the right:

$$HCl + AcO^- \rightleftharpoons Cl^- + HOAc$$

One can regard this as a competition between the two bases, acetate and chloride, for control of the proton; the acetate ion wins and is thus the stronger base. Conversely hydrogen chloride is more successful than acetic acid in shedding its proton and is therefore the stronger acid.

In this manner a series of acid–base equilibria could be used to establish a sequence of acid and base strengths. However, it is more convenient (within the limits of what is practicable; see below) to use

---

* Lewis acids are discussed in chapter 6.

water both as a solvent and as a reference base, and to relate the strengths of acids to the position of the general equilibrium:

$$HA + H_2O \rightleftharpoons H_3O^+ + A^-$$

Solutions are kept sufficiently dilute for activity coefficients to be taken as unity and for $[H_2O]$ to be regarded as constant. The strength of the acid HA is then measured by the acid dissociation constant or by its negative logarithm:

$$K_a = [H_3O^+][A^-]/[HA]$$
$$pK_a \equiv -\log K_a \equiv \log 1/K_a = pH - \log[A^-]/[HA] \qquad [4.1]$$

The hydronium ion concentration $[H_3O^+]$ can be measured with a pH meter. If the species $A^-$ and HA have significantly different u.v. or visible spectra, then the ratio $[A^-]/[HA]$ can be determined spectrophotometrically (figure 4.1). Alternatively it can be calculated from the requirements of charge balance ($[H_3O^+] = [A^-] + [HO^-]$), stoicheiometry ($[HA] + [A^-] = [HA]_{stoi}$), and the ionic product of water ($[H_3O^+][HO^-] = 10^{-14}$ mol$^2$ l$^{-2}$).

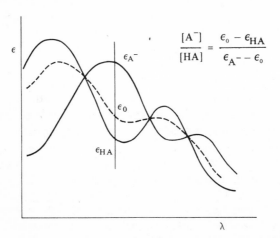

$$\frac{[A^-]}{[HA]} = \frac{\epsilon_0 - \epsilon_{HA}}{\epsilon_{A^-} - \epsilon_0}$$

Figure 4.1. Spectrophotometric determination of $[A^-]/[HA]$. The solid lines show the spectra of the acid and of the conjugate base, measured at least 2 pH units away from $pK_a$. The dotted line shows the spectrum of the equilibrium mixture at an intermediate pH. The ratio can be determined as shown from the absorption coefficients at any convenient wavelength.

Water itself is, of course, an acid as well as a base, and its $pK_a$ can be determined from its self-ionisation or autoprotolysis.

$$H_2O + H_2O \rightleftharpoons H_3O^+ + HO^-$$
$$K_a = [H_3O^+][HO^-]/[H_2O] = 10^{-14}/55.4 \text{ mol l}^{-1}$$
$$pK_a = 15.74$$

(Note that for comparison with other acids one molecule of water but not both must be included in the equation. The molar concentration of pure water at 25 °C is 55.4.)

Base strengths can be measured by reference to water as a standard acid.

$$B + H_2O \rightleftharpoons BH^+ + HO^-$$

$$K_b = [BH^+][HO^-]/[B]$$

$$pK_b \equiv -\log K_b \equiv \log 1/K_b$$

More commonly, however, the strength of a base is described in terms of the $pK_a$ of its conjugate acid.

$$BH^+ + H_2O \rightleftharpoons H_3O^+ + B$$

$$K_{BH^+} = [H_3O^+][B]/[BH^+]$$

$$= [H_3O^+][HO^-]/K_b$$

$$pK_{BH^+} = 14 - pK_b$$

The quantity here described as $pK_{BH^+}$ is sometimes called simply the $pK_a$ of the base – meaning, of course, that of its conjugate acid; the usage is extraordinarily confusing when discussing the behaviour of species that can be acids or bases and should be shunned.

A large positive value of $pK_a$ or $pK_{BH^+}$ describes a weak acid or a strong base, and a small or negative value describes the opposite. The value of $pK_{BH^+}$ for water ($pK_a$ for $H_3O^+$) is $-1.74$.

### 4.1.2    Solvent levelling

A direct competitive measurement of the relative acid strengths of perchloric acid and hydrogen chloride shows that perchloric acid is much the stronger:

$$HClO_4 + Cl^- \rightleftharpoons HCl + ClO_4^-$$

In aqueous solution, however, the difference between the two is not discernible. Both are so much stronger than the hydronium ion that within measurable limits they are totally ionised:

$$HX + H_2O \longrightarrow H_3O^+ + X^-$$

Water is said to exert a levelling effect on any acid which is stronger than hydronium. Consider the ionisation of a 0.1M aqueous solution of an acid with $pK_a = -2$:

$$K_a = [H_3O^+][A^-]/[HA] = 10^2$$

Charge balance requires that $[H_3O^+] = [A^-]$, since $[HO^-]$ will be utterly insignificant at this level of acidity; stoicheiometry requires $[A^-] + [HA]$ to be 0.1. Whence $[H_3O^+] = 0.0999$ and $[HA] = 0.0001$. Repeating the calculation for an acid a hundredfold stronger, with

$pK_a = -4$, gives for the corresponding concentrations 0.099999 and 0.000001. Within the limits of experimental accuracy both results are equivalent to $0.1 M\ H_3O^+$ with no detectable residue of undissociated HA.

In a solvent more basic than water the cut-off point occurs at weaker acid strengths. Thus in liquid ammonia even carboxylic acids are fully ionised and they therefore appear to be as strong as the mineral acids:

$$AcOH + NH_3 \longrightarrow NH_4^+ + AcO^-$$

Conversely in solvents of low basicity, such as anhydrous formic acid, it is possible to differentiate strong acids; perchloric acid is still fully ionised in this solvent, but hydrogen chloride acts as a weak acid.

At the other end of the scale bases that are stronger than hydroxide are levelled in aqueous solution because they are effectively fully protonated:

$$NH_2^- + H_2O \longrightarrow NH_3 + HO^-$$

If the solvent is more acidic than water it will level weaker bases; phenoxide and hydroxide appear equally strong in anhydrous formic acid, and even the carboxylic acids are strong bases when dissolved in sulphuric acid. Conversely, very weakly acidic solvents such as liquid ammonia can differentiate strong bases.

### 4.1.3    Buffers

Because a strong acid or base is fully ionised in aqueous solution a very small quantity can make an enormous difference to the acidity of a solution. The addition of 0.01 ml ($\simeq 0.0001$ mol) of concentrated hydrochloric acid to a litre of water reduces the pH from 7 to 4 (i.e. $-\log 0.0001$). A buffered solution is one that contains comparable concentrations of a relatively weak acid and of its conjugate base; it is much less susceptible to such pH changes. Rearranging eqn 4.1 gives:

$$pH = pK_a + \log[A^-]/[HA]$$

Thus a solution that contains equal concentrations of sodium acetate and acetic acid will have a pH equal to the $pK_a$ of acetic acid (4.75). Suppose $[A^-] = [HA] = 0.01M$. If we now make the same addition of concentrated hydrochloric acid it will protonate the acetate, reducing $[A^-]$ to 0.0099 and increasing $[HA]$ to 0.0101. The buffer ratio, $[A^-]/[HA]$, is barely changed: the pH alters by only $\log(99/101) = -0.01$.

## 4.2    Acid and base catalysis

A very large number of reactions are catalysed by acids or bases, and not a few are catalysed by both. Ester hydrolysis (chapter 12) is a typical example. The reaction entails an attack by the weakly nucleophilic water molecule on the weakly electrophilic carbonyl group of the ester. In acidic solution some of the ester molecules are protonated on the carbonyl oxygen atom, thus enhancing their electrophilicity. In base there is a significant concentration of hydroxide ion, much more nucleophilic than neutral water. (Strictly speaking the base is not a catalyst in this reaction since it is consumed with the formation of the carboxylate anion; the process is referred to as pseudo-catalysis.)

The way in which an acid or base influences a reaction can be illustrated by a rate profile, a plot of log $k$ against pH (or $H_0$ or other acidity function: see §4.3) as exemplified in figure 4.2. Curve (a) represents

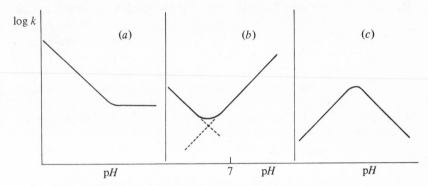

Figure 4.2. Rate profiles for various types of acid or base catalysis.

a reaction that is catalysed by acid but not by base; the horizontal line corresponds to the uncatalysed reaction, or more accurately, to the reaction as catalysed by water. Curve (b) shows a reaction catalysed by acid and by base, the observed solid line being the superimposition of the two straight line plots; in this example base catalysis is rather more effective than acid catalysis because it becomes dominant while the solution is still slightly acidic. Curve (c) is found in two-step reactions when one step is inhibited by acid and the other accelerated. For example, the formation of oximes from ketones with hydroxylamine is retarded by acid because it inactivates the hydroxylamine by protonating it, but the second step of the reaction is an acid catalysed dehydration which decelerates with increasing pH (§11.3). In all these rate profiles the slopes are generally $\pm 1$ since, as discussed below, the rates are usually proportional to $[H_3O^+]$ or $[HO^-]$.

### 4.2.1 *Specific acid or base catalysis*

Proton transfer between electronegative elements is normally a very fast process (but see §11.3.2, §13.1.1). Thus reactions such as acid catalysed ester hydrolysis commonly entail a rapid equilibrium between a reactant R and its conjugate acid, followed by a rate-limiting formation of products P. In aqueous solution this can be represented as:

$$R + H_3O^+ \underset{\text{Fast}}{\overset{K_1}{\rightleftharpoons}} RH^+ + H_2O$$

$$RH^+ \underset{\text{Slow}}{\overset{k_2}{\rightarrow}} P$$

Applying the steady-state approximation to $[RH^+]$ gives:

$$\text{Rate} = K_1 k_2 [R][H_3O^+]$$

If the acid is a true catalyst the proton consumed in the first step is regenerated in the second, and so $[H_3O^+]$ remains constant throughout the reaction and pseudo-order kinetics are obtained. However, the catalysis can be detected by measuring the kinetics in a series of solutions at different pH and finding that the observed pseudo-rate coefficient is proportional to $[H_3O^+]$.

This behaviour is described as *specific acid catalysis*. The reaction rate is unaffected by the nature of any acidic species present and depends specifically on the concentration of $H_3O^+$, or more generally of pro-tonated solvent: $NH_4^+$ for reactions in liquid ammonia, $HC(OH)_2^+$ for those in formic acid, and so on.

*Specific base catalysis* analogously describes reactions in which the rate depends specifically on the concentration of the conjugate base of the solvent: hydroxide ion in aqueous solution. Mechanistically it arises when the reaction is initiated by rapid proton abstraction:

$$RH + HO^- \underset{\text{Fast}}{\overset{K_1}{\rightleftharpoons}} R^- + H_2O$$

$$R^- \underset{\text{Slow}}{\overset{k_2}{\rightarrow}} P$$

$$\text{Rate} = K_1 k_2 [RH][HO^-]$$

### 4.2.2 General acid and base catalysis

Early workers in the field assumed that all acid and base catalysis was controlled only by the pH of the solution but this is not always true. The hydrolysis of ethyl orthoacetate can be carried out under conditions mild enough to exclude further hydrolysis of the ethyl acetate which is produced. The reaction is catalysed by acids but not by bases:

$$CH_3C(OEt)_3 + H_2O \overset{H^+}{\rightarrow} CH_3COOEt + EtOH$$

When the hydrolysis is carried out in a series of *m*-nitrophenoxide buffers the rate is found to vary with the concentration of the buffer even when the pH is kept constant. Since base catalysis has been excluded the only species that can be responsible for this effect is the undissociated *m*-nitrophenol which must be acting as a catalyst in addition to the hydronium ions. The overall reaction rate can thus be expressed as the sum of separate contributions from hydronium ion catalysis, *m*-nitrophenol catalysis, and an 'uncatalysed' reaction which is probably catalysed by undissociated water:

$$\text{Rate} = \{k_1[H_2O] + k_2[H_3O^+] + k_3[m\text{-}NO_2\text{-phenol}]\}[CH_3C(OEt)_3]$$

At 20 °C the values of the separate *catalytic coefficients* $k_1$, $k_2$ and $k_3$ are $10^{-7}$, $2.1 \times 10^4$ and $1.7 \times 10^{-3}$ l mol$^{-1}$ s$^{-1}$, respectively.

This is an example of *general acid catalysis*: independent catalysis by all the acidic species present and not only by the conjugate acid of the solvent. The general rate expression for such reactions is:

$$k_{\text{obs}} = \Sigma k_i[HA_i]$$

Similarly *general base catalysis* is revealed by a rate expression:

$$k_{\text{obs}} = \Sigma k_i[B_i]$$

General acid or base catalysis usually arises when proton transfer occurs in the rate-limiting step. In the hydrolysis of orthoesters this happens because the proton transfer is associated with C–O bond breaking:

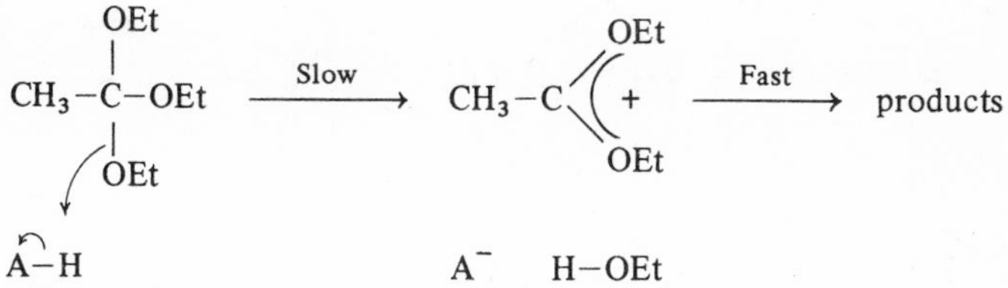

The usual source of general acid or base catalysis is proton transfer to or from a carbon atom; these are normally much slower than proton transfers between two electronegative elements. There are two mechanistically distinct reaction sequences.

(*a*) The C–H bond may be broken in a slow first step, generating a reactive species which decomposes rapidly to products. For example, the formation of an enolate anion, as in $CH_3CHO \rightarrow (CH_2CHO)^-$, is usually slow, leading to a mechanism of the type:

$$RH + B_i \underset{\text{Slow}}{\overset{k_i}{\rightarrow}} R^- + B_iH$$

$$R^- \underset{\text{Fast}}{\rightarrow} P$$

$$\text{Rate} = k_i[RH][B_i]$$

If there are several different bases, $B_i$, then each is independently capable of abstracting a proton from the substrate RH, and each such reaction will have its own characteristic rate coefficient $k_i$. The total reaction rate will be the sum of the rates of these separate reactions. The corresponding mechanism leading to general acid catalysis is:

$$R + HA_i \xrightarrow[\text{Slow}]{k_i} RH^+ + A_i^-$$

$$RH^+ \xrightarrow{} P$$
$$\text{Fast}$$

An example is the hydrolysis of enol ethers, where the rate-limiting step is C-protonation:

$$CH_2{=}CH{-}OEt \xrightarrow[\text{or HA}]{H_3O^+} CH_3{-}CH{=}\overset{+}{O}Et + H_2O \text{ or } A^-$$

(b) The same type of kinetic behaviour results from a mechanism in which a rapid equilibrium protonation is followed by a rate-limiting proton transfer in a subsequent step, as in acid catalysed enolisation:

$$CH_3{-}CHO \underset{\text{Fast}}{\rightleftharpoons} CH_3{-}CH{=}OH^+ \xrightarrow[\text{Slow}]{H_2O \text{ or } A^-} CH_2{=}CH{-}OH$$
$$+ H_3O^+ \text{ or } HA$$

In general terms such a mechanism is:

$$R + H_3O^+ \underset{\text{Fast}}{\overset{K_1}{\rightleftharpoons}} RH^+ + H_2O$$

$$RH^+ + A_i^- \xrightarrow[\text{Slow}]{k_2} P + HA_i$$

$$\text{Rate} = K_1 k_2 [R][H_3O^+][A_i^-]$$

Since the acid dissociation constant of $HA_i$ is

$$K_{a_i} = [H_3O^+][A_i^-]/[HA_i]$$

then

$$\text{Rate} = K_{a_i} K_1 k_2 [R][HA_i] \hspace{2cm} [4.2]$$

This corresponds to general acid catalysis by the species $HA_i$. It seems at first sight rather odd that a reaction in which the rate-limiting step involves attack by a base should be acid catalysed. However, one must remember that the concentration of undissociated $HA_i$ is related both to $[H_3O^+]$ and to $[A_i^-]$; depending on the buffering of the reaction mixture an increase in $[HA_i]$ may lead to an increase in $[H_3O^+]$, accelerating the reaction by increasing $[RH^+]$, or it may increase $[A_i^-]$, accelerating the rate-limiting step. This type of behaviour is sometimes called *general base-specific acid* catalysis, but it should be remembered that it is kinetically indistinguishable from general acid catalysis.

Again there is a corresponding mechanism for general base (*general acid-specific base*) catalysis:

$$RH + HO^- \underset{\text{Fast}}{\overset{K_1}{\rightleftharpoons}} R^- + H_2O$$

$$R^- + HB_i^+ \underset{\text{Slow}}{\overset{k_2}{\rightarrow}} P + B_i$$

$$\text{Rate} = K_{b_i}K_1k_2[RH][B_i]$$

(c) There is a third mechanistic possibility which can generate the kinetics characteristic of general acid catalysis but which may not necessarily involve proton transfer and so it falls outside the strict definition. An initial adduct (usually hydrogen bonded) is formed between substrate and acid, and this decomposes in the rate-limiting step:

$$R + HA_i \underset{\text{Fast}}{\overset{K_1}{\rightleftharpoons}} R\text{---}HA_i \underset{\text{Slow}}{\overset{k_2}{\rightarrow}} P$$

$$\text{Rate} = K_1k_2[R][HA_i]$$

Similar kinetics follow from the formation of an adduct between a substrate and a base, but in most reactions of this type (as, for example, the alkaline hydrolysis of esters: chapter 12) the base is acting as a nucleophile and there is no hint of proton transfer.

### 4.2.3 The Brønsted relations

The behaviour of an acid or base on ionisation (exchanging a proton with solvent) is closely related to its behaviour as a general acid or base catalyst (exchanging a proton with a substrate). The relation was quantified by Brønsted and Pederson [24ZP(108)185] following their development of the principles of general acid and base catalysis. Their equations may be expressed as:

$$\log k_a = \alpha \log K_a + c_a; \qquad \log k_b = -\beta \log K_{BH^+} + c_b$$

In these equations $k_a$ and $k_b$ are the catalytic coefficients for the acid and base catalysed reactions, respectively, $K_a$ and $K_{BH^+}$ are the acid dissociation constants for the catalysts, and $\alpha$, $\beta$ and $c$ are constants dependent on the reaction and its conditions but independent of the catalyst. For example, figure 4.3 shows the correlation between $\log k_a$ and $pK_a$ for the reaction: $CH_3CH(OH)_2 \rightarrow CH_3CHO + H_2O$ in acetone, which has been studied using 47 different carboxylic acid and phenolic catalysts.

The Brønsted equations are historically the earliest linear Gibbs energy relations: $\log k$ is proportional to the Gibbs energy of activation for the catalysed reaction, and $\log K$ to the Gibbs energy of ionisation of the catalyst. The close analogy with the Hammett equation (chapter 3) may be seen by comparing an experiment such as that illustrated in figure 4.3

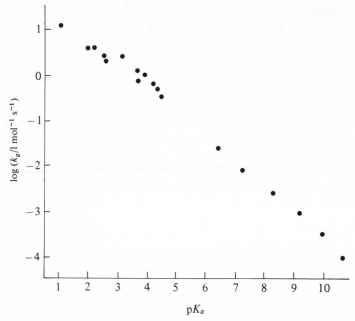

Figure 4.3. Sample points for Brønsted plot of acid catalysed dehydration of $CH_3CH(OH)_2$ in 92.5 per cent aqueous acetone at 25 °C [49PRS(A197)141].

with a Hammett analysis of the same reaction. The catalysts would then have to be restricted to *meta* and *para* substituted benzoic acids, and we should expect to find a linear correlation:

$$\log(k/k_0) = \rho\sigma = \rho \log(K_a/K_{a0})$$

since Hammett's $\sigma$ values are defined in terms of acid dissociation constants. Therefore

$$\log k = \rho \log K_a + (\log k_0 - \rho \log K_{a0})$$

The terms in parentheses are constant for a given reaction, and the equation thus has exactly the same form as Brønsted's, with $\rho = \alpha$.

The mechanistic significance of the Brønsted equations is easier to visualise if it is discussed in Hammett-like terms, comparing the behaviour of a catalyst with that of some arbitrary reference. Consider the reaction profile of a general base catalysed reaction controlled by the rate-limiting transfer of a proton from a reactant, RH, to a catalyst, $B_0$. We could approximate to this profile by superimposing the separate R−H and H−$B_0^+$ stretching curves (figure 4.4). (Bond stretching curves are strictly functions of potential energy, but within the limits of this approximation it is acceptable to use Gibbs energy; compare the discussion in [73MI*a*].)

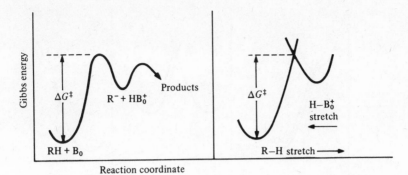

Figure 4.4. Proton transfer reaction, modelled by two overlapping bond stretching profiles.

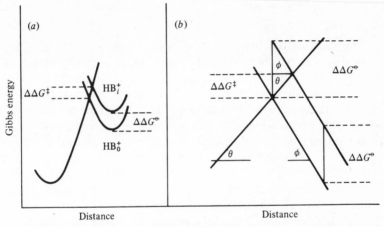

Figure 4.5. The origin of the Brønsted relation.

Now consider the effect of changing the catalyst to a slightly weaker base, $B_i$. The energy of the conjugate acid, $HB_i^+$, will be slightly higher than that of $HB_0^+$, but as long as the two bases are not of grossly different structural types the shapes of the two $H-B^+$ curves will be similar (figure 4.5(a)). The difference between the Gibbs energies of activation for the two reactions, $\Delta\Delta G^\ddagger$, is approximately proportional to the standard Gibbs energy difference, $\Delta\Delta G^\ominus$, between the two protonated bases, as shown by the geometrical construction of figure 4.5(b), and this proportionality is simply a restatement in Gibbs energy terms of the Brønsted equation:

$$\Delta\Delta G^\ddagger = \frac{\tan\theta}{\tan\theta + \tan\phi} \cdot \Delta\Delta G^\ominus$$

$$log\,(k/k_0) = \beta \log(K_b/K_{b0}) = -\beta \log(K_{BH^+}/K_{B_0H^+})$$

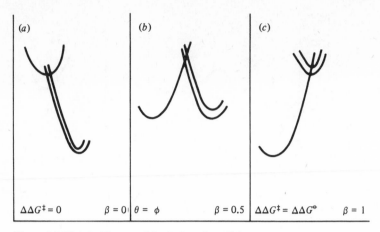

Figure 4.6. The significance of the Brønsted coefficient.

The value of the Brønsted coefficient $\beta$ (and similarly of $\alpha$ for acid catalysis) is thus seen to depend on the slopes of the R–H and H–B$^+$ curves at the point of intersection. Figure 4.6 illustrates the extremes and the midpoint, showing that $\beta$ can be interpreted in terms of the extent of proton transfer at the transition state. If the transition state is very reactant-like (figure 4.6(a)) the proton has barely moved and $\beta = 0$. At the other extreme, (c), the transition state is not reached till the proton is almost fully transferred to the catalyst and $\beta = 1$. When in the transition state the proton is just halfway from reactant to catalyst, (b), $\beta = 0.5$. Brønsted analysis thus provides a convenient parallel to the study of H/D kinetic isotope effects (§2.4).

This analysis is applicable to general acid or base catalysis of the first mechanistic type (§4.2.2(a)) where the reaction is initiated by a rate-limiting proton transfer. If the mechanism involves a rapid pre-equilibrium (§4.2.2(b)) the significance of the Brønsted coefficient is slightly different. For general acid catalysis we have:

$$R + H_3O^+ \underset{\text{Fast}}{\overset{K_1}{\rightleftharpoons}} RH^+ + H_2O$$

$$RH^+ + A^- \underset{\text{Slow}}{\overset{k_2}{\rightarrow}} P + HA$$

The rate equation [4.2] for such a process can be written:

$$k_{\text{obs}} = K_a K_1 k_2$$

or     $\log k_{\text{obs}} = \log K_a + \log K_1 + \log k_2$

Applying the Brønsted equation for base catalysis to the slow step:

$$\log k_2 = -\beta \log K_a + c$$

Therefore

$$\log k_{\text{obs}} = (1 - \beta)\log K_a + \log K_1 + c$$

$K_1$ is independent of the nature of the catalyst HA, so this expression has the form of a normal Brønsted acid catalysis equation with $\alpha = 1 - \beta$. Thus a low value of $\alpha$ arises from a high value of $\beta$ and implies that in the transition state there is considerable proton transfer from $RH^+$. The argument can be applied analogously to the corresponding general base catalysed mechanism.

The approach to the Brønsted relation illustrated in figures 4.5 and 4.6 also reveals the main reasons for departures from it. If the p$K$ range of the catalysts being used is very wide there will be changes in the slopes of the curves at the intersection and $\alpha$ or $\beta$ will not remain constant throughout the whole series of reactions. A plot of $\log k$ against p$K$ will give a curve rather than a straight line. This is indeed found; a hint of curvature can be seen in figure 4.3 where the acidity of the catalysts spans a range of $10^{10}$. In fact it is surprising that these plots show such little curvature; compare the discussion on the conflict between the Hammett and Hammond relations (§3.4). More serious discrepancies arise when the catalysts are of markedly different chemical types [82JO3224]. In this case the assumption that their dissociation curves have similar shapes is no longer valid and the slopes at intersection can be markedly different. If the data of figure 4.3 are extended by including carbon acids such as nitromethane the resulting points lie well off the line.

A more serious anomaly in Brønsted analyses has been the discovery of linear correlations with coefficients outside the range 0–1. The logarithmic rates of deprotonation of $ArCHMeNO_2$ and $ArCH_2CHMeNO_2$ by hydroxide ion in aqueous methanol correlate with their p$K_a$ values with $\alpha = 1.6$ and $1.4$ for the two series of compounds, respectively, whereas in the series $CH_3NO_2$, $MeCH_2NO_2$, $Me_2CHNO_2$ there is an inverse relation between the kinetics and thermodynamics of deprotonation corresponding to a negative value of $\alpha$ [70JA5926].

Such behaviour has been interpreted by Marcus [69JA7224], Kresge [70JA3210] and others in terms of specific effects of a substituent on the transition state, effects which may be quite different from those influencing the ground state. For example, in the transition state for the deprotonation of $ArCHMeNO_2$ (**4.1**) there can be an electrostatic interaction between the aryl group and the partial negative charge on the hydroxide oxygen atom. At either extreme of the reaction the hydroxide

$$\left[ \overset{\delta-}{HO}\cdots H \cdots \overset{\overset{\displaystyle Ar}{|}}{C}Me \overset{}{\cdots} NO_2^{\delta-} \right]^{\ddagger} \qquad (\textbf{4.1})$$

ion or the water molecule it generates is remote from the nitro compound and there can be no such interaction. Differential solvation of the ground and transition states may also be important (compare [77JA7653; 79JA1295]). Marcus analyses this in terms of two independent contributions to the Gibbs energy of activation: a thermodynamic term, $\Delta G^{\ominus}$, which influences $\Delta G^{\ddagger}$ in the manner represented in figures 4.5 and 4.6, and an 'intrinsic' term, $\Lambda (= \Delta G^{\ddagger}$ when $\Delta G^{\ominus} = 0)$, which relates specifically to the transition state. Figure 4.7 shows how this can lead to anomalous Brønsted coefficients.

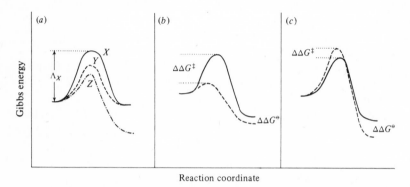

Figure 4.7. Marcus theory applied to anomalous Brønsted coefficients.

(a) The separation of thermodynamic and intrinsic contributions to the Gibbs energy of activation: curve $Y$ shows the effect of reducing $\Lambda$ while maintaining $\Delta G^{\ominus}$ constant (in this example, $\Delta G^{\ominus} = 0$); curve $Z$ shows the effect of reducing $\Delta G^{\ominus}$.

(b) If both terms contribute significantly, and work in the same direction, then $\Delta \Delta G^{\ddagger} > \Delta \Delta G^{\ominus}$ and the Brønsted coefficient is greater than 1.

(c) If both terms contribute significantly, but work in opposite directions, then $\Delta \Delta G^{\ddagger}$ and $\Delta \Delta G^{\ominus}$ have opposite signs and the Brønsted coefficient is negative.

Such behaviour is confined largely to carbon acids. Proton transfers between electronegative elements are usually very fast so that $\Lambda$ is small and the effects of structural changes are dominated by the thermodynamic term. The ionisation of a carbon atom is generally accompanied by considerable structural reorganisation (there is a great deal of C–N double bond character in a nitro anion, for example) and this retards the reaction. It follows that for such systems the value of the Brønsted coefficient may give little guidance towards the nature of the transition state [72JA3907; 74JA7222].

## 4.3 Acidity functions

### 4.3.1 The $H_0$ scale

In dilute aqueous solution the acidity is measured by pH. With a table of pK values we can use equation [4.1] to calculate, for example, that at pH = 7 aniline ($pK_{BH^+}$ = 4.63) is less than 0.5 per cent protonated ([B]/[BH$^+$] = 235), and that phenol ($pK_a$ = 9.89) is only about 0.1 per cent dissociated ([A$^-$]/[HA] = 0.0013).

In concentrated acid solutions equation [4.1] fails because the solutions can no longer be treated as ideal, and equilibrium constants cannot be measured simply in terms of concentrations. Moreover, the equation cannot be applied in solvents other than water. Nevertheless it would be very desirable to have some measure of the acidity of such media: a measure, that is, of the ability of the system to donate a proton to a base according to the equilibrium:

$$B + H^+ \rightleftharpoons BH^+$$

where H$^+$ represents protons in whatever form they occur in the solvent in question. The equilibrium constant needs to be expressed in terms of activities as:

$$K \ ( \equiv 1/K_{BH^+}) = a_{BH^+}/a_B \cdot a_{H^+}$$

The equilibrium constant can be related to the degree of protonation of the base, [BH$^+$]/[B], by introducing the appropriate activity coefficients:

$$K = \left( \frac{1}{a_{H^+}} \cdot \frac{y_{BH^+}}{y_B} \right) \cdot \frac{[BH^+]}{[B]}$$

Expressed logarithmically:

$$pK_{BH^+} = \log\left( \frac{1}{a_{H^+}} \cdot \frac{y_{BH^+}}{y_B} \right) - \log[B]/[BH^+] \qquad [4.3]$$

Comparison with eqn 4.1 shows that the term $\log(y_{BH^+}/a_{H^+} y_B)$ is the equivalent of pH – but how is it to be measured? Hammett and Deyrup [32JA2721] overcame the problem by using a series of indicator bases; that is, bases whose electronic spectra are markedly different from those of their conjugate acids so that the indicator ratio [B]/[BH$^+$] can be measured spectrophotometrically or colorimetrically (compare figure 4.1). This ratio is conveniently given the symbol $I$.

Consider the behaviour of two different indicator bases, X and Y, in solution in sulphuric acid of the same concentration, so that $a_{H^+}$ for the two measurements is the same. The indicators are added in very low concentrations, so their influences on the medium are negligible. In practice they need to have fairly similar pK values so that the indicator ratio for each can simultaneously lie in the accurately measurable range of about 0.1 to 10. Subtracting equation [4.3] for Y from that for X:

$$\Delta pK_{BH^+} = \log(y_{XH^+}/y_X) - \log(y_{YH^+}/y_Y) + \log I_Y - \log I_X$$

Using weakly basic anilines as indicators, Hammett found that plots of $\log I_X$ and $\log I_Y$ against acid concentration were virtually parallel over the whole range for which both could be accurately measured: in other words, $\log I_Y - \log I_X$ is independent of $a_{H^+}$, and therefore so is $\log(y_{XH^+}/y_X) - \log(y_{YH^+}/y_Y)$.

This series of pairwise comparisons of indicators could be extended over a very wide range of acid concentrations, from very dilute with *p*-nitroaniline to 98% sulphuric acid with the extremely weakly basic 2,4,6-trinitroaniline. In dilute acid the activity coefficients approach unity, so the term $\log(y_{XH^+}/y_X) - \log(y_{YH^+}/y_Y)$ must be zero. From the parallel plots it follows that, within the limits of experimental error, it is zero at all concentrations: that is, the ratio $y_{BH^+}/y_B$ depends only on the acidity of the medium and not on the nature of the indicator base. This is not unreasonable; it means that although the actual activity coefficients may change substantially as the medium is altered, the relative behaviour of structurally similar indicators remains the same.

Equation [4.3] can now be expressed as:

$$pK_{BH^+} = -\log h_0 - \log I \qquad [4.4]$$
$$\text{or} \quad pK_{BH^+} = H_0 - \log I$$

where $h_0 = a_{H^+}y_B/y_{BH^+}$ and $H_0 = \log(1/h_0)$. These terms provide the information we seek: they measure the proton-donating ability of the acid medium and are therefore called *acidity functions*.

The pairwise comparison of two indicators at the same acid concentration gives the difference between their $pK_{BH^+}$ values:

$$pK_{XH^+} - pK_{YH^+} = \log I_Y - \log I_X \qquad [4.5]$$

Individual $pK_{BH^+}$ values can thus be found by starting the comparisons with an indicator that is sufficiently basic for its $pK_{BH^+}$ to be measured in dilute aqueous solution. Hammett used *p*-nitroaniline ($pK_{BH^+} = 1.00$). These values can then be substituted into equation [4.4] to construct the $H_0$ scale. The method is illustrated schematically in figure 4.8, and the resulting $H_0$ scale is plotted in figure 4.9.

In dilute aqueous solutions equation [4.4] reduces to the form of [4.1], with $h_0 = [H_3O^+]$ and $H_0 = pH$. In concentrated acid, however, $h_0$ differs markedly from $[H^+]$: in 7.5M (52 per cent) sulphuric acid $[H^+] \simeq 7.5 \text{ mol l}^{-1}$ but $h_0 = 4000$ ($H_0 = -3.6$). This apparently greatly enhanced acidity of concentrated acid solutions is in reality a reflection of the reduced availability of the proton in dilute solution. The hydrated proton is conventionally written $H_3O^+$, but there are actually further water molecules tightly bound in solvation (§5.1.4). In concentrated acid

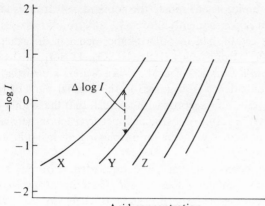

Figure 4.8. Construction of an acidity function. Indicator X is sufficiently basic for $pK_{XH^+}$ to be 'anchored' in dilute acid solution, so that over the range in which this indicator can be used $H_0 = pK_{XH^+} + \log I$ (equation [4.4]). In somewhat more acid solution indicator Y is used; $\Delta \log I$, empirically found to be constant gives $pK_{YH^+}$ from equation [4.5], and thus allows $H_0$ to be determined. Effectively the whole $H_0$ scale is established by moving all the individual indicator curves vertically upwards so that they overlap and generate a continuous line as in figure 4.9.

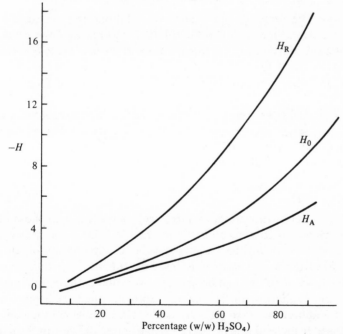

Figure 4.9. Acidity functions in aqueous sulphuric acid. The curves show the variations in three different acidity functions with percentage acid concentration [$H_R$: 55JA3044; $H_0$: 69JA6654; $H_A$: 64CJ1957; 67J(B)1235].

there are too few water molecules to satisfy the demand, solvation is less effective, and the proton is consequently more available to other bases.

The $H_0$ scale can be established in other acidic media in an exactly similar manner. For very strong acids (HBr, $HClO_4$, $H_2SO_4$, etc.) the plots of $H_0$ against *molality* (moles of acid per kg water) are virtually identical, though this coincidence is disguised in the normal plots of $H_0$ against molarity or per cent. This accords with the idea that the acidity of these solutions is largely the result of a stoicheiometric reaction between the proton and water molecules with the anion playing an insignificant role.

In weaker acids the behaviour can be complicated. For a given concentration they all have smaller values of $-H_0$ than the strong acids, attributable to incomplete dissociation. But some of them do reach $H_0$ values that are much more negative than would be expected from their dissociation constants in water. For HF, for example, $pK_a = 3.2$, but for a 10M solution $H_0 = -0.4$. This must arise from the specific behaviour of the fluoride anion, generating such species as $HF_2^-$: in other words, concentrated HF behaves as if it were a solution of the strong acid $H_2F_2$.

### 4.3.2 Other acidity functions

Within the limits of dilute aqueous solutions the pH scale is universal; it indicates the acidity of the solution towards any base which may be present. Unfortunately the same is not true for the $H_0$ scale which applies only to bases for which changes in $y_{BH^+}/y_B$ parallel those for primary anilines. Bases which do behave this way are called Hammett bases.

Structurally dissimilar bases may follow very different protonation behaviour. If a series of amides is used as indicators a self-consistent acidity function (designated $H_A$) can be set up in a manner precisely corresponding to that used for $H_0$ [64CJ1957; 67J(B)1235]. The defining equation is:

$$pK_{BH^+} = H_A - \log I \qquad [4.6]$$

Numerically, however, the values of $-H_A$ are always less than those of $-H_0$ for a given acid concentration (figure 4.9); in 7.5M sulphuric acid $H_A$ is only $-2.6$, showing that sulphuric acid is a less powerful proton donor towards amides than it is towards anilines. The difference undoubtedly arises largely from differences in hydration of the cations and the implication is that amide cations are normally more strongly hydrated than anilinium ions. In strongly acidic solution, where the competition for water molecules is fierce, it becomes more difficult to form a cation which seeks considerable stabilisation from hydration. Amides protonate on oxygen:

$$R-C\overset{\diagup O}{\underset{\diagdown NH_2}{}} \xrightarrow{H^+} \left[ R-C\overset{\diagup OH^+}{\underset{\diagdown NH_2}{}} \longleftrightarrow R-C\overset{\diagup OH}{\underset{\diagdown NH_2^+}{}} \right]$$

This leads to a very different charge distribution from that found in anilinium ions, and to different patterns of hydrogen bonding and ion association. Pyridine $N$-oxides, which likewise protonate on oxygen ($py^+-O^- \rightarrow py^+-OH$), also follow the $H_A$ function and have been used to extend it over a wider range of acidities.

A variety of other acidity functions have been defined in a similar manner using other types of base.

A somewhat different class of acidity function is derived from the ionisation of alcohols [55JA3044]:

$$ROH + H^+ \rightleftharpoons R^+ + H_2O$$

where R is commonly a triarylmethyl group, $Ar_3C$. The acidity function, designated $H_R$ ($C_0$ or $J_0$ in early literature) is defined as:

$$H_R = \log \frac{a_{H_2O} \cdot y_{R^+}}{a_{H^+} \cdot y_{ROH}}$$

or $\qquad H_R = pK_{R^+} - \log[R^+]/[ROH]$

where $K_{R^+}$ is the equilibrium constant for the reverse of the defining equilibrium. Water appears as a product of the ionisation, which is therefore greatly encouraged in strongly acidic water-seeking solutions (figure 4.9).

In principle any of these acidity functions can be set up in any solvent system. However, in media with low dielectric constants, especially nonaqueous solutions, the association of ions in ion-pairs and other aggregates can become dominant. Even small structural differences may make themselves felt, and the value of $y_{BH^+}/y_B$ may not remain constant even over a closely related series of indicators.

### 4.3.3 Relations between acidity functions

How is one to cope with the seeming vagaries of acid–base behaviour in strongly acidic media? A clue can be found in the form of figure 4.9 and similar plots of other acidity functions; the lines fan out from a point, never crossing each other, and maintaining an approximate proportionality. In fact the plot of one acidity function against another over a range of acid concentrations gives a reasonably good straight line except near the origin, and it is not a bad approximation to relate any acidity function to $H_0$ by the expression:

$$H_X = nH_0 + \text{constant} \qquad\qquad [4.7]$$

This is widely used to investigate the protonation behaviour of bases that are not obvious candidates for one of the established acidity functions; one can plot $\log I$ against different functions looking for a slope of unity. For example, for the tautomeric benzotriazole derivative (**4.2**) the slope of the $\log I$ plot against $H_0$ is 1.55 whereas against $H_A$ it is 0.97. This suggests that the compound protonates on oxygen rather than nitrogen and implies that the $N$-oxide tautomer (**4.2a**) predominates [73J(P2)160].

(**4.2a**)                                                   (**4.2b**)

A more rigorous analysis by Bunnett and Olsen [66CJ1899] came from the observation that for a number of acid–base equilibria, regardless of the nature of the base, there exist linear relations of the type:

$$\log \frac{[B][H^+]}{[BH^+]} = n \log \frac{[X][H^+]}{[XH^+]} + c$$

$$\therefore \log I_B + \log [H^+] = n(\log I_X + \log [H^+]) + c$$

Bunnett and Olsen imagined such a relation applying when X is a hypothetical Hammett base that has $pK_{BH^+} = 0$. From equation [4.4] it follows that $\log I_X = H_0$. Therefore:

$$\log I_B + \log [H^+] = n(H_0 + \log [H^+]) + c$$

In dilute acid solutions $H_0 = -\log [H^+]$ and, from equation [4.1], $\log I_B + \log [H^+] = -pK_{BH^+}$. Whence $c = -pK_{BH^+}$. Therefore:

$$\log I_B + \log [H^+] = n(H_0 + \log [H^+]) - pK_{BH^+} \qquad [4.8]$$

which allows $pK_{BH^+}$ to be determined for any base, no matter what its protonation behaviour, from a plot of $(\log I + \log [H^+])$ against $(H_0 + \log [H^+])$.

The Bunnett analysis also suggests that all acid–base equilibria can be described in terms of three parameters: an acidity function to describe the proton-donating power of the solvent ($H_0$ is used above, but in principle any acidity function would do); the basicity parameter, $pK_{BH^+}$; and a sensitivity parameter $n$, analogous to Hammett $\rho$, which depends mainly on solvation (see below), and which measures the way the base responds to a protonating medium. There is clearly a continuum of acid–base behaviour, with $n$ a continuous variable. However, within a series of structurally related bases it is likely that the $n$ values will all be closely similar, giving rise to the established acidity function scales. This can be seen by combining equation [4.8] with the generalised version of equation [4.4] or [4.6] to give:

$$H_B + \log[H^+] = n(H_0 + \log[H^+]) \qquad [4.9]$$

where $H_B$ is an acidity function applying to all bases in the same class as B. Empirically it is found that mutual plots of $(H_B + \log[H^+])$ are indeed more accurately linear than direct plots of one acidity function against another.

Equation [4.9] can be simplified into a relation between activity coefficients, because the term $(H_B + \log[H^+])$ is equivalent to $\log(y_{BH^+}/y_B y_{H^+})$. Therefore:

$$\log(y_{BH^+}/y_B y_{H^+}) = n\log(y_{XH^+}/y_X y_{H^+}) \qquad [4.10]$$

where X is a Hammett base.

By *assuming* that such a linear relation is valid between all bases, Marziano [73J(P2)1915; 77J(P2)306, 309, 845; 81J(P2)1070], followed by Cox and Yates [78JA3861; 81CJ2116], modified the approach of Bunnett and Olsen. Using Cox and Yates' symbolism, equation [4.8] is written as:

$$\log I + \log[H^+] = -m^*X - pK_{BH^+} \qquad [4.11]$$

where $X$ is the activity coefficient function $\log(y_X y_{H^+}/y_{XH^+})$ for an arbitrary reference base. However, unlike the Bunnett and Olsen reference, this does not have any predetermined value of $pK_{BH^+}$, nor does it have to follow $H_0$ or any other specific acidity function. Values of $X$ are determined by analysing experimental values of $\log I$ and $\log[H^+]$ for a variety of bases of different types in acid solutions of varying concentrations. The sensitivity and basicity parameters $m^*$ and $pK_{BH^+}$ are taken to be unknown constants characteristic of each base, and $X$ is allowed to vary with acid concentration but applies equally to all bases. An iterative procedure then gives the values of all three parameters that best fit all the data. The function $X$ is called the 'excess acidity' because it measures the difference between the acidity of a solution and that of an ideal solution of the same concentration (see below).

The advantages claimed for this approach are that it avoids the multiple overlap measurements needed to define conventional acidity functions (figure 4.8) and that there is no difficulty with anchoring the scale in dilute acid solutions because $X$ is, by definition, zero in the standard state. In practice the two approaches have been shown to be effectively equivalent [82JA1958]. As expected from a comparison of equations [4.8] and [4.11] it is found that $-X \simeq H_0 + \log[H^+]$ and that $m^* \simeq n$. Values of $pK_{BH^+}$ determined from equations such as [4.4] or [4.6] do not in general differ greatly from those obtained statistically. The main discrepancies occur with extremely weak bases ($pK_{BH^+} < -8$): assertions that one method or the other gives superior results await substantiation.

The significance of the sensitivity parameter, $n$ or $m^*$, may be seen by considering the protonation of a base B in dilute aqueous solution and as a dilute solution in a mineral acid. The species B, $BH^+$, and $H^+$ are all hydrated, but the average degree of hydration will be less in the acid than in water (§4.3.1).

In water:

$$B.a + H^+.b \rightleftharpoons BH^+.c + (a + b - c)H_2O$$

$$pK_{BH^+} = \log([BH^+]/[B][H^+])$$

where $a$, $b$, and $c$ represent the average number of hydrating water molecules. These do not appear in the expression for $pK_{BH^+}$ because they are deliberately excluded from its definition.

In acid the *difference* between the number of hydrating water molecules and the number allowed for in water must be taken into account:

$$B.x + H^+.y \rightleftharpoons BH^+.z + (x + y - z)H_2O$$

$$K_{BH^+} = \frac{[B][BH^+]}{[BH^+](a_{H_2O})^r}$$

$$pK_{BH^+} = \log([BH^+]/[B][H^+]) + r \log a_{H_2O}$$

where $r = (x + y - z) - (a + b - c)$. The species B, $BH^+$, and $H^+$ are all present in low concentrations and to a first approximation their concentrations are an adequate measure of their activities. For the water $a_{H_2O}$ must be used rather than $[H_2O]$. When $pK_{BH^+}$ values are actually measured, however, the water of hydration is not considered explicitly; the deviation from ideal behaviour, conventionally dealt with by using activities in place of concentrations, is therefore mainly due to the $r \log a_{H_2O}$ term:

$$pK_{BH^+} = \log \frac{y_{BH^+}[BH^+]}{y_B[B] \cdot y_{H^+}[H^+]} = \log \frac{[BH^+]}{[B][H^+]} + \log \frac{y_{BH^+}}{y_B y_{H^+}}$$

$$\therefore \log(y_{BH^+}/y_B y_{H^+}) = r \log a_{H_2O}$$

Comparison with equation [4.10] gives $n = r_B/r_X$, and the sensitivity parameter is seen to be a measure of the difference in hydration between the protonation of B and of X.

### 4.3.4 Acidity functions and reaction rates

When transition state theory (§2.3.3) is applied to non-ideal solutions the pseudo-equilibrium between reactants and transition state must be described in terms of activities not concentrations. The effectiveness of an acid as a catalyst depends not on $[H^+]$ but on $a_{H^+} y_B/y_{\ddagger}$ – a term similar to that which defines $h_0$, but which incorporates $y_{\ddagger}$, the activity coefficient of the transition state, in place of $y_{BH^+}$. Johnson has used such an approach to define $H_c^{\ddagger}$, a kinetic acidity function for the protonation of carbon bases. It was established from the rates of H/D exchange at the *ortho* positions of *p*-nitro-*NN*-dimethylaniline in sulphuric acid, and it correlates the reactions of many aromatic, heteroaromatic, olefinic and acetylenic substrates in acid [79JO745, 753; 82J(P2)1025]. Using the same principle, and assuming a linear relation between $y_{H^+}y_B/y_{\ddagger}$ and $y_{H^+}y_B/y_{BH^+}$ (compare equation [4.10]), Cox and Yates have extended their $X$-function approach to reaction rates [79CJ2944]; an example may be found in §12.2.1.

Bunnett and Olsen [66CJ1917] found that linear correlations similar to those observed for acid–base equilibria also applied to the rates of acid catalysed reactions in strongly acidic solutions:

$$\log k = -n(H_0 + \log[H^+]) + \log k^0$$

Here $k$ is the rate coefficient for the catalysed reaction and $k^0$ is the value extrapolated to infinite dilution (when $H_0 + \log[H^+] = 0$). In practice one normally observes the pseudo-order rate coefficient $k_\psi = k[H^+]$, whence:

$$\log k_\psi - \log[H^+] = -n(H_0 + \log[H^+]) + \log k^0$$

or $\quad \log k_\psi + H_0 = \phi(H_0 + \log[H^+]) + \log k^0 \qquad [4.12]$

where $\phi = 1 - n$. Experimentally one plots $(\log k_\psi + H_0)$ against $(H_0 + \log[H^+])$. As with equilibria the values of $n$, and hence of $\phi$, are related to changes in hydration. By comparing the measured $\phi$ values for a number of reactions whose mechanisms were generally accepted, Bunnett and Olsen suggested that for substrates that are more or less fully protonated during the reaction the following correlations could be made between $\phi$ and the involvement of water in the reaction: if water is a nucleophile in the rate-limiting step ($H_2O + RH^+ \rightarrow H_2O \cdot RH^+$), $\phi$ should lie in the range 0.22 to 0.56; if water acts as a base ($RH^+ + H_2O \rightarrow R + H_3O^+$), then $\phi$ is greater than 0.58; if water is not involved at all ($RH^+ \rightarrow P$), $\phi$ is negative. [For an extension of this treatment, see 77JA3387.]

### 4.3.5 *Strongly basic media*

When Hammett first developed the $H_0$ scale he also envisaged the existence of other functions, which he designated $H_+$ and $H_-$, to correlate the protonation of cationic and anionic bases in the same way that $H_0$ correlated the neutral anilines. Some work on $H_+$ has been carried out using, for example, the second protonation of aminopyridines as the indicator equilibrium. Studies on $H_-$ have been limited by the lack of adequate indicators – organic anions that are such weak bases that they are protonated only at high concentrations of acid. Some cyanocarbons have been found suitable: for $(NC)_2C=C(CN)-CH(CN)_2$

$pK_a = -8.5$ because it ionises to give the highly stabilised $C_3(CN)_5^-$ anion [63JPC737].

However, another aspect of the $H_-$ acidity function is more important. The protonation of very *strongly* basic anions – or in other words the abstraction of a proton from very weak acids – can be used to provide a measure of the basicity of strongly basic media such as aqueous dimethyl sulphoxide containing $Me_4N^+OH^-$. Nitroanilines and nitrodiphenyl-amines have been used as acid indicators (HA), so that the equilibria being studied are:

$$HO^- + HA \rightleftharpoons H_2O + A^-$$

The $H_-$ scale is then established in exactly the same way as other acidity functions, from pairwise comparison of indicators using the equation:

$$pK_a = H_- - \log I$$

and anchoring the scale in dilute aqueous solution where $H_- = pH$ [64CJ1681; 67CJ911; compare also 76JA488].

Figure 4.10 illustrates the way in which $H_-$ changes with the ratio of water to DMSO. Note the dramatic increase in $H_-$ at low water concentrations where the hydroxyl ion is stripped of solvating water molecules.

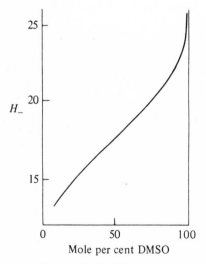

Figure 4.10. The $H_-$ function in aqueous dimethyl sulphoxide; $[Me_4N^+OH^-] = 0.01M$.

## 4.4 Further reading

### General:

R. P. Bell, *The Proton in Chemistry*, 2nd edn, Chapman & Hall, London, 1973.
R. P. Bell, *Quart. Rev.*, 1947, **1**, 113 (historical).
H. L. Finston & A. C. Rychtman, *A New View of Current Acid–Base Theories*, Wiley, New York, 1982.
E. F. Caldin & V. Gold (eds), *Proton-Transfer Reactions*, Chapman & Hall, London, 1975.
M. Eigen, *Angew. Chem. Internat. Ed.*, 1964, **3**, 1.

### p$K$ measurements:

A. Albert & E. P. Serjeant, *Determination of Ionisation Constants*, 2nd edn, Chapman & Hall, London, 1973 (includes p$K$ tables).
R. F. Cookson, *Chem. Rev.*, 1974, **74**, 5.
G. Kortum, W. Vogel & K. Andrussow, *Pure Appl. Chem.*, 1960, **1**, 187 (tables of p$K_a$ values for organic acids).
E. P. Serjeant & B. Dempsey, *Ionisation Constants of Organic Acids in Aqueous Solution*, Pergamon, Oxford, 1979.
D. D. Perrin, *Dissociation Constants of Organic Bases in Aqueous Solution*, Butterworths, London, 1965, and *Supplement* 1972.

### Acid and base catalysis:

R. P. Bell, *Acid–Base Catalysis*, Oxford University Press, Oxford, 1941.
W. P. Jencks, *Chem. Rev.*, 1972, **72**, 705; *Acc. Chem. Res.*, 1976, **9**, 425.
A. V. Willi, in *Comprehensive Chemical Kinetics*, ed. C. H. Bamford & C. F. H. Tipper, vol. 8, Elsevier, Amsterdam, 1977.
K. Yates & T. A. Modro, *Acc. Chem. Res.*, 1978, **11**, 190.
R. Stewart & R. Srinivasan, *Acc. Chem. Res.*, 1978, **11**, 271.
R. P. Bell, The Brønsted equation, in *Correlation Analysis in Chemistry* (N. B. Chapman and J. Shorter, eds), Plenum, New York, 1978.

### Acidity functions:

C. H. Rochester, *Acidity Functions*, Academic, London, 1970.
M. A. Paul & F. A. Long, *Chem. Rev.*, 1957, **57**, 1.
F. A. Long & M. A. Paul, *Chem. Rev.*, 1957, **57**, 935.
K. Yates & R. A. McClelland, *Progr. Phys. Org. Chem.*, 1974, **11**, 323.
K. Bowden, *Chem. Rev.*, 1966, **66**, 119 (strongly basic media).
E. M. Arnett & G. Scorrano, *Adv. Phys. Org. Chem.*, 1976, **13**, 83.

# 5 The reaction medium

## 5.1 Solvation

Most discussions of reactivity and of reaction mechanisms are couched in terms of the behaviour of isolated molecules; the solvent is generally treated as an inert support which can occasionally participate in the reaction in a rather passive way, for example as a convenient source of protons. The reality is often far removed from this picture. Interactions between solvent and solute molecules can profoundly influence – in some cases dominate – the course of a reaction. Nevertheless our understanding of solvent effects remains limited. We still know surprisingly little about the detailed structure of the liquid state even for a pure substance. How much more difficult it is to discuss the perturbed structure in the vicinity of an intruding solute molecule. It is possible to build up pictorial descriptions of the ways in which solvation can occur, and also to carry out quantitative studies of the effects of solvents on reaction rates and equilibria. What is still lacking is a broadly based theory that ties the two together and that is capable of general quantitative predictions.

### 5.1.1 Intermolecular forces

The only attractive interactions between non-polar molecules are the weak but ubiquitous *London* or *dispersion* forces. They arise from the motion of the electrons within a molecule, producing a transient electric dipole; this can polarise a neighbouring molecule and momentarily the two are held together. Looking at it in a slightly different way, one can say that intermolecular electrostatic interactions synchronise the movements of the electrons in such a way as to favour attraction between molecules, minimising the energy of the system as a whole. The strength of the attraction depends on the square of the molecular polarisability $\alpha$. This in turn relates to the number of electrons in the molecule and to their mobility. Pentane is more polarisable than butane; it has more electrons. But butadiene is also more polarisable than butane even though it has four electrons fewer; the weakly bound $\pi$ electrons are more mobile, freer to follow a changing electric field, than are $\sigma$ electrons. (The refractive index, $n$, of a liquid provides a convenient measure or polarisability since it is related to the ease with which the electrons can follow the oscillating electric field of a light wave. The relation is

$$\alpha = (3V_m/4\pi L)(n^2 - 1)/(n^2 + 2)$$

where $V_m$ is molar volume and $L$ the Avogadro constant.)

Molecules that possess a permanent dipole moment can also attract their neighbours by *induction*, the polarisation of one molecule by the dipole of another, and by direct *dipole–dipole attraction*. These interactions are confined to relatively small areas of contact between molecules; the dipole–dipole attractions are further limited because they require an alignment of dipoles and the molecules cannot all simultaneously adopt suitable orientations. Dispersion forces, on the other hand, operate at all points of contact between molecules throughout the whole bulk of the liquid. They therefore dominate intermolecular attractions even in polar compounds except where the molecules are small and strongly dipolar, or where hydrogen bonding can occur.

*Hydrogen bonding* plays a special role in solvent–solute interactions, particularly in hydroxylic solvents. First, the hydrogen atom at the positive end of the O–H dipole protrudes from the molecule giving easy access to anions and other centres of negative charge. By contrast the positively charged end of most other polar groups is embedded inside the molecule; the sulphur atom of dimethyl sulphoxide is an example. Secondly, the hydrogen bonded interaction can be exceptionally strong; this is partly a consequence of its unencumbered steric environment, but there is also a measure of covalency: the two bonding electrons of the 'normal' O–H bond and a lone pair from the hydrogen bond acceptor are to some extent delocalised over the three nuclei. The oxygen–oxygen distance in an O–H$\cdots$O hydrogen bond is normally considerably less than the equilibrium van der Waals separation of two oxygen atoms; the interatomic repulsions must therefore be considerable, but the bonding forces more than outweigh them. Finally one must remember the importance of hydrogen bonding in the structure of liquid water (and to a lesser extent in that of the lower alcohols), and that this must be partially disrupted by the intrusion of molecules of solute.

### 5.1.2 Solvation of ions

A strong electrolyte can dissolve only if the solvent interacts with the ions in such a way as to overcome their natural tendency to aggregate. This calls first for a solvent of high dielectric constant (relative permittivity, $\varepsilon_r$) because the force between charges is inversely proportional to the dielectric constant of the medium. The solvent should also be able to solvate the ions, surrounding them with a shell of solvent molecules. In order to come together they would have to shed some of these attendants, and if the energy needed to do this were greater than could be recouped by aggregation then the ions would remain separate.

For small ions the forces of solvation are almost entirely electrostatic, each ion attracting to itself dipolar solvent molecules. Water is the solvent *par excellence*: it has a high dielectric constant; the anions are solvated through hydrogen bonds, and the cations cling to the oxygen atoms (**5.1**). Those water molecules which are directly in contact with the ions are thereby polarised and can themselves attract other water molecules the more strongly. In this way a solvation shell or *cybotactic region* can build up to a thickness of several molecules.

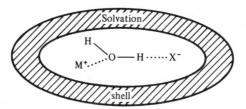

(5.1)

In more concentrated solutions, and particularly when multiply charged ions are present, the forces of solvation are unable to keep all the ions completely free. They begin to cluster together in *ion-pairs* and higher aggregates. These still retain a surrounding solvent shell, and there may also be water molecules separating the ions, but the ions belonging to an aggregate are no longer free to move independently through the solution; rather the whole aggregate behaves as a single entity (figure 5.1). In the limit, when the concentration of ions is too high for solvation to cope any further, the solution reaches saturation.

In non-aqueous solutions ion association generally occurs at lower concentrations. Few solvents other than water combine a high dielectric constant with the ability to solvate both anions and cations.

Figure 5.1. A solvent-separated ion-pair.

For larger ions, those with substantial hydrocarbon moieties or the more polarisable inorganic ions, the picture is more complex because solvation via dispersion forces becomes significant. Only in the most extreme cases are non-polar solvents, with their low dielectric constants, able to dissolve electrolytes, but the balance between other solvents becomes much finer. Water has a very low polarisability; the alcohols are better in this respect but have low dielectric constants; some of the dipolar aprotic solvents, for example acetone or dimethyl sulphoxide, have high dielectric constants, but they are not hydrogen bond donors.

### 5.1.3 Solvation of electrically neutral molecules

Non-polar solutes dissolve readily in non-polar solvents because the dispersion forces within the solution are little different from those in the separate components. One cannot point to any specific solute–solvent interactions, and so the term 'solvation' is perhaps inappropriate in this context.

With more polar solutes the complexities of solvation are akin to those of large ions: dispersion forces are always present but their importance depends on the polarisability of both solvent and solute, whereas hydrogen bonding and dipole–dipole interactions are strongly influenced by the extent of charge localisation within the solute and by the geometry of the interactions. The overall extent of solvation depends on the balance between these solute–solvent interactions and the disruption of intermolecular attractions within the unmixed solvent and solute.

### 5.1.4 Thermodynamics of solvation

Solvation is necessarily a stabilising process: if it were not, it would not occur. It is therefore always accompanied by a decrease in Gibbs energy. However, this represents a balance between enthalpy and entropy effects. Solvation is generated by attractive forces between solvent and solute, so the enthalpy of solvation is negative. But as the solute gathers around itself a cluster of solvent molecules it restricts their mobility and reduces their entropy, so in general the entropy of solvation is also negative – and the stronger the solvation, the greater the entropy effect.

This is well illustrated by the hydration of the proton. In aqueous solution the hydrated proton is conventionally represented as $H_3O^+$, but most of the evidence suggests that there are actually four water molecules tightly bound to each proton (**5.2**); there will also be others more loosely associated in outer solvation layers. Thus when strong acids are extracted from aqueous solution into organic solvents they frequently bring four molecules of water with them, and anhydrous acidic ion exchange resins readily take up four molecules of water per acid residue.

$$
\begin{array}{c}
\overset{+}{\text{O}}\text{----H}\cdots\text{OH}_2 \\
\text{H}\diagup\quad\diagdown \\
\text{H}_2\text{O}\cdots\quad\quad\text{H} \\
\diagdown \\
\text{OH}_2
\end{array}
\qquad\qquad (5.2)
$$

This behaviour is reflected in the enormously large enthalpy and entropy of hydration for the proton, estimated at $-1129\ \text{kJ mol}^{-1}$ and $-131\ \text{JK}^{-1}\ \text{mol}^{-1}$, respectively [73MI*b*].

The balance between entropy and enthalpy effects is often disguised in the overall Gibbs energy term. Aqueous trichloroacetic acid is, as the inductive effect of chlorine predicts, a stronger acid than acetic acid, but the enthalpy term actually favours the ionisation of acetic acid; it is the entropy which tips the balance in the expected direction [67JA213; 69JA6057]:

| (At 25 °C) | $\Delta G^{\ominus}/\text{kJ mol}^{-1}$ | $\Delta H^{\ominus}/\text{kJ mol}^{-1}$ | $\Delta S^{\ominus}/\text{J K}^{-1}\,\text{mol}^{-1}$ |
|---|---|---|---|
| $CH_3COOH$ | $+27.2$ | $-0.1$ | $-91.6$ |
| $CCl_3COOH$ | $+2.9$ | $+1.2$ | $-5.9$ |

This is quite reasonable if we consider the hydration of the two anions. Acetate is a small ion with its negative charge concentrated on the two oxygen atoms. In aqueous solution it is very strongly hydrogen bonded, so contributing to the negative enthalpy of ionisation; but this also generates a thick hydration shell of ordered water molecules and leads to the large negative entropy term. The trichloroacetate ion, on the other hand, has a much more disperse negative charge, a considerable portion being borne by the chlorine atoms. It is less strongly hydrated and the enthalpy, and especially the entropy of ionisation are more positive.

The thermodynamic parameters quoted above are, of course, an amalgam containing contributions from the fundamental ionisation process itself and from the solvation of the various species present. With the advent of mass spectrometric techniques for studying ion–molecule reactions in the gas phase it has become possible to separate the various terms. Consider, for example, the protonation of amines [e.g. 76JA318], as shown in the following scheme.

$$
\begin{array}{ccccc}
\text{B(aq)} + \text{H}^+\text{(aq)} & \xrightarrow{\ 1\ } & \text{BH}^+\text{(aq)} \\
\Big\uparrow{\scriptstyle 3} & & \Big\uparrow{\scriptstyle 4} & & \Big\uparrow{\scriptstyle 5} \\
\text{B(gas)} + \text{H}^+\text{(gas)} & \xrightarrow{\ 2\ } & \text{BH}^+\text{(gas)}
\end{array}
$$

In this scheme the thermodynamic parameters for reaction 1 are derived from the normal protonation equilibrium in water, those for reaction 2 from gas phase measurements, and those for hydration of the neutral base (step 3) by direct calorimetry and solubility measurements. The thermody-

Table 5.1. *Thermodynamics of the protonation and hydration of methylamine at 25°C* [76JA318]

| Reaction step | $\Delta G^{\ominus}/kJ\,mol^{-1}$ | $\Delta H^{\ominus}/kJ\,mol^{-1}$ | $\Delta S^{\ominus}/J\,K^{-1}\,mol^{-1}$ |
|---|---|---|---|
| 1: Protonation in water | −61 | −55 | +19 |
| 2: Gas phase protonation | −879 | −914 | −118 |
| 3: Hydration of $MeNH_2$ | −11 | −45 | −114 |
| 4: Hydration of $H^+$ | −1090 | −1129 | −131 |
| 5: Hydration of $MeNH_3^+$ | −283 | −315 | −108 |

Table 5.2. *Free energies* $(-\Delta G^{\ominus}/kJ\,mol^{-1})$ *of protonation and hydration for the series* $Me_nNH_{3-n}$ *at 25°C* [76JA318]

| Reaction step | $NH_3$ | $MeNH_2$ | $Me_2NH$ | $Me_3N$ |
|---|---|---|---|---|
| 1: Protonation in water | 53 | 61 | 62 | 56 |
| 2: Gas phase protonation | 828 | 879 | 906 | 924 |
| 3: Hydration of free base | 10 | 11 | 10 | 6 |
| 4: Hydration of $H^+$ | 1090 | 1090 | 1090 | 1090 |
| 5: Hydration of ammonium ion | 326 | 283 | 255 | 228 |

namics of hydration of the proton (step 4) cannot be directly measured but the values quoted above and used in the following tables are fairly reliable estimates [73MI*b*]. The parameters for the remaining step, the hydration of the ammonium cation, are obtained from the other four steps by completing the thermodynamic cycle.

Such an analysis reveals the overwhelming importance of the solvation terms; for methylamine the various thermodynamic parameters for each step are listed in table 5.1. Table 5.2 records the various values of $\Delta G^{\ominus}$ for the series $NH_3$, $MeNH_2$, $Me_2NH$, $Me_3N$. The irregular order of basicities of primary, secondary, and tertiary amines (line 1) has long been a source of speculation, but now it can be seen as merely a consequence of the balance between regular trends in the inherent basicities of the isolated molecules (line 2) and in the solvation of the cations (line 5). The effects on these two properties of substituting methyl groups for hydrogen are large, but they almost cancel out.

### 5.1.5 The effects of solvation on reaction rates

Gas phase studies have also been used to probe the effects of solvents on the rates of reactions, especially nucleophilic substitutions [80Acc76]. These effects can be dramatic. The rate coefficient for the room-temperature reaction between hydroxide ion and bromomethane in aqueous solution is $1.4 \times 10^{-4}\,l\,mol^{-1}\,s^{-1}$: in the gas phase it is

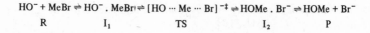

$$HO^- + MeBr \rightleftharpoons HO^-.MeBr \rightleftharpoons [HO \cdots Me \cdots Br]^{-\ddagger} \rightleftharpoons HOMe.Br^- \rightleftharpoons HOMe + Br^-$$

R           $I_1$           TS           $I_2$           P

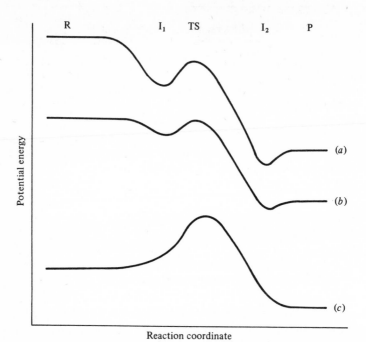

Reaction coordinate

Figure 5.2. Qualitative representation of energy changes during the reaction between hydroxide ion and bromomethane: (*a*) in the gas phase with unsolvated $HO^-$, (*b*) in the gas phase with $HO^-.(H_2O)_2$, (*c*) in aqueous solution.

$6 \times 10^{+11} \, l \, mol^{-1} \, s^{-1}$ [81JA978]. In this study it was possible to investigate the transition from gas phase to solution kinetics because the specifically hydrated nucleophilic species $HO^-.H_2O$, $HO^-.(H_2O)_2$, and $HO^-.(H_2O)_3$ could be monitored separately. (The model is imperfect because the nucleofugal $Br^-$ is not similarly hydrated [83JA5509].) The results could be interpreted in terms of a model (figure 5.2) in which in the gas phase the reacting entities first come together in an activation-less process to form a stabilised adduct in which the negative charge is more dispersed than in the isolated hydroxide ion. This then rearranges through a conventional $S_N2$ transition state. In water the hydroxide ion is greatly stabilised by solvation and it cannot approach the bromomethane without the energy-demanding necessity of shedding some of its attendant water molecules.

The use of such gas phase experiments will no doubt lead in due course to accurate quantitative analyses of the changing enthalpy and entropy of solvation throughout a reaction. Such analyses have already been attempted in studies of *differential* solvation, the effect of transferring a reaction from one solvent to another.

Much of our knowledge of solvent effects comes from qualitative rationalisations of the ways in which different solvents influence reaction rates, and various empirical scales have been developed in attempting to quantify these effects. Some of this work is described in §5.2. More sophisticated analyses of differential solvation seek to distinguish the various thermodynamic contributions to the overall solvent effect. To avoid using a plethora of $\Delta$s, let us write $\S G^{\ominus}$ to mean the change in standard Gibbs energy that accompanies the transfer of a solute from one solvent to another. For neutral molecules $\S G^{\ominus}$ can be obtained from the distribution coefficient if the two solvents are immiscible; if not it can be calculated from relative solubilities or from the distribution coefficients relative to a third mutually immiscible solvent. For single ions the calculations are less reliable: one cannot dissolve a cation without taking the accompanying anion into solution as well. How does one assess their relative contributions? Thermodynamics cannot answer the question; one has to make some additional assumption. A common one, probably fairly reliable, is to assume that the charge plays an insignificant role in solvation if it is embedded deep inside the hydrocarbon part of a large, polarisable organic ion. Thus for the salt $Ph_4As^+ Ph_4B^-$ (in which the ions are structurally identical and of closely similar size) the observed value of $\S G^{\ominus}$ is divided into equal contributions from each ion. Then one can obtain $\S G^{\ominus}$ values for any other ion, say $X^-$, by subtracting the value for $Ph_4As^+$ from the measured value for $Ph_4As^+ X^-$. In this way values of $\S G^{\ominus}$ for the transfer of many single ions from one solvent to another have been established, notably through the work of Parker [69CR1; 72JA1148; 78JO1843]. (Parker reports his earlier results as relative activity coefficients, $\gamma$; the relation is $\S G^{\ominus} = RT \ln \gamma$.)

Finally, if the reaction is measured in each solvent to obtain $\Delta G^{\ddagger}$ values, one can obtain $\S G^{\ddagger}$, the Gibbs energy of solvent transfer for the transition state, by completing the cycle illustrated in figure 5.3.

The application of this technique is illustrated in table 5.3 for three $S_N2$ reactions: $MeI + Cl^-$, $MeI + SCN^-$, and $Me_3S^+ + SCN^-$.

The sulphonium ion, and to a small extent the methyl iodide molecule, are stabilised by transfer to DMF ($G^{\ominus}$ lowered); polarisability and more particularly the ability of the amide oxygen to solvate cationic charge are important here. On the other hand the anions are destabilised in DMF because of the lack of hydrogen bonding; the less polarisable chloride ion feels the difference more sharply. This is clearly the origin of the enormous difference in differential solvent effects on the rates of the first two reactions, for the transition state stabilities are barely altered. Both

Table 5.3. *Thermodynamic analysis of differential solvation at 25 °C in* $S_N2$ *reactions transferred from methanol to dimethylformamide* [68JA5049]

| Reactant $§G^{\ominus}$/kJ mol$^{-1}$ | MeI $-2.9$ | Me$_3$S$^+$ $-17.7$ | Cl$^-$ $+37.1$ | SCN$^-$ $+15.4$ |
|---|---|---|---|---|
| Reaction | MeI + Cl$^-$ | MeI + SCN$^-$ | Me$_3$S$^+$ + SCN$^-$ | |
| $k_{\text{DMF}}/k_{\text{MeOH}}$ | $8 \times 10^5$ | 160 | 16 | |
| $\Delta G^{\ddagger}_{\text{DMF}} - \Delta G^{\ddagger}_{\text{MeOH}}$ | $-33.7$ | $-12.6$ | $-6.8$ | |
| $§G^{\ddagger}$ | $+0.5$ | $-0.1$ | $-9.1$ | |

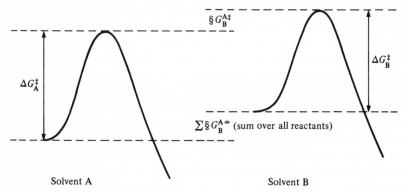

Figure 5.3. Calculating $§G^{A\ddagger}_B$, the Gibbs energy of transfer of transition state from solvent A to solvent B.

$$-RT \ln(k_B/k_A) = \Delta G^{\ddagger}_B - \Delta G^{\ddagger}_A$$
$$§G^{A\ddagger}_B = \sum §G^{A\ominus}_B + \Delta G^{\ddagger}_B - \Delta G^{\ddagger}_A$$

transition states are negatively charged, $(X\cdots Me\cdots I)^-$, but both are also fairly polarisable and presumably the hydrogen bonding and van der Waals interactions are finely balanced. By contrast in the third reaction the solvent transfer effects on the reactant ions almost cancel out and the relatively small difference in reaction rates is largely due to the differential solvation of the transition state. This is overall electrically neutral, $(NCS\cdots Me\cdots SMe_2)$, less susceptible to hydrogen bonding, and stabilised more by DMF than by methanol.

Ideally one would like to dissect $§G$ into $§H$ and $§S$ components. $§H^{\ominus}$ values can be obtained as the difference between heats of solvation in the two solvents, and $§S^{\ominus}$ then follows as $(§H^{\ominus} - §G^{\ominus})/T$. Unfortunately many of the published heats of solvation for electrolytes are somewhat unreliable, and their separation into single ion enthalpies has been based on a variety of different extra-thermodynamic assumptions, so that direct comparison with $§G^{\ominus}$ is often a dubious exercise. Where it does seem to be justified the results can be surprising. The azide ion, $N_3^-$, is markedly destabilised on transfer from methanol to dimethyl sulphoxide ($\S G^{\ominus}$ = $+17.6$ kJ mol$^{-1}$), but the transfer is actually exothermic with $\S H^{\ominus}$ = $-4.2$ kJ mol$^{-1}$. The DMSO must be more strongly bound to azide than is methanol, despite the absence of hydrogen bonding, but in the process it becomes highly oriented so that $\S S^{\ominus}$ is $-73$ J K$^{-1}$ mol$^{-1}$ [73JA408].

To complete a full thermodynamic analysis of a reaction we need $\S H^{\ddagger}$ and $\S S^{\ddagger}$, which can be derived after the manner of figure 5.3 from measurements of the reaction rates in each solvent over a range of temperatures.

## 5.2 Empirical correlations of solvent effects

The sophisticated analyses of solvent effects on reaction rates described in the previous section are recent developments. Qualitative and empirical quantitative correlations have been with us for a much longer time.

If non-polar reactants generate a non-polar transition state then any solvation is likely to be weak and will have little influence on the reaction rate. For example, the rate of the Diels–Alder dimerisation of cyclopentadiene at a given temperature is much the same in the gas phase, in neat liquid, or in solution in solvents as disparate as liquid paraffin and acetic acid.

On the other hand if there are significant changes in charge distribution during the reaction the solvent plays a more important role. In qualitative terms, if the transition state is more polar than the reactants it will be stabilised by polar solvents thus accelerating the reaction, and vice versa.

Hughes and Ingold in the 1930s discussed the basic principles with reference to nucleophilic substitution reactions in which a variety of different patterns of charge distribution are possible (table 5.4). In general the reactions that start with relatively non-polar reactants (reactions 1 and 4) lead to transition states with considerable charge separation. In reactions 2, 3, and 6 one of the reactant species is ionic and its charge is dispersed in the transition state. In reaction 5, between oppositely charged ions, there is a marked reduction in charge separation; the transition state, though dipolar, is electrically neutral.

Ingold quotes many examples of solvent effects that accord with this pattern. The $S_N2$ hydrolysis of methyl or ethyl iodides with hydroxide ion in aqueous ethanol (type 3) are retarded by increasing the proportion of water, but the $S_N1$ solvolysis of t-butyl chloride (type 1) is accelerated. With positively charged leaving groups the pattern is different, as predicted. The hydrolysis of $Me_3S^+$ with hydroxide ion (type 5) and the

Table 5.4. *Solvent effects on* $S_N$ *reactions of different charge types* (after Ingold [69MI])

| Reaction type (see text) | | Ground state charge distribution | Transition state charge distribution | Effect of increasing solvent polarity |
|---|---|---|---|---|
| $S_N1$ | 1 | R–Z | $\overset{\delta+}{R}\cdots\overset{\delta-}{Z}$ | Large acceleration |
| | 2 | R–Z$^+$ | $\overset{\delta+}{R}\cdots\overset{\delta+}{Z}$ | Small deceleration |
| $S_N2$ | 3 | X$^-$ + RZ | $\overset{\delta-}{X}\cdots R\cdots\overset{\delta-}{Z}$ | Small deceleration |
| | 4 | X + RZ | $\overset{\delta+}{X}\cdots R\cdots\overset{\delta-}{Z}$ | Large acceleration |
| | 5 | X$^-$ + RZ$^+$ | $\overset{\delta-}{X}\cdots R\cdots\overset{\delta+}{Z}$ | Large deceleration |
| | 6 | X + RZ$^+$ | $\overset{\delta+}{X}\cdots R\cdots\overset{\delta+}{Z}$ | Small deceleration |

solvolysis of t-BuSMe$_2^+$ (type 2) in aqueous ethanol are both retarded by adding more water. The Menschutkin reaction (Et$_3$N + EtI; type 4) is accelerated by increasing solvent polarity in the order hexane < benzene < acetone < ethanol < methanol, but the reaction of Me$_3$N with Me$_3$S$^+$ (type 6) is retarded by polar solvents: water < methanol < ethanol < nitromethane.

The predictions also hold in a semi-quantitative manner. For example, the reaction of Me$_3$S$^+$ with HO$^-$ is accelerated some 20 000 times when the solvent is changed from water to ethanol, but for Me$_3$S$^+$ with Me$_3$N the acceleration is about tenfold. However, the data of table 5.3 illustrate how difficult it is to make simple comparisons on this basis between protic and dipolar aprotic solvents.

Grunwald and Winstein [48JA846] used the same basic idea to establish an empirical scale of solvent polarity. Supposing the ionisation of t-butyl chloride to be a true $S_N1$ process, they suggested that the rate of the reaction in different solvents would depend solely on the 'solvent ionising power', which they designated $Y$:

$$\log(k_s/k_0) = Y$$

Their standard solvent ($k_0$) was 80 per cent ethanol in water (by volume). Much of their work was carried out in mixed solvents of various proportions; in aqueous ethanol the $Y$ value changes from +3.5 for pure water, through +1.7 for a 1:1 mixture, 0.0 (by definition) for 80 per cent, to −2.0 for pure ethanol. Some other $Y$ values for pure solvents are:

|  HCOOH | HCONH$_2$ | MeOH | AcOH | i-PrOH | t-BuOH |
|---|---|---|---|---|---|
| +2.1 | +0.6 | −1.1 | −1.6 | −2.7 | −3.3 |

If $Y$ is truly a property of the solvent alone, then one should be able to use it to measure the solvent effect on $S_N1$ solvolyses of other substrates,

RX, according to the equation:

$$\log(k_s^{RX}/k_0^{RX}) = m\,Y$$

where $m$ is a substrate parameter depending only on RX. For a particular pair of mixed solvents, say aqueous ethanol, this equation is entirely consistent, giving excellent straight line plots; indeed, it even correlates a range of $S_N2$ substrates such as methyl tosylate. But if the solvent system is changed, say to aqueous dioxan, then the value of $m$ for a particular substrate also changes, and so does the intercept at $Y = 0$. In other words the $Y$ values are not uniquely a solvent parameter, but depend on the substrate as well. They are nonetheless still widely quoted as giving at least a semi-quantitative idea of how effectively different solvent systems can solvate ions.

Several other parameters have been proposed for characterising solvent properties. A detailed statistical survey of correlations between no fewer than 54 different chemical and spectroscopic properties and ten different solvent parameters has been carried out. The study covered 57 different solvents, though data were not available for each property in every solvent [71J(B)460]. The single parameter giving the best overall correlation for all properties (though still with several major discrepancies) is the so-called Dimroth parameter, $E$ [63LA(661)1; see also 82T1615], which is the transition energy associated with the electronic absorption maximum of the zwitterion:

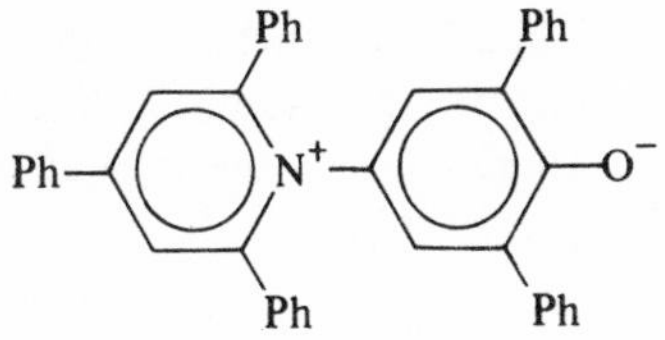

This is markedly dependent on solvent, varying between 450 and 1000 nm. The ground state of the molecule is more polar and more susceptible to hydrogen bonding than the first excited state, so both protic and dipolar solvation preferentially stabilise the ground state and raise the transition energy. The $E$ value may be calculated as:

$$E = 119\,600/\lambda$$

where $\lambda$ is measured in nm; the resulting $E$ values are in kJ mol$^{-1}$. Values of $\Delta E$, referred to hexane as a standard solvent, are given in table 5.5, which also lists dielectric constants. It can be seen that both polarity and hydrogen bonding play their part: acetic acid has a dielectric constant similar to that of ethyl acetate, but its hydrogen bonding ability is reflected in its larger $\Delta E$ value; the aprotic but much more polar dimethylformamide also has a high $\Delta E$ value.

Table 5.5. *Dimroth solvent parameters (relative to hexane; $\Delta E/kJ\ mol^{-1}$). Dielectric constants (relative permittivities) $\varepsilon_r$ at $25\,°C$*

| | $\varepsilon_r$ | $\Delta E$ | | $\varepsilon_r$ | $\Delta E$ |
|---|---|---|---|---|---|
| Hexane | 1.9 | 0.0 | Dimethyl formamide | 36.7 | 54.0 |
| Cyclohexane | 2.0 | 1.3 | t-Butanol | 12.2 | 54.4 |
| Carbon tetrachloride | 2.2 | 6.7 | Sulpholan | 44.0 | 54.8 |
| Carbon disulphide | 2.6 | 7.1 | Dimethyl sulphoxide | 48.9 | 59.0 |
| Benzene | 2.3 | 15.1 | Acetonitrile | 37.5 | 63.2 |
| Diethyl ether | 4.2 | 15.5 | Nitromethane | 38.6 | 64.4 |
| 1,4-Dioxan | 2.2 | 21.3 | Isopropanol | 18.3 | 74.1 |
| Tetrahydrofuran | 7.4 | 27.2 | Propanol | 20.1 | 82.8 |
| Ethyl acetate | 6.0 | 30.1 | Acetic acid | 6.2 | 84.5 |
| 1,2-Dimethoxyethane | 7.0 | 30.5 | Ethanol | 24.3 | 87.9 |
| Chloroform | 4.7 | 34.3 | N-Methylacetamide | 175.7 | 88.3 |
| Pyridine | 12.3 | 38.9 | 2-Methoxyethanol | 15.9 | 89.5 |
| Hexamethylphosphoramide | 29.6 | 41.8 | N-Methylformamide | 182.4 | 97.1 |
| Dichloromethane | 8.9 | 42.7 | Methanol | 32.6 | 103 |
| Nitrobenzene | 34.8 | 46.4 | Ethylene glycol | 29.4 | 106 |
| Acetone | 20.7 | 47.3 | Formamide | 109.5 | 108 |
| Dimethylacetamide | 37.8 | 53.6 | Water | 78.5 | 135 |

It is hardly surprising that a single parameter fails to sum up the complexities of solvation. We can distinguish six possible modes of interaction: dispersion, solvent–solute and solute–solvent induction, dipole–dipole interaction, and solvent–solute and solute–solvent hydrogen bonding. Not all of these will be equally important, and moreover different phenomena will exhibit different sensitivities towards these various modes. This line of thought suggests that we should invoke a multiparameter equation of the type:

$$\Delta P = r_1 s_1 + r_2 s_2 + \ldots$$

where $\Delta P$ represents the change in some solvent-dependent property: for kinetic studies $P = \log k$, but it could also represent $\log K$ for an equilibrium, or a spectroscopic transition frequency.

Kamlet and Taft have recently used such an approach by seeking to measure separately the hydrogen bond acceptor, hydrogen bond donor, and polarity/polarisability properties of solvents [81PPO485]. At first they followed Dimroth in basing their measurements on solvent-induced changes in electronic spectra. Thus the hydrogen bond acceptor contributions, denoted $\beta$, were measured by comparing the shifts of *p*-nitrophenol with those of *p*-nitroanisole. In non-hydrogen bonding solvents there is a good linear correlation between the two; in hydrogen bond-accepting solvents there is a deviation which is assigned to hydrogen bonding between the solvent and the phenolic proton. Similar behaviour is found for *p*-nitroaniline and *NN*-

Table 5.6. *Kamlet–Taft solvent parameters:* α, *hydrogen bond donation;* β, *hydrogen bond acceptance;* π* *or* π, *polarity/polarisability* [81JA6924; 81JO661; 81PPO485; 83JO2877]

|  | α | β | π* |  | α | β | π |
|---|---|---|---|---|---|---|---|
| Cyclohexane | 0 | 0 | 0 | Acetone | 0.10 | 0.48 | 0.71 |
| Carbon tetrachloride | 0 | 0 | 0.27 | Acetonitrile | 0.22 | 0.31 | 0.85 |
| Benzene | 0 | 0.10 | 0.59 | t-Butanol | 0.62 | 1.01 | 0.40 |
| Diethyl ether | 0 | 0.47 | 0.27 | i-Propanol | 0.77 | 0.95 | 0.48 |
| Triethylamine | 0 | 0.71 | 0.14 | Ethanol | 0.85 | 0.77 | 0.54 |
| Pyridine | 0 | 0.64 | 0.87 | Methanol | 0.98 | 0.62 | 0.60 |
| Ethyl acetate | 0 | 0.45 | 0.55 | Acetic acid | (1.01) | ? | 0.64 |
| Dimethyl sulphoxide | 0 | 0.76 | 1.00 | Water | 1.10 | (0.18) | 1.09 |

Values in parentheses are uncertain.

dimethyl-*p*-nitroaniline, and the differential shifts also correlate with some earlier estimates of hydrogen bond acceptance. All are averaged to produce the β scale.

It proved more difficult to disentangle the hydrogen bond donor and polarity/polarisability terms. The latter, designated π*, was originally derived from the effect of the solvent on the $n \rightarrow \pi^*$ and $\pi \rightarrow \pi^*$ transitions of a variety of solutes, omitting all systems in which hydrogen bonded solvent–solute interactions might be significant. In protic solvents, however, hydrogen bond donation is so ubiquitous that its effect on the electronic transitions of aromatic π orbitals cannot be ignored, and for these solvents a scale (designated π) based on the $^{13}$C n.m.r. chemical shifts of the *para* carbon atoms of $PhCF_3$ and $PhSF_5$ was developed [81JA6924]. Moreover, it became clear that the balance between polarity and polarisability differs for different solvents, and a correction factor δ was introduced to distinguish three classes of solvents: aromatic, polychlorinated aliphatic, and other aliphatic. [For discussion of the significance of π* see also 82J(P2)923.]

Finally, the α scale of hydrogen bond donation was established, originally using electronic spectral shifts such as those of Dimroth's zwitterion, and later extended to include properties such as $^{13}$C n.m.r. chemical shifts in PhSOMe. In each case the observed solvent effect had to be statistically corrected to allow for the polarity/polarisability contribution.

Sample values of α, β, and π* or π are recorded in table 5.6. Kamlet and Taft have used them to analyse a wide variety of solvent-dependent properties, thermodynamic, kinetic, and spectroscopic, by means of the equation:

$$\Delta P = a\alpha + b\beta + s(\pi^* + d\delta)$$

In practice it has been difficult to use this equation when $P$ is markedly affected by all four solvent parameters (a common problem with multi-parameter equations), but if the number of variables can be reduced the equation can provide interesting information about the role of the solvent. This may be accomplished deliberately, for example by using only a single class of solvents so that the $\delta$ term can be omitted, or it may happen that the property $P$ is unresponsive to one or other of the parameters. Thus the logarithms of the rate coefficients for the unimolecular decomposition of t-butyl halides in non-chlorinated aliphatic solvents correlate with only two of the parameters [81JO3053]:

$$\log(k/k_0) = a\alpha + s\pi^*$$

The lack of dependence on $\beta$ is surprising: although there is little potential for hydrogen bonding from reactants to solvent one might have expected the incipient t-butyl cation to have sought stabilisation from nucleophilic solvents (compare the discussion of nucleophilic solvation in §7.3.1). The value of $s$ is much the same ($7.0 \pm 0.2$) for t-butyl chlorides, bromides, and iodides. The values of $a$, however, change progressively: I, 2.7; Br, 4.0; Cl, 5.3. The importance of hydrogen bonding in stabilising the incipient halide anion in the transition state clearly increases through the series.

A further example is seen in the correlation of electronic transition frequencies of *o*-nitrophenol and *o*-nitro-*N*-methylaniline in non-hydrogen bonding and in hydrogen bond-accepting solvents [82JO1734]. For the amines there is a good correlation for all solvents with $\pi^*$ alone: hydrogen bonding from solute to solvent is clearly insignificant so that the amino hydrogen must be tightly held by intramolecular hydrogen bonding to the nitro group. The phenol, however, requires a dual parameter correlation with $\beta$ and $\pi^*$, showing that the more acidic phenol hydrogen atom is readily diverted from intramolecular to intermolecular hydrogen bonding with acceptor solvents:

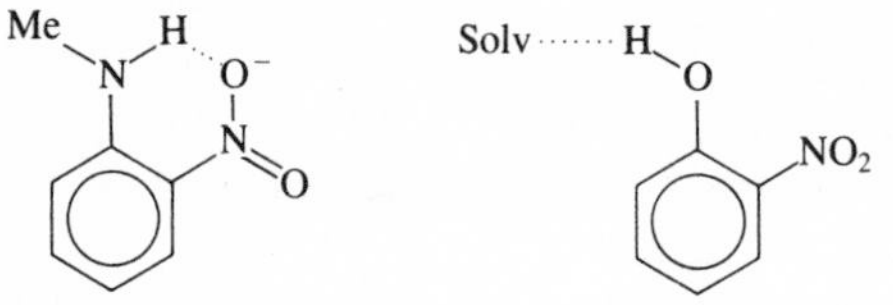

### 5.3 Solvent isotope effects

The rates of reactions carried out in heavy water or in deuteriated alcoholic solution frequently differ from the rates in the normal solvent. This may be the consequence of a normal primary isotope effect: a proton, either coming directly from the solvent or from some other species which exchanges rapidly with the solvent, is transferred in the rate-limiting step. The magnitude of this *primary solvent kinetic isotope effect* can often be calculated with reasonable accuracy from infrared stretching frequencies and from estimates of the extent of hydron* transfer at the transition state derived, for example, from Brønsted coefficients (§2.4; §4.2.3).

*Secondary solvent isotope effects* are also important. These arise from two main sources. First, an isotopically distinguished species, solvent or solute, may participate directly in the reaction without transferring a hydron. For example, the nucleophilicity of $HO^-$ (reacting via the oxygen atom) is not necessarily the same as that of $DO^-$. The second contribution comes from changes in solvation that arise because $D_2O$ is more structured than $H_2O$ (witness its lower molar volume and the reduced solubility of sparingly soluble compounds in $D_2O$).

Secondary effects can be measured empirically from equilibria such as:

$$XH + ROD \rightleftharpoons XD + ROH$$

where ROH is a hydroxylic solvent. The equilibrium constant is called the *isotopic fractionation factor* and measures the extent to which secondary effects favour the formation of XD over XH relative to ROD/ROH:

$$\phi_X = \frac{[XD]/[XH]}{[ROD]/[ROH]}$$

In practice it is found that $\phi$ depends mainly on the nature of the atom to which the mobile hydrogen is attached and on the charge it bears. For example:

$$\phi_O = 1.0 \qquad \phi_{O^+} = 0.69 \qquad \phi_{O^-} = 0.5$$

$$\phi_N = 0.9 \qquad \phi_{N^+} = 1.0 \qquad \phi_S = 0.4$$

These fractionation factors can be used to estimate *equilibrium* solvent isotope effects. For an equilibrium between two systems XH and YH we have:

$$XH \rightleftharpoons YH \qquad\qquad XD \rightleftharpoons YD$$

$$K_H = [YH]/[XH] \qquad\qquad K_D = [YD]/[XD]$$

$$\frac{K_H}{K_D} = \frac{[YH]/[YD]}{[XH]/[XD]} = \frac{\phi_X}{\phi_Y}$$

Strictly speaking this ratio is applicable to equilibria measured in the same ROH/ROD mixture, but within the range of accuracy of $\phi$ values it can be used to compare equilibria measured in pure $H_2O$ and pure $D_2O$.

For equilibria involving several isotopically differentiated sites the $\phi$

---

\* Hydron: a term proposed by J. F. Bunnett to describe $H^+$, $D^+$ or $T^+$ without implying any particular isotope.

factors are multiplied. Consider the dissociation of an oxy-acid XOH:

$$XOH + H_2O \rightleftharpoons XO^- + H_3O^+$$

In the products there are three $H-O^+$ bonds, in the reactants three $H-O$ bonds. If all the labile protons are changed to deuterons we have:

$$K_H/K_D = (\phi_O)^3/(\phi_{O^+})^3 = 1/(0.69)^3 = 3$$

This implies that oxy-acids are about three times stronger in $H_2O$ than in $D_2O$ or, looking at it in another way, that $D_3O^+$ is about three times stronger an acid than $H_3O^+$ because it is better at hydronating $XO^-$. Likewise, $DO^-$ is a stronger base than $HO^-$:

$$XO^- + H_2O \rightleftharpoons XOH + HO^-$$

$$K_H/K_D = (\phi_O)^2/\phi_O \cdot \phi_{O^-} = 2$$

An analysis of *kinetic* solvent isotope effects requires an estimate of the secondary effects on the transition state. A simple approach is to assume that secondary effects (unlike primary) change regularly with bond breaking and bond making, so that the kinetic effect will be intermediate between the calculated equilibrium effects for the systems either side of the rate-limiting step. For example, the acid catalysed cleavage of ethylene oxide could be $S_N1$ or $S_N2$:

$$
\begin{array}{ccc}
& & \overset{+}{C}H_2 \\
& & | \\
& & CH_2OH \quad + H_2O \longrightarrow \text{products} \\
& S_N1 \nearrow & 0.33 \\
CH_2 & CH_2 & \\
| \rangle O \ + H_3O^+ \rightleftharpoons & | \rangle OH^+ + H_2O & \\
CH_2 & CH_2 & \\
& 0.48 & H_2O^+ - CH_2 \\
& S_N2 \searrow & | \\
& & CH_2OH \\
& & 0.69
\end{array}
$$

The italic numbers show the $K_H/K_D$ values relating the reactants to the various other stages of the reaction. For $S_N1$ the transition state structure should be close to that of the high-energy carbenium ion intermediate (according to the Hammond postulate §3.4) and $K_H/K_D$ should be close to 0.33. For $S_N2$ the transition state should be closer to the cyclic oxonium ion and $K_H/K_D$ would be expected to be a little above 0.48. The observed value is 0.45, not accurately in accord with either mechanism (reflecting the relatively crude approximations in this approach) but pointing more towards $S_N2$ than $S_N1$ [56JA6008].

In reactions in which a proton is transferred in the rate-limiting step the observed kinetic solvent isotope effect is the product of the primary and secondary terms.

## 5.4 Kinetic electrolyte effects

The rates of many reactions, particularly those involving ions, are markedly influenced by the addition of electrolytes even though the added ions do not themselves participate in the reaction. The origin of these effects, commonly called *salt effects*, lies in the electrostatic interactions between ions, which introduce deviations from ideal behaviour even at low concentrations. These deviations depend on, and apply to, all the ions in solution, so changes in the concentrations of non-reacting ions can alter the activities of reacting ones.

Consider a reaction in which two reactant species come together in the transition state: $A + B \rightarrow AB^{\ddagger} \rightarrow$ products. Transition state theory (§2.3.3) gives us an expression for the rate of reaction:

$$\text{Rate} = v^{\ddagger}[AB^{\ddagger}]$$

where $v^{\ddagger}$ is the rate of decomposition of the transition state. In non-ideal conditions the pseudo-equilibrium constant for the formation of the transition state must be expressed in activities:

$$K^{\ddagger} = a_{AB\ddagger}/a_A a_B = y^{\ddagger}[AB^{\ddagger}]/y_A y_B[A][B]$$

$$\text{Rate} = k[A][B] = v^{\ddagger}K^{\ddagger}\frac{y_A y_B}{y^{\ddagger}}[A][B]$$

$$k = v^{\ddagger}K^{\ddagger}y_A y_B/y^{\ddagger} \qquad [5.1]$$

We now invoke the Debye–Hückel theory, which predicts the activity coefficient of an ion in terms of its electrostatic interactions with other ions in the neighbourhood:

$$\log y = -Dz^2 I^{\frac{1}{2}} \qquad [5.2]$$

$D$ is a constant for a given solvent at a given temperature (for water at 25 °C it has the value of 0.511 $l^{\frac{1}{2}}$ mol$^{-\frac{1}{2}}$), $z$ is the charge number of the ion; $I$ is the ionic strength of the solution, defined as:

$$I = \tfrac{1}{2}\Sigma c_i z_i$$

where $c_i$ is the concentration and the sum is taken over all the ions in solution.

Now the charge on the transition state, $z^{\ddagger}$, must be the sum of the charges on its precursors, $z_A + z_B$, so if we combine equations [5.1] and [5.2] we can write:

$$\log k = \log v^{\ddagger}K^{\ddagger} - D\{z_A^2 + z_B^2 - (z_A + z_B)^2\}I^{\frac{1}{2}}$$
$$= \log v^{\ddagger}K^{\ddagger} + 2Dz_A z_B I^{\frac{1}{2}}$$

In the limit when $I = 0$, the right hand term vanishes, whence:

$$\log(k/k_0) = 2Dz_A z_B I^{\frac{1}{2}} \qquad [5.3]$$

where $k_0$ is the rate coefficient extrapolated to infinite dilution. In water at 25 °C, $2D \simeq 1$, and the equation simplifies further to:

$$\log(k/k_0) = z_A z_B \, I^{\frac{1}{2}}$$

In dilute solutions (the simple Debye–Hückel relation does not apply at concentrations where ion association becomes important), equation [5.3] holds with remarkable accuracy. Its conceptual significance is closely related to the ideas of Hughes and Ingold as expressed in table 5.4, since increasing the ionic strength of a solution increases its polarity. Reactions between ions of like charge ($z_A z_B$ positive) are accelerated by the addition of electrolytes; those between ions of opposite charge ($z_A z_B$ negative) are retarded.

The Hughes–Ingold analysis also points to the reason for deviations from equation [5.3]. These are most commonly encountered in reactions involving neutral molecules ($z_A z_B$ zero), which nevertheless frequently do exhibit electrolyte effects. Many reactions between neutral, relatively non-polar reactants pass through a transition state which, though it has no net charge, is very dipolar (e.g. type 4, table 5.4); it can be stabilised by an ionic medium, and a large positive electrolyte effect is found. Conversely reactions between ions and neutral molecules frequently show a small negative electrolyte effect because the charge in the transition state is more dispersed than in the ionic reactant.

The existence of electrolyte effects has an obvious significance in the conduct of kinetic experiments, since it will not always be apparent whether a change of rate is due to the actual participation of added ions in the reaction, or only to an electrolyte effect. It is best to carry out control experiments using an inert electrolyte to compare with the behaviour of the potentially reactive species. Alternatively the reactions can be carried out in the presence of a large excess of an inert electrolyte so that the ionic strength remains effectively constant.

The effects described above are called *primary* kinetic electrolyte effects because they directly influence the rate-limiting step. Ions can also modify reaction rates by changing the positions of pre-equilibria and this is called a *secondary* kinetic electrolyte effect. It is commonly encountered in reactions subject to specific acid catalysis (§4.2.1) where the rate is proportional to $[H_3O^+]$. If the catalyst, HA, is a relatively weak acid then $[H_3O^+]$ can be altered by the influence of electrolytes on the dissociation:

$$HA + H_2O \rightleftharpoons H_3O^+ + A^-$$

$$K_a = \frac{[H_3O^+][A^-]}{[HA]} \cdot \frac{y_{H_3O^+} y_{A^-}}{y_{HA}} \qquad [5.4]$$

Rearranging this equation, taking logarithms, and applying the Debye–Hückel relation [5.2] with the approximation that $D = \frac{1}{2}$ gives:

$$\log[H_3O^+] = \log K_a - \log[A^-]/[HA] + I^{\frac{1}{2}}$$

This shows that increasing the ionic strength accelerates the reaction by shifting the protonation equilibrium.

Note that this type of electrolyte effect is not generated by acids of the charge type $HA^+$; now the activity coefficient term in equation [5.4] takes the form $y_{H_3O^+}y_A/y_{HA^+}$, and the charge contributions cancel out. This also applies to catalysis by strong acids which are fully ionised to $H_3O^+$.

Electrolyte effects in specific base catalysis are similar: catalysis by relatively weak neutral bases such as ammonia is enhanced by increasing ionic strength because this increases $[HO^-]$. For bases of the type $B^-$ the effect is small.

The investigation of general acid or base catalysis (§4.2.2) is hampered by the intrusion of secondary electrolyte effects. The method requires that the *concentration* of a buffer is varied while the buffer *ratio*, $[A^-]/[HA]$, is held constant to maintain a constant pH. It is therefore necessary to balance the buffer concentration with the complementary addition of an inert salt in order that the ionic strength is kept constant. If this is not done the rate changes due to the electrolyte effect may be misinterpreted as evidence for general acid or base catalysis.

## 5.5    Further reading

J. H. Hildebrand, J. M. Prausnitz, & R. L. Scott, *Regular and Related Solutions*, Van Nostrand Reinhold, New York, 1970.

J. F. Coetzee & C. D. Ritchie (eds), *Solute–Solvent Interactions*, Marcel Dekker, New York, 1969.

F. Franks (ed.), *Water: a Comprehensive Treatise*, vol. 3, Plenum, New York, 1973.

C. Reichardt, *Solvent Effects in Organic Chemistry*, Verlag Chemie, Weinheim, 1979.

J. B. F. N. Engberts, in *Water: a Comprehensive Treatise*, ed. F. Franks, vol. 6, Plenum, New York, 1979.

R. W. Taft, *Progr. Phys. Org. Chem.*, 1983, **14**, 247 (gas phase acidities and basicities).

E. M. Arnett, *Acc. Chem., Res.*, 1973, **7**, 404 (gas phase proton transfer).

F. M. Jones & E. M. Arnett, *Progr. Phys. Org. Chem.*, 1974, **11**, 263 (ionisation and hydration of amines).

E. M. Arnett, in *Proton-Transfer Reactions*, ed. E. F. Caldin & V. Gold, Chapman & Hall, London, 1975 (ionisation and hydration of amines).

J. T. Edward, *J. Chem. Educ.*, 1982, **59**, 354 (ionisation and hydration of acids).

M. H. Abraham, *J. Amer. Chem. Soc.*, 1982, **104**, 2085 (thermodynamics of solvation of gaseous non-electrolytes).

T. H. Morton, *Tetrahedron*, 1982, **38**, 3195 (gas phase reactions).

A. J. Parker, *Chem. Rev.*, 1969, **69**, 1 (reactions in dipolar aprotic solvents; differential solvation).

E. Buncel & H. Wilson, *Adv. Phys. Org. Chem.*, 1977, **14**, 133 (reactions in dimethyl sulphoxide).

E. Buncel & H. Wilson, *Acc. Chem. Res.*, 1979, **12**, 42 (differential solvation).

M. H. Abraham, *Progr. Phys. Org. Chem.*, 1974, **11**, 1 (differential solvation; solvent polarity parameters).

C. Reichardt, *Angew. Chem. Internat. Ed.*, 1965, **4**, 29; 1979, **18**, 98 (solvent polarity parameters).

M. J. Kamlet, J. L. M. Abboud, and R. W. Taft, *Progr. Phys. Org. Chem.*, 1981, **13**, 485 (solvent polarity parameters).

H. Langhals, *Angew. Chem. Internat. Ed.*, 1982, **21**, 724 (solvent polarity parameters for mixed solvents).

V. Gold, *Adv. Phys. Org. Chem.*, 1969, **7**, 259 (solvent isotope effects).

R. L. Schowen, *Progr. Phys. Org. Chem.*, 1972, **9**, 275 (solvent isotope effects).

W. J. Albery, in *Proton-Transfer Reactions*, eds E. F. Caldin & V. Gold, Chapman & Hall, London, 1975 (solvent isotope effects).

M. C. R. Symons, *Acc. Chem. Res.*, 1981, **14**, 179 (water structure and reactivity).

W. A. P. Luck, *Angew. Chem. Internat. Ed.*, 1979, **18**, 350; 1980, **19**, 28 (structures of liquids).

M. D. Zeidler, *Angew. Chem. Internat. Ed.*, 1980, **19**, 697 (structures of liquids).

J. A. Barker and D. Henderson, *Scientific American*, 1981, **245**, No. 5, 94 (structures of liquids).

# 6    Molecular orbital methods

## 6.1    Perturbation theory

### 6.1.1    Two-orbital interactions

The description of molecular orbitals in terms of interacting atomic orbitals is called *perturbation theory*. We imagine two atomic orbitals coming together: as they do so they perturb each other generating two molecular orbitals, one of higher energy than either atomic orbital, one of lower energy. The extent of their interaction can be described using three parameters. The *Coulomb integral*, $\alpha$, is the energy of the isolated atomic orbital. The *overlap integral*, $S$, is a measure of the extent of the overlap between the interacting atomic orbitals, and the *resonance* or *bond integral*, $\beta$, relates to the energy of the interaction. Energies are customarily measured relative to the energy of an electron at infinity, so $\alpha$ and $\beta$ are negative. In terms of these parameters the energy, $E$, of the molecular orbitals formed from atomic orbitals A and B is given by:

$$(\alpha_A - E)(\alpha_B - E) = (\beta - SE)^2 \qquad [6.1]$$

If the two atomic orbitals are degenerate (for example, the hydrogen $1s$ orbitals in the construction of hydrogen molecular orbitals), $\alpha_A = \alpha_B = \alpha$:

$$(\alpha - E) = \pm(\beta - SE)$$

Therefore:

$$E = \frac{\alpha + \beta}{1 + S} \quad \text{or} \quad \frac{\alpha - \beta}{1 - S}$$

This is illustrated in figure 6.1(*a*). The frequently used simplification of neglecting overlap ($S = 0$) leads to:

$$E = \alpha \pm \beta \quad \text{(figure 6.1(}b\text{))} \qquad [6.2]$$

The overlap term introduces asymmetry into the orbital splitting pattern; the $\sigma$ molecular orbital is less bonding than $\sigma^*$ is antibonding. Thus whereas the hydrogen molecule is stable, with both electrons in the $\sigma$ orbital, the interaction of two helium atoms, which would fill both $\sigma$

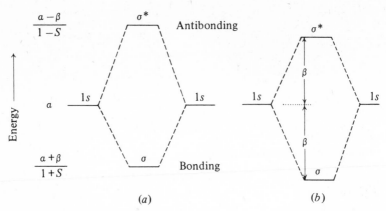

Figure 6.1. Mutual perturbation of two $1s$ orbitals, generating $\sigma$ bonding and $\sigma^*$ antibonding molecular orbitals: (a) including overlap; (b) neglecting overlap.

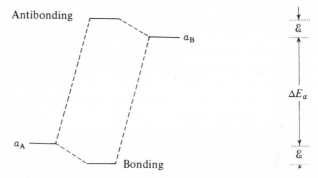

Figure 6.2. Second order perturbation between orbitals of markedly different energies.

and $\sigma^*$ orbitals, is not just energetically neutral as the simplified picture of figure 6.1(b) suggests, but is actually repulsive.

Figure 6.2 illustrates the other extreme – the interaction of two orbitals whose energies, $\alpha_A$ and $\alpha_B$, are very different. To a first approximation $(\alpha_B - E)$ for the bonding molecular orbital can be replaced by $(\alpha_B - \alpha_A)$, and likewise for the antibonding orbital $(\alpha_A - E)$ approaches $(\alpha_A - \alpha_B)$. As a further simplification we may neglect $S$; this is a less drastic assumption than in the previous case because overlap between very different orbitals is likely to be small. Thus equation [6.1] becomes:

$$(\alpha_A - E)(\alpha_B - \alpha_A) = \beta^2$$
$$(\alpha_A - \alpha_B)(\alpha_B - E) = \beta^2$$

The energy differences $\mathscr{E}$ (figure 6.2) are related to the separation between the atomic orbitals, $\Delta E_\alpha$, by the expression:

$$\mathscr{E} = \beta^2/\Delta E_\alpha \qquad\qquad [6.3]$$

The interaction between degenerate (or closely similar) orbitals, which to a first approximation is measured directly by $\beta$ (equation [6.2]), is called a *first order* perturbation. That between widely different orbitals depends on $\beta^2$ and is called *second order*. There are clearly intermediate stages where the splitting depends on $\beta$ in a more complex manner.

The wave functions for the molecular orbitals, $\Psi$, can be represented as combinations of the atomic orbitals, $\psi$, according to the equations:

$$\Psi_+ = c_A\psi_A + c_B\psi_B \qquad \text{(bonding)}$$

$$\Psi_- = c_A^*\psi_A - c_B^*\psi_B \qquad \text{(antibonding)}$$

where the coefficients $c$ measure the relative contribution of each atomic orbital to the molecular orbital. For normalised orbitals ($\oint\Psi^2 = 1$; $\oint\psi^2 = 1$) the sums of the squares of the coefficients – that is, $c_A{}^2 + c_B{}^2$, $c_A^{*2} + c_B^{*2}$, $c_A{}^2 + c_A^{*2}$, and $c_B{}^2 + c_B^{*2}$ – are all unity. If the atomic orbitals are identical then their coefficients must all be the same: $c_A = c_B = c_A^* = c_B^* = 1/\sqrt{2}$. If the atomic orbitals are widely different the bonding molecular orbital closely resembles the lower energy atomic orbital, $A$, and the antibonding molecular orbital resembles the higher energy atomic orbital, $B$: $c_A \gg c_B$ and $c_A^* \ll c_B^*$ (compare §6.3).

### 6.1.2 Localised two-centre bonds

It requires an enormous conceptual jump to go from the simple interaction of two atomic orbitals to the much more complex pattern of interactions in multi-electron, multi-orbital systems. Fortunately, for most purposes of organic chemistry the description of saturated compounds and of those containing unconjugated multiple bonds can be reduced to a very simple level.

The fact is that such compounds behave in almost every respect as a collection of atoms held together by localised bonds – bonds that have clearly defined properties of length, strength, polarity, polarisability, vibrational frequency, and so on, which can be transferred from one molecule to another. The heats of atomisation of alkanes can be accurately expressed as the sum of constituent C–C and C–H bond energies, and dipole moments can be calculated by vector addition of individual bond moments. It is this behaviour which leads to the familiar concept of the localised two-centre molecular orbital constructed from hybridised atomic orbitals on each carbon atom.

This is not to say that the localised bond is 'right'. It provides a simple model that is adequate for most chemical purposes and that is readily visualised. It fails – at least in its simplest form – to predict that there are two different kinds of bonding molecular orbital in methane as revealed

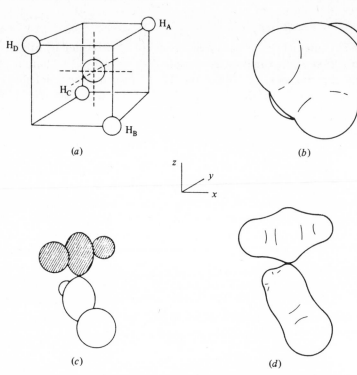

Figure 6.3. A delocalised molecular orbital treatment of the bonding in methane. Diagram (*a*) shows the geometrical relation between the tetrahedral symmetry of the methane molecule and Cartesian coordinates. The spherical $2s$ orbital of carbon can interact in phase with all four hydrogen $1s$ orbitals to produce a delocalised bonding molecular orbital (*b*) with tetrahedral symmetry. A carbon $2p$ orbital interacts with the hydrogen $1s$ orbitals as shown in (*c*), where one lobe of $2p_z$ interacts with $H_A$ and $H_D$ and the other lobe with $H_B$ and $H_C$. This generates a delocalised molecular orbital (*d*), of higher energy than (*b*), with a nodal plane through the carbon atom. Two similar orbitals are formed by the analogous interactions of $2p_x$ and $2p_y$.

by photoelectron spectroscopy and as predicted by the delocalised molecular orbital treatment illustrated in figure 6.3. On the other hand, the delocalised picture does not readily relate to the chemically satisfying concept of discrete carbon–hydrogen bonds and it is difficult to extend to large molecules: the pattern of delocalised molecular orbitals becomes rapidly more complex as one has to handle the simultaneous interactions of more and more atomic orbitals.

Unconjugated $\pi$ bonds, too, can be treated in isolation as the interaction of two unhybridised $p$ orbitals, independent of the $\sigma$ framework which lies in a plane orthogonal to them. Their mutual perturbation generates bonding and antibonding $\pi$ orbitals (figure 6.4) with energies, to

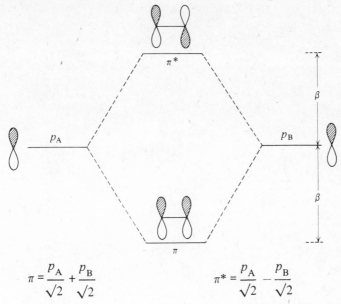

$$\pi = \frac{p_A}{\sqrt{2}} + \frac{p_B}{\sqrt{2}} \qquad\qquad \pi^* = \frac{p_A}{\sqrt{2}} - \frac{p_B}{\sqrt{2}}$$

Figure 6.4. Mutual perturbation of two atomic $p$ orbitals. (Note the conventional representation of molecular orbitals in terms of their constituent atomic orbitals.)

a first approximation, given by the simple first order expression: $E = \alpha \pm \beta$ (equation [6.2]).

### 6.1.3 Conjugation and delocalisation

In conjugated systems, with three or more adjacent atoms involved in $\pi$ bonding, one must consider the extended overlap of the $p$ orbitals. In butadiene, for example, there is bound to be significant interaction between the two inner $p$ orbitals. The butadiene $\pi$ system can be visualised as a hypothetical construct of two molecules of ethylene and the effect of delocalisation can then be estimated by seeing how the two localised $\pi$ systems perturb each other.

There are two differences between intermolecular perturbation and the interatomic perturbations discussed previously. An atomic orbital is centred entirely on the atom; the ethylene $\pi$ orbital is shared between the two carbons, but only one of them is directly involved in the perturbation (figure 6.5) so the overlap is incomplete. The extent to which each orbital participates is given by the coefficients $c_x$ and $c_y$ which measure the contribution of the separate atomic $p$ orbitals in the total $\pi$ molecular orbital. In the present example $c_x = c_y = 1/\sqrt{2}$ (figure 6.4). We must also take into account that there are *two* orbitals, $\pi$ and $\pi^*$, on each ethylene fragment and all four mutual perturbations must be considered. The

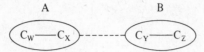

Figure 6.5. Hypothetical construct of butadiene from two ethylene fragments, A and B, linked via atoms $x$ and $y$.

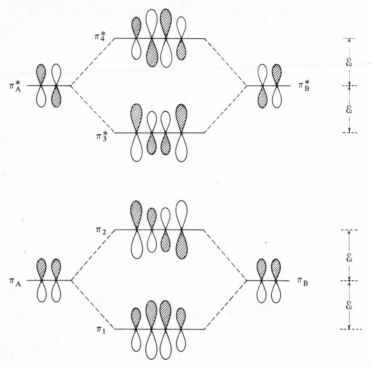

Figure 6.6. First order perturbation of ethylene $\pi$ orbitals in construction of butadiene $\pi$ system.

interaction between the two bonding orbitals is first order, generating the molecular orbitals $\pi_1$ and $\pi_2$ (figure 6.6). The corresponding interaction between the antibonding orbitals gives $\pi_3^*$ and $\pi_4^*$. The energy changes that accompany these perturbations are given by introducing the coefficients $c_x$ and $c_y$ into equation [6.2]:

$$E = \alpha \pm c_x c_y \beta \qquad [6.4]$$

whence $\mathscr{E} = \beta/2$. However, the total energy of the four bonding electrons is not altered by this interaction; the lowering of $\pi_1$ is counterbalanced by the raising of $\pi_2$. Indeed, if overlap is taken into account (equation [6.1]), the first order perturbation is repulsive.

The interaction of bonding with antibonding orbitals is second order.

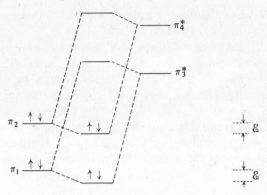

Figure 6.7. Second order perturbation between bonding and antibonding orbitals in construction of butadiene $\pi$ system.

It is conveniently treated as a subsidiary perturbation of the first order molecular orbitals (figure 6.7). Orbitals of opposite symmetry cannot mix with each other. For example, if $\pi_1$ were to interact with $\pi_4^*$ the in-phase overlap on carbon atoms 1 and 3 would be exactly cancelled by the corresponding out-of-phase overlap on carbon atoms 2 and 4. However, $\pi_2$ and $\pi_4^*$, which are both symmetrical about the centre of the molecule, can mix with each other, and likewise $\pi_1$ and $\pi_3^*$, which are both antisymmetrical about the same point. The magnitude of this second order perturbation, by analogy with equation [6.3], is:

$$\mathscr{E} = (c_x c_y \beta)^2 / \Delta E \qquad [6.5]$$

and since $\Delta E$ in this case is $2\beta$ (figure 6.4), $\mathscr{E}$ is only $\beta/8$. However, unlike the first order perturbation this lowers the energy of all the bonding electrons; the delocalised four-centre $\pi$ system is thus more stable than the isolated $\pi$ bonds by $4\mathscr{E} = \beta/2$.

It is not usually easy to calculate the coefficients, $c$, that measure the contributions of each $p$ atomic orbital to a delocalised molecular orbital. For straight chain polyenes with $n$ carbon atoms they are given by the following expression:

$$c_{ij} = \{2/(n + 1)\}^{\frac{1}{2}} \sin \{ij\pi/(n + 1)\}$$

Here $i$ is the number of the molecular orbital in question (starting with $i = 1$ for the lowest), and $j$ is the position of the particular carbon atom in the chain. For butadiene this gives the following values (compare the qualitative representations of orbital size and phase in figure 6.6).

$$\left.\begin{array}{l} \Psi_1 = 0.37\,\psi_1 + 0.60\,\psi_2 + 0.60\,\psi_3 + 0.37\,\psi_4 \\ \Psi_2 = 0.60\,\psi_1 + 0.37\,\psi_2 - 0.37\,\psi_3 - 0.60\,\psi_4 \\ \Psi_3 = 0.60\,\psi_1 - 0.37\,\psi_2 - 0.37\,\psi_3 + 0.60\,\psi_4 \\ \Psi_4 = 0.37\,\psi_1 - 0.60\,\psi_2 + 0.60\,\psi_3 - 0.37\,\psi_4 \end{array}\right\} \qquad [6.6]$$

There are two features about these orbitals which are noteworthy and to which we shall return later (§6.2.2):

(*a*) For each individual molecular orbital $\Psi$ the sum of $c^2$ over all the constituent atomic orbitals is unity, and for each individual atomic orbital $\psi$ the sum of $c^2$ over all the molecular orbitals to which it contributes is also unity. This is a universal requirement for normalised orbitals since $\int \Psi^2 = 1$ and $\int \psi^2 = 1$.

(*b*) The orbitals are arranged in pairs, $\Psi_1$ and $\Psi_4$, $\Psi_2$ and $\Psi_3$, and the coefficients within each pair differ only in the sign on alternate atoms. This is a general property of so-called even alternant hydrocarbons (§6.2.1).

## 6.2 Alternant hydrocarbons

### 6.2.1 The pairing of orbitals

In constructing the $\pi$ system of butadiene from two localised $\pi$ bonds (§6.1.3) the pattern of orbitals remains throughout symmetrical about the energy level of an atomic $p$ orbital (compare figures 6.6 and 6.7). The same behaviour is found if we bring together two butadiene fragments to make an octatetraene or if butadiene interacts at each end with an ethylene fragment to generate the $\pi$ system of benzene. In all such cases we start with two symmetrically distributed sets of orbitals and their mutual perturbation is also symmetrical. For each bonding orbital of energy $\alpha - \mathscr{E}$ there is a corresponding antibonding orbital of energy $\alpha + \mathscr{E}$, where $\alpha$ is the energy of a single $p$ orbital.

A similar pattern appears if we bring together a single $p$ orbital, A, and an ethylene fragment, B, to generate the three-centre allyl $\pi$ system (figure 6.8). Bearing in mind the requirements of symmetry we see that $p_A$ and $\pi_B$ can interact in a bonding way to produce $\pi_1$, and $p_A$ and $\pi_B^*$ can interact

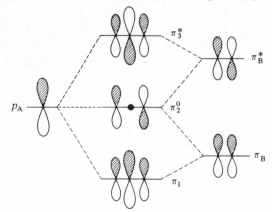

Figure 6.8. Perturbations between a single $p$ orbital and ethylene in construction of the allyl $\pi$ system.

in an antibonding way to produce $\pi_3^*$. The third molecular orbital can only be obtained by mixing all three contributory orbitals together, $(p_A - \pi_B) + (p_A + \pi_B^*)$, to give an orbital in which bonding and antibonding contributions cancel out. The result is the nonbonding orbital $\pi_2^0$, which has a node at the central carbon atom. Thus the symmetrical pattern is retained – bonding and antibonding orbitals running in pairs.

The only exception to this occurs with odd-membered rings. Figure 6.9 shows the interactions of a $p$ orbital and an ethylene fragment to generate cyclopropenyl. The perturbations now involve both ends of the ethylene

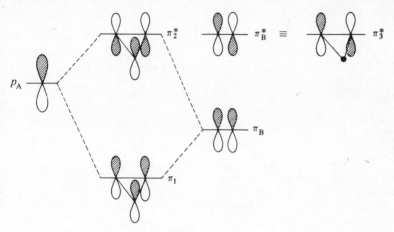

Figure 6.9. Perturbations between a single $p$ orbital and ethylene in construction of the cyclopropenyl $\pi$ system.

orbitals, and $p_A$ can interact with $\pi_B$ in either a bonding or antibonding way to produce $\pi_1$ and $\pi_2^*$. But there is no way that $p_A$ and $\pi_B^*$ can perturb each other; any interaction at one end is cancelled by the other. The third molecular orbital, shown as $\pi_3^*$, is in fact equivalent to the unperturbed $\pi_B^*$, so the cyclopropenyl $\pi$ system has one strongly bonding orbital $(E = \alpha + 2\beta)$ and two less strongly antibonding $(E = \alpha - \beta)$.

A $\pi$ system that has no odd-membered rings is called alternant, because the conjugated atoms can be divided into two groups – conventionally designated on formulae by starring one set – so that each atom is flanked only by atoms of the opposite group. Figure 6.10 gives some examples of odd and even alternant systems, and of two non-alternant systems containing odd-membered rings.

### 6.2.2 *Properties of alternant hydrocarbons*

Alternant systems are important because their structure and reactivity are amenable to very simple calculations – quite literally of the 'back-of-an-envelope' kind.

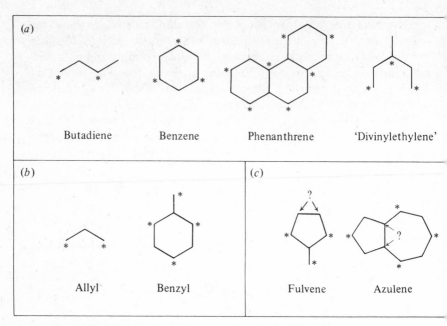

Figure 6.10. $\pi$ systems of conjugated hydrocarbons: (a) even alternant; (b) odd alternant; (c) non-alternant.

(a) *Electron distribution in even alternant hydrocarbons* (*EAH*). The orbitals of butadiene exemplify the general pattern for all EAH (compare equations [6.6] and discussion of them). For each atomic orbital, $\Sigma c^2$ over all the molecular orbitals is unity and, since the molecular orbitals are arranged in symmetrical bonding and antibonding pairs differing only in the sign of alternate coefficients, it follows that $\Sigma c^2$ over the bonding orbitals is 0.5. Each bonding orbital contains two electrons, so the total $\pi$ electron density on each atom is $2\Sigma c^2 = 1$; that is, there is an even distribution of $\pi$ electrons across the whole conjugated system.

(b) *Electron distribution in odd alternant hydrocarbons* (*OAH*). The key to the behaviour of OAH lies in the properties of the non-bonding molecular orbital (NBMO), for which the coefficient, $c_0$, can be found by a very simple procedure due originally to Longuet-Higgins and developed widely by Dewar [50JCP(18)265; 52JA3341]:
  (i) Star the larger group of alternate positions, as in figure 6.10.
  (ii) Set the coefficients for the non-starred positions equal to zero (compare the allyl system, figure 6.8).
  (iii) Set the sum of the coefficients of all starred atoms adjacent to any unstarred one equal to zero.

(iv) Apply the condition of normalisation for the whole orbital $\Sigma c_0^2 = 1$.

To illustrate these rules, consider how they apply to the benzyl system.

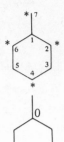

(i) Star four positions, leaving only three unstarred.

(ii) The coefficients at the unstarred positions are all zero.

(iii) Simple algebra gives the relative values of the coefficients for the starred positions. Start arbitrarily with a value of $x$ for $c_4$; then $c_2$ must be $-x$ so that the sum around position 3 is zero. Work through the molecule in this way. In larger systems one may have to introduce further arbitrary values, but in the end there are always enough equations to give the coefficients in terms of a single variable $x$.

(iv) $\Sigma c_0^2 = 7x^2 = 1$
$$x = \pm 1/\sqrt{7}$$

Note that though the relative signs of $c$ matter, their absolute signs do not.

The equation for the NMBO is therefore:

$$\Psi^0 = (1/\sqrt{7})(-\psi_2 + \psi_4 - \psi_6 + 2\psi_7)$$

The neutral benzyl radical has a single electron in the NBMO (figure 6.11(*a*)) and, as with the EAH, the total $\pi$ electron density on each carbon atom is unity. (Summing vertically for any particular atomic orbital: $\Sigma c^2 = \Sigma c^2$ (bonding) $+ \Sigma c^2$ (antibonding) $+ c^2$(NMBO) $= 1$. Symmetry requires: $\Sigma c^2$ (bonding) $= \Sigma c^2$ (antibonding). Electron density $= 2\Sigma c^2$ (bonding) $+ c^2$ (NBMO) $= 1$.)

Ionisation of the benzyl radical by abstracting an electron leaves a positive 'hole' where the non-bonding electron had been (figure 6.11(*b*));

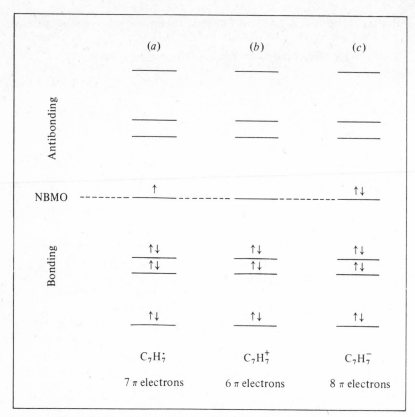

Figure 6.11. $\pi$ molecular orbitals of benzyl system.

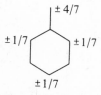

Figure 6.12. Charge distribution in benzyl cation $(+)$ and anion $(-)$.

the charge is distributed around the ring according to the value of $c_0^2$ at each atom. Likewise in the benzyl anion (figure 6.11($c$)) it is the NBMO that accommodates the extra electron and the charge again follows $c_0^2$ (figure 6.12).

(*c*) *Delocalisation energies and aromaticity.* The energy change caused by the mutual perturbation of two EAH $\pi$ systems – as in the construction of butadiene from two ethylene fragments, §6.1.3 – is not, in general, easily calculated. The first order effects are self-cancelling, involving the fully occupied bonding orbitals, and the interaction energy comes from the relatively small second order perturbations between bonding and anti-bonding orbitals. To a first approximation the magnitude of this second order effect, for acyclic compounds, is independent of the actual EAH involved; one can keep on adding ethylene fragments and the total energy change is about the same each time. Thus acyclic conjugated polyenes *behave like* localised systems and this is reflected in their alternating bond lengths. The double bond (in the classical representation) is about the same length as the double bond of ethylene, but the 'single' bond is rather shorter than the single bond of ethane, about 148 pm instead of 154 pm. This is a consequence of the second order $\pi$ contribution. However, since this contribution is effectively constant the heats of atomisation can be calculated simply as the sum of individual bond energies, using a special C–C 'single' bond value (383 instead of 346 kJ mol$^{-1}$) to allow for it. As with the alkanes the use of a working model based on a localised picture is not the same as asserting that the system really is localised. The pattern of orbitals in butadiene (figure 6.6) shows that delocalisation is indeed important, but its effect on the total energy is small and can be built into the localised picture.

The coming together of two OAH fragments, on the other hand, is dominated by a large first order interaction between their NBMOs, which are necessarily degenerate (figure 6.13). The total change in $\pi$ energy, $\delta E_\pi$, follows from equation [6.4]:

$$\delta E_\pi = 2\mathscr{E} = 2 c_x c_y \beta \qquad [6.7]$$

Here $c_x$ and $c_y$ are the NBMO coefficients for the atoms $x$ and $y$ through which the interaction occurs.

Two allyl fragments can unite end-to-end to form hexatriene or they can complete the circle to form benzene (figure 6.14). Since first order perturbations are additive we merely apply equation [6.7] twice to calculate the value of $\delta E_\pi$ for the formation of benzene, which is seen to be more stable than the open chain triene: more stable, in other words, than would be calculated from tables of bond energies. Benzene, we say, is aromatic.

Figure 6.15 illustrates the corresponding interactions of allyl with a single $p$ orbital (which can be regarded as the limiting case of an OAH). Here the ring closing is energetically neutral because $c_x c_y$ and $c_{x'} c_y$ cancel out. A delocalised cyclobutadiene is not merely not aromatic, it is actually antiaromatic, less stable than its open chain analogue or than tabulated bond energies predict.

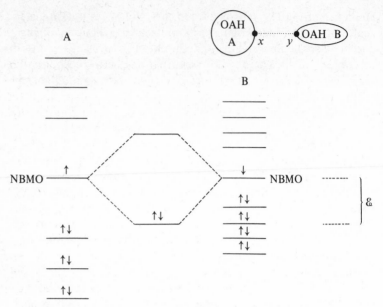

Figure 6.13. First order perturbation between NBMOs of two OAH fragments, A and B.

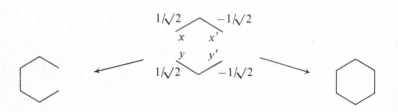

$$\delta E_\pi = 2c_x c_y \beta = \beta \qquad\qquad \delta E_\pi = 2(c_x c_y \beta + c_{x'} c_{y'} \beta) = 2\beta$$

Figure 6.14. Construction of $\pi$ systems of hexatriene and of benzene from allyl fragments.

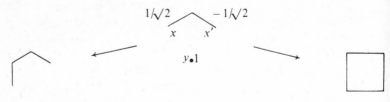

$$\delta E_\pi = 2c_x c_y \beta = \sqrt{2}\beta \qquad\qquad \delta E_\pi = 2(c_x c_y \beta + c_{x'} c_{y'} \beta) = 0$$

Figure 6.15. Construction of $\pi$ systems of butadiene and of cyclobutadiene from an allyl fragment and a single $p$ orbital.

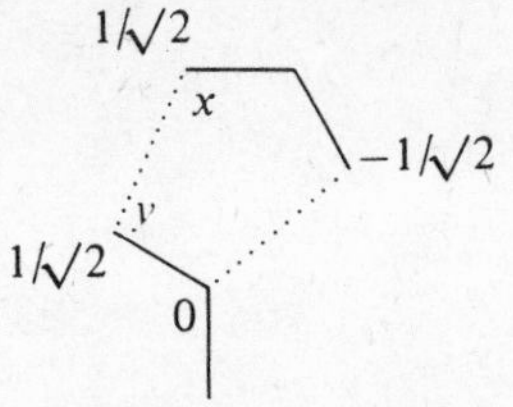

$$\delta E_\pi = 2c_x c_y \beta + 0 = \beta$$

Figure 6.16. Construction of $\pi$ system of fulvene from allyl fragments.

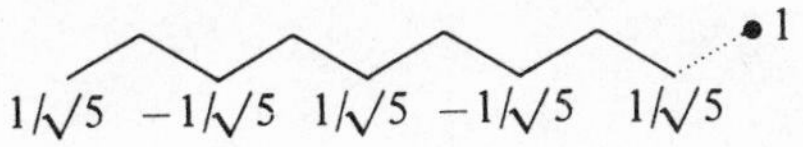

$$\delta E_\pi = 2\beta/\sqrt{5}$$

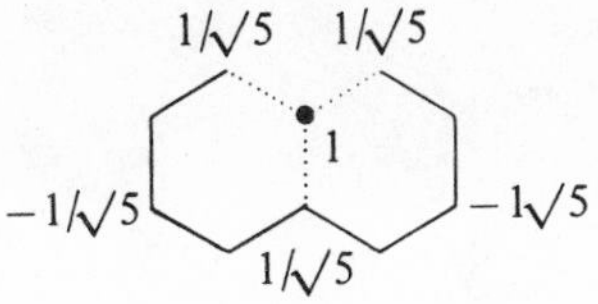

$$\delta E_\pi = 3 \times 2\beta/\sqrt{5} = 6\beta/\sqrt{5}$$

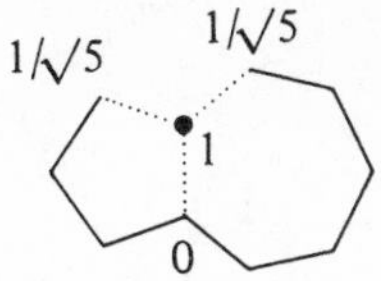

$$\delta E_\pi = 2 \times 2\beta/\sqrt{5} = 4\beta/\sqrt{5}$$

Figure 6.17. Construction of $\pi$ systems of decapentaene, naphthalene and azulene from a 9-$\pi$ OAH and a single $p$ orbital.

(*d*) *Extension to non-alternant hydrocarbons.* The method described in the previous paragraphs can be used to estimate the delocalisation energy of even non-alternant hydrocarbons. Figure 6.16 illustrates the formation of fulvene from two allyl residues. It is calculated to be no more stable than the open chain triene; that is, it is non-aromatic.

Azulene, on the other hand, though less stable than naphthalene, is nevertheless aromatic (figure 6.17).

## 6.3 Heteroatomic systems

When two orbitals of different energies perturb each other the resulting bonding orbital resembles the lower energy contributor and the antibonding orbital resembles the higher. This is reflected in the coefficients of the molecular orbitals and in the electron distribution. Figure 6.18 illustrates this for the localised carbon–oxygen $\sigma$ and $\pi$ bonds.

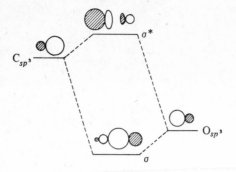

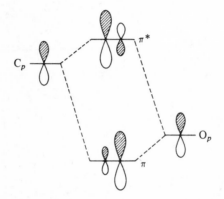

In each case: $\Psi_+ = c_C \psi_C + c_O \psi_O \quad (c_C \ll c_O)$

and: $\Psi_- = c_C^* \psi_C - c_O^* \psi_O \quad (c_C^* \gg c_O^*)$

$$\frac{c_C}{c_O} = \frac{c_O^*}{c_C^*} = 1 + \left(\frac{\Delta E}{4\beta^2}\right)^{\frac{1}{2}} - \frac{\Delta E}{2\beta}$$

Figure 6.18. Polarisation of carbon-oxygen bonds. Bonding electrons are concentrated near the oxygen atom.

The same principle applies to delocalised orbitals, but the pattern of changes introduced when a carbon atom is replaced by something else is qualitatively a little more complex, and calculating the coefficients usually calls for a computer.

The $\pi$ system of an enolate anion is related to that of the allyl anion (figure 6.19). By comparison with figure 6.8 we should expect the oxygen to play a major role in the lowest energy molecular orbital, a minor one in $\pi_2$, and an insignificant one in $\pi_3^*$. Thus oxygen bears the bulk of the negative charge (aided by the polarisation of the $\sigma$ bond). However, in the $\pi_2$ orbital the largest single contribution comes from the $p$ orbital of $C_2$, and this has important consequences (§6.4.1).

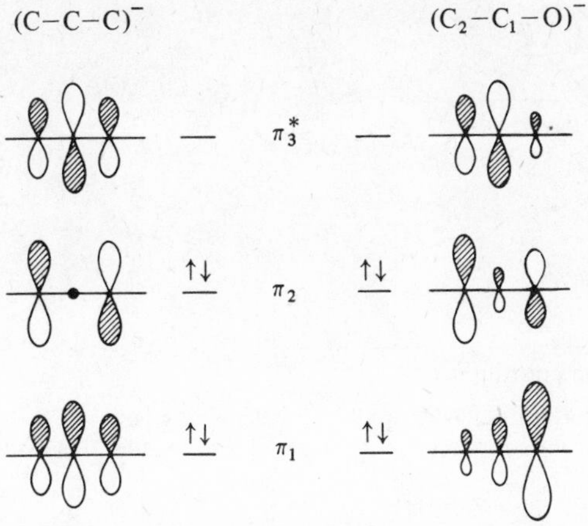

$(C-C-C)^-$ $\qquad\qquad$ $(C_2-C_1-O)^-$

Figure 6.19. Comparison between $\pi$ orbitals of allyl and enolate anions. (It is not intended to suggest that the orbitals of the two systems have the same energies.)

## 6.4 Perturbation theory and reactivity

### 6.4.1 *Frontier molecular orbitals and Coulombic effects*

Another type of system that cannot be adequately described in terms of localised two-centre bonds is the partially bonded structure of a transition state. In an $S_N2$ reaction, for example, the central carbon atom is approximately $sp^2$ hybridised and the remaining $p$ orbital is simultaneously interacting with both nucleophile and leaving group. Advanced molecular orbital techniques, aided by powerful computers, are beginning to make progress in calculating the energies of reactants and of various possible transition states and hence to predict the course and rate of a reaction. The problems, however, are immense. The total energy of the benzene molecule relative to its isolated nuclei and electrons is some $600\,000$ kJ mol$^{-1}$; to estimate the rate of a reaction even within a factor of ten requires that the energy difference between reactants and transition state be known to within 6 kJ mol$^{-1}$.

Perturbation theory provides a very simple way, without elaborate calculations, of gaining insight into reactivity, particularly the relative reactivities of two similar systems or of two alternative modes of reaction of a particular substrate. In the previous sections we have considered the formation of molecular orbitals in terms of a hypothetical coming together of atomic orbitals or molecular orbitals from smaller fragments. The same approach can be applied to the real coming together of two reactants: how does the energy of the system change as the orbitals begin to perturb each other? In general it is not possible to follow the change right through to the transition state. This involves considerable changes

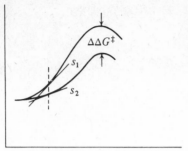

Reaction coordinate

Figure 6.20. Perturbation theory applied to reactivity: $s_1$ and $s_2$ are the slopes of the reaction profiles at similar early stages of the reaction. It is assumed that $\Delta\Delta G^{\ddagger} \propto \Delta s$.

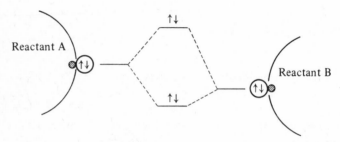

Figure 6.21. Closed shell repulsion. The total effect is the sum of all such repulsions between interacting orbitals.

in the positions of the nuclei and is clearly more than just an electronic perturbation. What we can do is to see how the reaction begins. This is usually an indication of how it will continue; unless some unlooked-for factor intervenes – a steric interaction, for example – the reaction coordinates for two similar reactions tend to keep in step with each other (figure 6.20). If it were not so it is hard to see how linear Gibbs energy relations could be so common. Perturbation theory, as applied to chemical reactivity, therefore postulates that the initial perturbation governs the course of the subsequent reaction.

When two reactants approach each other the total change in energy can be divided into three contributory parts:

(a) *Interactions between fully occupied orbitals* (figure 6.21). To a first approximation these are self-cancelling (compare figures 6.1(*b*) and 6.2). When overlap is taken into account the net effect is repulsive (as in $He_2$), and these repulsions are the main contributors to the activation energy for a reaction. Nevertheless, in comparing *relative* reactivities the *differ-*

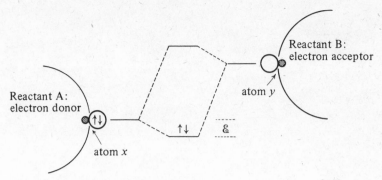

Figure 6.22. The covalent term.

*ences* in repulsion terms are usually small and will be ignored in future discussion. There is in fact little justification for this, but as a first approximation it gives useful results.

(*b*) *Interactions between filled orbitals of one reactant and vacant orbitals of the other* (figure 6.22). In general the interacting orbitals have significantly different energies, so this is a second order perturbation. It is a stabilising interaction; the energy of the occupied orbital is lowered and the corresponding increase in the energy of the other orbital has no effect since it is empty of electrons. The stabilisation generated by a single interaction is $2\mathscr{E}$ since two electrons are involved; from equation [6.5] this equals $2(c_x c_y \beta)^2/\Delta E$, where $c_x$ and $c_y$ are the coefficients for atoms $x$ and $y$, the sites of reaction, in the interacting molecular orbitals. In principle one should consider all the interactions between occupied orbitals of A and vacant orbitals of B with the same symmetry, and vice versa. Notice, however, the term $\Delta E$ in the energy expression: the major contribution to the second order perturbation comes from the interactions between the occupied and unoccupied orbitals that are closest in energy. For an essentially qualitative approach to reactivity we may confine our attention to this one set of interactions, between the so-called *frontier molecular orbitals* (FMOs); figure 6.23 illustrates this.

Molecule A has a relatively high energy HOMO (highest occupied molecular orbital) and B has a relatively low energy LUMO (lowest unoccupied molecular orbital). The interaction between them is large (figure 6.23(*a*)). For the interaction of a lower lying occupied orbital of A with a higher unoccupied orbital of B, $\Delta E$ is much larger and the interaction is small (figure 6.23(*b*)). In figure 6.23(*c*) the interaction is between the HOMO of B and the LUMO of A, but this too is relatively unimportant. Clearly in this reaction A is a nucleophile, donating electrons from its high level HOMO to electrophilic B. (The discussion in

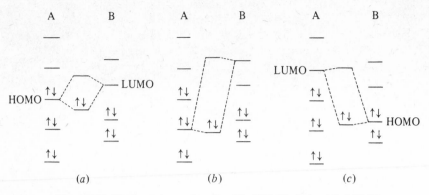

Figure 6.23. The importance of nucleophile HOMO-electrophile LUMO interaction.

this section is couched exclusively in terms of electron pairs, but it can obviously be extended to radical processes.)

(c) *Electrostatic interactions.* In reactions involving ions or dipolar molecules there is a simple Coulombic attraction or repulsion between the interacting atoms. Its magnitude is $Q_x Q_y/r$, where $Q_x$ and $Q_y$ are the charges on the interacting atoms and $r$ their separation. (The effective dielectric constant of the 'space' within a molecule is probably about 1.)

In short, a comparison between reactivities in similar systems can be made on the basis of only two criteria: the major HOMO–LUMO interaction, and the contribution from electrostatic attraction or repulsion. These can be summarised in the equation:

$$P = 2(c_x c_y \beta)^2/\Delta E_{\text{FMO}} - Q_x Q_y/r \qquad [6.8]$$

where $P$ is the fall in energy as the reactants approach each other (ignoring closed shell repulsions), and $\Delta E_{\text{FMO}}$ is the energy difference between the nucleophile HOMO and the electrophile LUMO.

Consider, for example, the alternative modes of reaction of an enolate anion with an electrophile:

$$C=C-O^- \quad E^+ \rightarrow C=C-O-E$$

$$E^+ \quad C=C-O^- \rightarrow E-C-C=O$$

Attack at oxygen tends to be charge-controlled, dominated by the Coulombic term of equation [6.8], whereas attack at carbon is orbital-controlled. The oxygen bears the bulk of the negative charge, but in the HOMO, which governs the covalent term, the carbon $p$ orbital is the major contributor (figure 6.19). Thus the proton – positively charged, small so that $r$ can be small, and with a relatively high LUMO – reacts more rapidly at oxygen. Bromine, on the other hand – uncharged, large, and with low-lying vacant orbitals – reacts much faster at carbon.

### 6.4.2  SHAB: soft and hard acids and bases

The Lewis definition of acids and bases extends beyond that of Brønsted and Lowry: a base is an electron-pair donor, an acid an electron-pair acceptor. For bases the two definitions are effectively synonymous, but Lewis acids include a wide range of species other than proton donors, in particular metal cations, carbenium ions and other electron-deficient compounds such as $BF_3$ and $CH_2$, and compounds with low-lying vacant $d$ orbitals such as the higher halogens.

One of the difficulties with Lewis' definition was that it could not be quantified; there are no self-consistent scales of Lewis acid strengths analogous to the $pK_a$ scale for proton acids. It had been clear to coordination chemists for many years that metal–ligand interactions fell into two fairly distinct classes. Ammonia, water and fluoride ion complex strongly with the cations of the alkali and alkaline earth metals but only weakly with those of the heavier transition metals such as $Hg^{2+}$ or $Pt^{2+}$. The pattern is reversed for complexes of phosphines, sulphides and the heavier halogen ions. (The old name for thiol, mercaptan, comes from *corpus mercurium captans*: mercury-catching.) In other words, $Hg^{2+}$ is a weaker Lewis acid than $Mg^{2+}$ if $H_2O$ is the reference base, but stronger if it is $H_2S$.

In 1963 Pearson brough a degree of rationalisation into the situation with the concept of hard and soft acids and bases. A soft species is fairly large, polarisable, and with relatively small charge; a hard one is small, of low polarisability, and usually charged. The SHAB principle states that hard acids react preferentially (both kinetically and thermodynamically) with hard bases, and soft with soft. Thus one could establish a rough scale of softness/hardness by setting up an equilibrium such as:

$$BH + MeHg^+ \rightleftharpoons MeHgB + H^+$$

A hard base, $B^-$, reacts preferentially with the hard acid $H^+$ and the equilibrium lies to the left; a soft base will move the equilibrium to the right. Table 6.1(a) lists some acids and bases in categories of hard, borderline and soft.

Pearson's original approach was entirely empirical, but it can be readily understood in terms of perturbation theory. The soft–soft interaction is largely covalent. Soft bases are of low electronegativity so their electron pairs are readily available (high energy HOMO); soft acids have low-lying vacant orbitals (low energy LUMO). The orbital contribution to their mutual perturbation is large and the interaction favour-

Table 6.1. *Soft and hard acids and bases*

(*a*) Qualitative division into hard, borderline, soft (after Pearson: see references in §6.5)

| Acids | Bases |
|---|---|
| | *Hard* |
| $H^+$ $Li^+$ $Na^+$ $K^+$ $Mg^{2+}$ $Ca^{2+}$ | $H_2O$ $HO^-$ $F^-$ $Cl^-$ $AcO^-$ |
| $Al^{3+}$ $Cr^{3+}$ $Fe^{3+}$ | $PO_4^{3-}$ $SO_4^{2-}$ $ClO_4^-$ $NO_3^-$ |
| $BF_3$ $AlMe_3$ $AlCl_3$ | $ROH$ $R_2O$ $RO^-$ |
| $SO_3$ $RCO^+$ $CO_2$ | $NH_3$ $RNH_2$ $N_2H_4$ |
| | *Borderline* |
| $Fe^{2+}$ $Cu^{2+}$ $Zn^{2+}$ | $PhNH_2$ $N_3^-$ $Br^-$ $SO_3^{2-}$ |
| $SO_2$ $R_3C^+$ | |
| | *Soft* |
| $RHal$ $ROTos$ | $RSH$ $R_2S$ $RS^-$ $HS^-$ |
| $Cu^+$ $Ag^+$ $Hg^{2+}$ $MeHg^+$ | $I^-$ $SCN^-$ $CN^-$ |
| $BH_3$ | $R_3P$ |
| $Br^+$ $I^+$ $HO^+$ $Br_2$ $I_2$ | $C_2H_4$ $C_6H_6$ $R^-$ $H^-$ |

(*b*) Order of softness for some inorganic ions derived from 'orbital electro-negativities' (after Klopman [68JA223])

| | *Hard* | *Soft* |
|---|---|---|
| Acids | $Al^{3+}$ $Mg^{2+}$ $Ca^{2+}$ $Fe^{3+}$ $Fe^{2+}$ $Li^+$ $H^+$ $Na^+$ $Cu^{2+}$ $Cu^+$ $Ag^+$ $Hg^{2+}$ | |
| Bases | $F^-$ $HO^-$ $Cl^-$ $Br^-$ $CN^-$ $SH^-$ $I^-$ $H^-$ | |

able. The hard–hard interaction is largely electrostatic, between ions or strongly dipolar molecules; they tend to be small so that they can approach each other closely, increasing the Coulombic attraction. A high charge similarly encourages hardness: $Fe^{3+}$, for example, is harder than $Fe^{2+}$. Hard acids tend to have high energy, inaccessible LUMOs and hard bases, clinging tightly to their electron pairs, have low energy HOMOs. Hard–soft interactions are weak: the FMO of the hard com-ponent is too high or too low for strong covalent interaction, and the size and low polarity of the soft component minimises electrostatic attraction. Table 6.1(*b*) sets down an order of softness for some inorganic ions based on this sort of argument and derived from FMO energies.

Most of the early work in the field was based on equilibria but the relevance of equation [6.8], which was discussed in terms of reaction rates, is apparent. More recently the SHAB principle has been extended to kinetic studies, particularly in organic chemistry where rates are often relatively slow and easily measured. The words acid and base in this context can be replaced by electrophile and nucleophile. Thus ethylene, a soft nucleophile because it lacks charge and has a fairly high energy $\pi$ orbital, reacts faster with bromine than with aqueous acid ($H^+$). The example discussed in §6.4.1 of the reactivity of the enolate anion shows that an ambident species may be hard at one reaction site (oxygen in this case) and soft at another (carbon).

### 6.4.3 Stereoelectronic effects

In orbital-controlled reactions the approaching frontier orbitals must be able to overlap effectively if the reaction is to proceed readily. A well-known example is the rear-side approach of the nucleophile in $S_N2$ reactions, governed by the need for the nucleophile HOMO to interact with the $\sigma^*$ LUMO of the substrate (§7.1.1). Likewise the preferred stereochemistry for elimination reactions (§8.2.1), with the two leaving groups coplanar, derives from the overlap between the frontier orbitals that generate the newly forming $\pi$ bond:

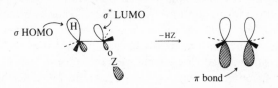

During cyclisation reactions the presence of the ring may prevent the approaching orbitals from following trajectories that would lead to effective overlap; the reaction may then be inhibited altogether or else it may follow an alternative pathway. In acyclic systems favourable inter-action between an oxygen lone pair (HOMO) of alcohols and the $\pi^*$ (LUMO) orbital on the $\beta$ carbon atom of a methacrylic ester leads to Michael addition in preference to transesterification:

$$MeOH + CH_2{=}CMe{-}COOEt$$

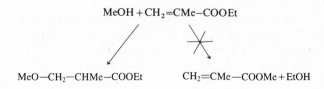

$$MeO{-}CH_2{-}CHMe{-}COOEt \qquad CH_2{=}CMe{-}COOMe + EtOH$$

However, in the intramolecular reaction of (**6.1**) the product is the lactone and not the intramolecular Michael adduct, because the hydroxyl HOMO cannot readily approach the LUMO at the terminal methylene group [76CC736]:

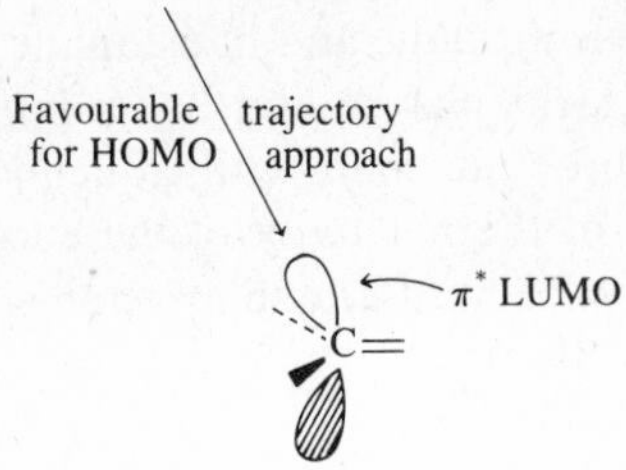

Favourable trajectory for HOMO approach

$\pi^*$ LUMO

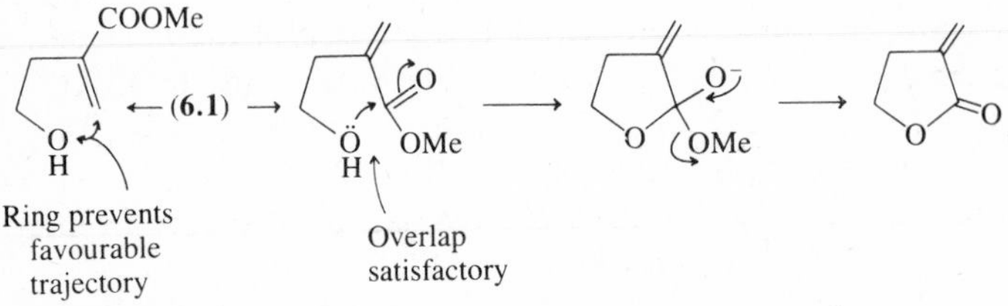

COOMe

← (6.1) →

Ring prevents favourable trajectory

Overlap satisfactory

The lactone formation is an example of what Baldwin [76CC734] calls a 5-*exo-trig* process: the formation of a five-membered ring by attack on a trigonal atom with the breaking bond exocyclic to the newly forming ring. The disfavoured intramolecular Michael addition would be 5-*endo-trig*. Baldwin has summarised those cyclisations that are relatively more or relatively less favoured as follows (where *tet* and *dig* describe ring closures to tetrahedral and digonal atoms, respectively):

| Favoured | Disfavoured |
|---|---|
| 3 to 7-*exo-tet* | 5 to 6-*endo-tet* |
| 3 to 7-*exo-trig* | |
| 6 to 7-*endo-trig* | 3 to 5-*endo-trig* |
| 5 to 7-*exo-dig* | 3 to 4-*exo-dig* |
| 3 to 7-*endo-dig* | |

Johnson has pointed out that similar considerations must apply to the reverse ring openings [83TL2851]. Cleavage of the ring C–O bond of the intermediate (**6.2**) is facilitated because there are *anti*-periplanar electrons on the exocyclic oxygen atoms that can overlap effectively with the rear lobe of the antibonding LUMO [75T2463]. In the reverse Michael addition, however, the HOMO of the anionic fragment in (**6.3**) is at right angles to the ring C–O bond, and the cleavage does not occur [76CC736].

(**6.2**)

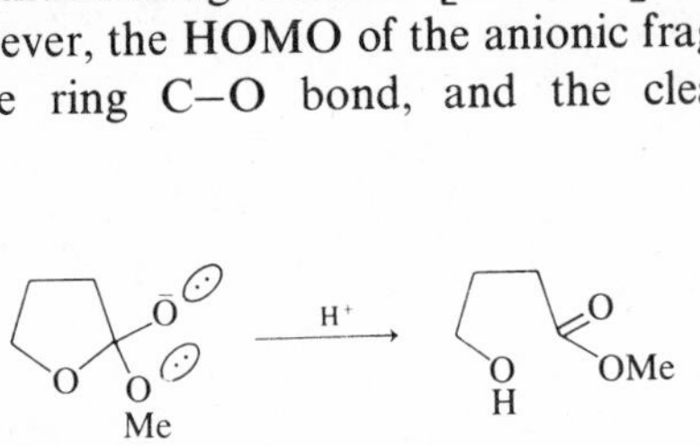

$\xrightarrow{\text{H}^+}$

(6.3)

## 6.5    Further reading

Molecular orbital theory in general and perturbation theory in particular:

A Streitwieser, *Molecular Orbital Theory for Organic Chemists*, Wiley, New York, 1961.

W. T. Borden, *Modern Molecular Orbital Theory for Organic Chemists*, Prentice-Hall, Englewood Cliffs, 1975.

M. J. S. Dewar, *The Molecular Orbital Theory of Organic Chemistry*, McGraw-Hill, New York, 1969.

M. J. S. Dewar & R. C. Dougherty, *The PMO Theory of Organic Chemistry*, Plenum, New York, 1975.

K. Fukui, *Angew. Chem. Internat. Ed.*, 1982, **21**, 801 (frontier orbitals).

Chemical reactivity in terms of perturbation theory:

G. Klopman (ed.), *Chemical Reactivity and Reaction Paths*, Wiley-Interscience, New York, 1974.

I. Fleming, *Frontier Orbitals and Organic Chemical Reactions*, Wiley-Interscience, London, 1976.

R. F. Hudson, *Angew. Chem. Internat. Ed.*, 1973, **12**, 36.

For an alternative theoretical approach to reactivity see:

A. Pross & S. S. Shaik, *Acc. Chem. Res.*, 1983, **16**, 10.

SHAB:

R. G. Pearson, *J. Chem. Educ.*, 1968, **45**, 581, 643; *Survey Progr. Chem.*, 1969, **5**, 1.

T. L. Ho, *Chem. Rev.*, 1975, **75**, 1.

Stereoelectronic effects:

F. M. Menger, *Tetrahedron*, 1983, **39**, 1013.

P. Deslongchamps, *Stereoelectronic Effects in Organic Chemistry*, Pergamon, Oxford, 1983.

Quantitative use of simple resonance theory (not discussed in this chapter, but related to NBMO calculations):

W. C. Herndon, *Tetrahedron*, 1973, **29**, 3; *J. Chem. Educ.*, 1974, **51**, 10; *Israel J. Chem.*, 1980, **20**, 270.

# PART 2

---

## 7     Aliphatic nucleophilic substitution

---

### 7.1     The $S_N1$ and $S_N2$ mechanisms

Our understanding of aliphatic nucleophilic substitution, perhaps the most thoroughly studied of all reaction types, is due primarily to a brilliant series of investigations by Hughes and Ingold dating from the mid-1930s. It was they who established the well-known mechanistic duality: the one-step concerted $S_N2$ mechanism and the dissociation mechanism, $S_N1$:

$$Nu^- \; R{-}Z \longrightarrow Nu{-}R + Z^- \qquad\qquad S_N2$$

$$R{-}Z \rightleftharpoons Z^- + R^+ \xrightarrow{\;Nu^-\;} R{-}Nu \qquad\qquad S_N1$$

Their ideas have been modified and extended over the years but they still provide the most convenient basis for discussion. Ingold's own book [69MI] provides a detailed survey of the field.

#### 7.1.1     Evidence for the $S_N2$ mechanism

Second order kinetics, showing dependence both on the nucleophile and on the substrate, are widely observed in nucleophilic substitutions. This is clearly a necessary criterion for the $S_N2$ mechanism, but it is not of itself sufficient proof. More convincing evidence comes from the stereochemistry of the reaction. The inversion of configuration that accompanies an $S_N2$ reaction, turning the substrate molecule inside-out like a joke umbrella, means that the nucleophile must approach the site of substitution from the side opposite the leaving group; stereochemical correlation between the two shows that both of them are associated in the transition state:

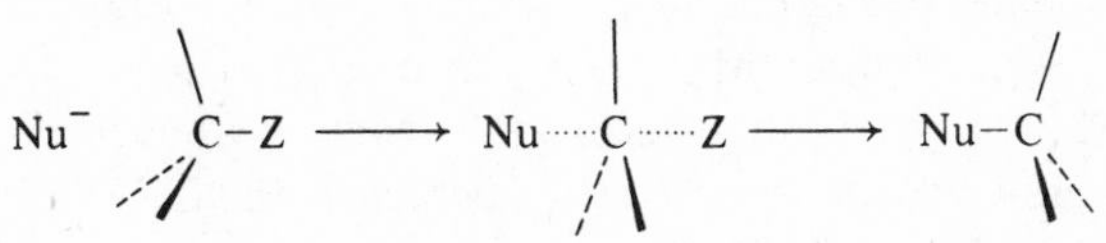

$$Nu^- \quad \overset{\diagdown}{\underset{\diagup}{C}}{-}Z \longrightarrow Nu\cdots\overset{\mid}{\underset{\diagup \diagdown}{C}}\cdots Z \longrightarrow Nu{-}\overset{\diagup}{\underset{\diagdown}{C}}$$

This stereochemical inversion was known long before Hughes and Ingold formulated the $S_N2$ mechanism; it was first noticed in the 1890s by Paul Walden, after whom it is named, and it was convincingly established by Phillips and Kenyon in the late 1920s and early 1930s. A typical experiment [23J(123)44] demonstrated the formation of the alternative enantiomeric ethers from a chiral alcohol in the following series of reactions. (R = MeC̊HCH₂Ph; $[\alpha]_D$ in parentheses.)

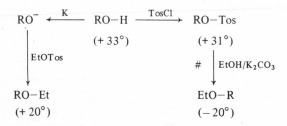

In these reactions the R—O bond is retained in every step except that marked #. A single $S_N2$ displacement is thus associated with one inversion of configuration. Similar results were obtained in a wide variety of reaction sequences and were extended to primary compounds using specifically deuterated substrates RC̊HDZ.

A particularly important experiment by Hughes [35J1525] concerned the reaction of chiral (+)-2-iodooctane with radioactive iodide ion. With iodide acting as both nucleophile and leaving group there is no net chemical change, but the reaction can be followed by measuring the rate of incorporation of radioactivity; at 30 °C in acetone the rate coefficient for this is $1.36 \, (\pm 0.11) \times 10^{-3} \, \mathrm{l \, mol^{-1} \, s^{-1}}$. In the same reaction the rate of stereochemical inversion (half the observed rate of racemisation, since inversion of one molecule generates a racemic pair) is $1.31 \, (\pm 0.01) \times 10^{-3} \, \mathrm{l \, mol^{-1} \, s^{-1}}$. The two rates are the same within experimental error, showing that inversion is not just an occasional event but that each and every nucleophilic attack is accompanied by inversion.

The original rationalisation of this behaviour was in terms of an electrostatic repulsion between the nucleophile and the leaving group. However, even a reaction such as $N_3^- + R{-}SMe_2^+$ occurs with inversion; one might have expected Coulombic attraction to encourage front-side attack in this case. Frontier MO theory provides a more satisfactory picture. The orbitals chiefly involved in the reaction are the HOMO of the nucleophile and the substrate LUMO which is the C—Z $\sigma^*$ orbital:

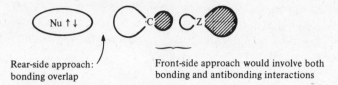

Rear-side approach: bonding overlap

Front-side approach would involve both bonding and antibonding interactions

### 7.1.2 Evidence for the S$_N$1 mechanism

S$_N$1 reactions are much more difficult to characterise than S$_N$2; the mechanistic criteria are seldom clear cut. If we apply the steady-state approximation to the S$_N$1 reaction scheme:

$$R-Z \underset{k_{-1}}{\overset{k_1}{\rightleftharpoons}} Z^- + R^+ \underset{Nu^-}{\overset{k_2}{\rightarrow}} RNu$$

we can derive the following expression.

$$\text{Rate} = \frac{k_1 k_2 [RZ][Nu^-]}{k_{-1}[Z^-] + k_2[Nu^-]} \qquad [7.1]$$

In the early stages of the reaction $[Z^-]$ is small and the kinetics simplify to first order with the rate equal to $k_1[RZ]$, independent of the concentration or even the nature of the nucleophile. Hughes and Ingold demonstrated this with the reactions of benzhydryl chloride (Ph$_2$CHCl) with different nucleophiles in liquid sulphur dioxide. As the reaction proceeds and the concentration of Z$^-$ builds up, the rate should slow down according to equation [7.1]. Moreover, the addition of extra Z$^-$ should also retard the reaction – the *common ion* effect – and both of these phenomena were observed [40J1011, 1017].

However, most S$_N$1 reactions are not carried out under these contrived conditions. They are usually solvolyses, where a good ionising solvent facilitates the ionisation and then also acts as the nucleophile in the second step of the reaction. The nucleophile concentration is thus effectively constant so even a bimolecular process would exhibit pseudo-first order kinetics. Significant changes in concentration of a nucleophilic solvent must inevitably entail fundamental changes to the whole system so that rate changes are ambiguous. For example, the hydrolysis of t-butyl bromide in aqueous acetone is accelerated 40 times as the solvent mix is changed from 10 per cent water to 30 per cent [40J960]. How can one decide whether this is a second order process accompanied by a 13-fold acceleration due to solvation changes, or a first order one with a 40-fold solvent effect?

The effects of added salts can shed more light on these solvolyses. Table 7.1 shows that the common ion effect can still be a useful criterion. The rates marked § show the normal kinetic electrolyte effect; the reaction is accelerated because an increase in ionic strength stabilises a polar transition state. This must still be present in the reactions marked †, but

Table 7.1. *First order rate coefficients* $(k_1/10^{-5} \ s^{-1})$ *for solvolyses in aqueous acetone (80 per cent)*[a] *at 25°C* [52J2488]

|  | Ph$_2$CHBr | Ph$_2$CHCl |
|---|---|---|
| Solvent only | 153 | 7.00 |
| + 0.1 M LiBr | 133† | 8.16§ |
| + 0.1 M LiCl | 194§ | 6.09† |

[a]80 per cent acetone, and similar expressions, are widely used to mean a mixture containing 80 per cent acetone and 20 per cent water by volume.

nevertheless the net effect of the added common ion is to retard the reaction as predicted by equation [7.1].

With trityl compounds (Ph$_3$CZ) the common ion effect is even more marked: 0.01M sodium chloride reduces four-fold the rate of solvolysis of trityl chloride in aqueous acetone. But t-butyl halides, on the other hand, exhibit no common ion effect, nor are their solvolyses retarded by build-up of the halide ion during reaction. This makes sense in terms of the $S_N1$ carbenium ion mechanism: the energetic t-butyl cation is highly reactive and relatively unselective, so $k_{-1}$ and $k_2$ are comparable in magnitude and the overwhelming concentration of the solvent is dominant. A more stable carbenium ion can discriminate between the concentrated but weakly nucleophilic solvent and the dilute but more nucleophilic common ion. Equation [7.1] can be rearranged to give [7.2], writing S (solvent) for Nu$^-$:

$$\text{Rate} = \frac{k_1[\text{RZ}]}{\left\{\dfrac{k_{-1}}{k_2[\text{S}]}\right\}[\text{Z}^-] + 1} \qquad [7.2]$$

The term $\{k_{-1}/k_2[\text{S}]\}$, calculated from the extent to which added Z$^-$ retards the reaction, is a measure of the selectivity between solvent and common ion. For various substrates in 80–90 per cent acetone its value is as follows.

Ph$_3$CCl  p-Tol$_2$CHCl  Ph$_2$CHCl  t-BuBr
$\sim$400     69–89          10–16       <2

The t-butyl value is too small to be numerically significant; allowances are made in the calculations for the ordinary kinetic electrolyte effects and these could well account for the whole of the very small rate changes observed with t-butyl bromide. The order of selectivity accords well with the expected order of stability of the carbenium ion intermediates [40J979; 53JA136].

The addition of a non-common ion that is itself strongly nucleophilic (azide is often used) can provide a convenient probe for distinguishing $S_N1$ and $S_N2$ behaviour. If the azide reacts directly with the substrate in an $S_N2$ mechanism then (after allowing for the electrolyte effect) it will accelerate the overall reaction in direct proportion to the amount of alkyl azide produced. In an $S_N1$ reaction, however, there is no correlation between the rate-limiting and product-forming steps. Benzhydryl bromide is solvolysed 34 times as fast as the chloride in 90 per cent acetone at 50 °C but in the presence of 0.1M sodium azide both substrates give identical mixtures of alcohol and alkyl azide (2.1:1) [40J979].

## 7.2 Ion-pairs

Many reactions that appear to involve $S_N1$-like ionisation nevertheless exhibit behaviour incompatible with the existence of a 'free' carbenium ion. Consider first the data of table 7.2.

A carbenium ion is a planar species, equally susceptible to attack from either side, so that an $S_N1$ reaction should lead to total racemisation, as in the last example in the table. $S_N2$ reactions, exemplified by the deuteriated n-butyl brosylate, give complete inversion. There is no clean division between the two extremes; how can one account for partial racemisation?

It could be that the two mechanisms are running in parallel, some molecules dissociating to ions while others undergo direct $S_N2$ inversion. This cannot always be the answer: the acetolysis of 1-phenylethylchloride is not accelerated by the addition of sodium acetate nor does this increase the proportion of inversion, so that it is unlikely that the net inversion in this case comes from an $S_N2$ process.

Table 7.2. *Stereochemistry of solvolyses*

| Substrate | Conditions | Inversion (per cent) | Retention (per cent) |
|---|---|---|---|
| n-Pr—$\overset{*}{C}$HD—OBs | AcOH, 99 °C | 98 ± 4 | — |
| n-$C_6H_{13}$—$\overset{*}{C}$HMe—OTos | AcOH, 75 °C | 96 | 4 |
| i-BuEtMe$\overset{*}{C}$—Ohph | MeOH, reflux | 77 | 23 |
| Ph—$\overset{*}{C}$HD—OTos | AcOH, 25 °C | 90 | 10 |
| Ph—$\overset{*}{C}$HMe—Cl | AcOH, 50 °C | 56 | 44 |
| p-MeOC$_6$H$_4$—$\overset{*}{C}$HPh—Ohph | MeOH, 'warm' | 50 | 50 |

Ohph = hydrogen phthalate.

The explanation originally put forward, and still widely quoted, asserts that partial racemisation is another manifestation of the differing stabilities of carbenium ions: a high-energy ion reacts very rapidly while one side is still shielded by the leaving group, so that attack from the back is preferred.

This argument merits careful analysis. Complete ionisation of a chiral substrate to a planar carbenium ion must destroy all chirality. The two alternative product-forming transition states (front-side and rear-side attack) are enantiomeric and of equal energy. Partial racemisation implies that there are two transition states of *different energies*. The argument invoking 'shielding-by-leaving-group' must mean that the leaving group is still associated with these transition states. The rate-limiting ionisation, therefore, cannot give a free carbenium ion because the leaving group is still associated with the subsequent transition state (figure 7.1).

What is the nature of the intermediate I? Before seeking an answer to this question, consider another situation that seems to demand the same sort of mechanism.

Rate-limiting transition state

I

$\overset{*}{R}-Z$

Product-forming transition states, differentiated by mutual dispositions of nucleophile and leaving group

$\overset{*}{R}-Nu$ (retained configuration)

$Nu-\overset{*}{R}$ (inverted configuration)

Figure 7.1

The solvolysis of 2-adamantyl sulphonates in aqueous ethanol gives a mixture of the adamantanol and adamantyl ethyl ether. Their ratio depends on the nature of the sulphonate group [74JA4484]:

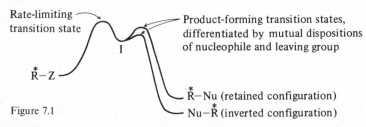

$p\text{-}X-C_6H_4SO_2O$ — $\xrightarrow[\text{70 per cent}]{\text{EtOH}:H_2O}$ Ad$-$OEt $+$ Ad$-$OH

| X | OMe | Me | Br | H | $NO_2$ |
|---|---|---|---|---|---|
| [AdOH]/[AdOEt] | 1.9 | 2.6 | 3.1 | 3.3 | 5.2 |

There is considerable evidence (§7.3.1) to show that there is no $S_N2$ contribution to the reactions of 2-adamantyl systems, so again we are forced to consider a reaction profile like that in figure 7.1, but this time with the products chemically rather than stereochemically distinct. The question again poses itself: what is the nature of the intermediate I?

Hammett proposed in 1940 that it was a partly dissociated system in which the covalent bonding was largely replaced by electrostatic attraction but with the two ions still held intimately together within a solvent cage; he called such a structure an *ion-pair*.

Saul Winstein confirmed the validity of this idea and extended it to show that, in some cases at least, there must be two different types of ion-pair on the path between the covalent compound and the fully disso-ciated ions. His key experiment [58JA169] was the acetolysis of (+)-*threo*-3-anisylbut-2-yl brosylate (**7.1**). The reactant becomes racemised during the reaction at a rate faster than the rate of solvolysis. The free anisylbutyl cation has a bridged structure (**7.2**; see §7.5.3) and is achiral, so

(7.1)          (7.2)

one might be tempted to explain fast racemisation as an $S_N1$-like process with a fast reversible dissociation followed by a slower solvolytic attack. This cannot be true, however, because the solvolysis is not subject to common ion retardation as it should be if, in equation [7.2], $k_{-1}$ is large.

An ion-pair mechanism offers a better alternative:

$$RZ \underset{k_{-1}}{\overset{k_1}{\rightleftharpoons}} R^+Z^- \rightarrow \text{Solvolysis products}$$
$$\text{Ion-pair}$$

The racemisation occurs via rapid equilibrium with the ion-pair; in the reverse step, 'internal return', the brosylate anion can easily move across to the carbon atom adjacent to the one it just left. However, this is an intramolecular process, with the anion never becoming free, so the addition of more anion does not interfere with the kinetics.

Both racemisation and solvolysis are accelerated somewhat by the normal electrolyte effect when lithium perchlorate is added to the solution, but the solvolysis shows an unusual and very large 'special electrolyte effect' at low concentrations of perchlorate (figure 7.2).

The perchlorate ion is negligibly nucleophilic, so this special effect can only arise from some electrostatic interaction. Winstein suggested that perchlorate intervenes by exchanging with the brosylate anion in an ion-pair, thus cutting out a rate-retarding return process. But this cannot be the same ion-pair that he invoked to explain fast racemisation: if there

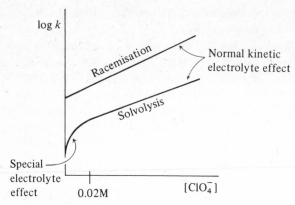

Figure 7.2. Effect of lithium perchlorate on acetolysis of (+)-*threo*-3-anisylbut-2-yl brosylate.

were efficient competition for $k_{-1}$ the racemisation would be slowed as well as the solvolysis, but racemisation is not subject to the special electrolyte effect. Winstein therefore proposed the scheme:

$$RZ \underset{k_{-1}}{\overset{k_1}{\rightleftarrows}} R^+Z^- \underset{k_{-2}}{\overset{k_2}{\rightleftarrows}} R^+ \| Z^- \underset{k_{-3}}{\overset{k_3}{\rightleftarrows}} R^+ + Z^-$$

Intimate    Solvent-
or contact   separated
ion-pair     ion-pair     AcOH

$k_{-4} \| ClO_4^-$
$k_4$      Products

$R^+ \| ClO_4^-$   AcOH

The addition of the perchlorate provides an extra product-forming route, $k_4$, which accelerates the reaction. With $[ClO_4^-] = 0.02$ M this route is so efficient that virtually every solvent-separated ion-pair is whisked through to products as soon as it is formed; further addition of perchlorate now merely induces the normal linear electrolyte effect. Fast racemisation $(k_{-1} > k_2)$ continues regardless of interference later in the reaction path. In agreement with this scheme lithium brosylate, which as already mentioned exerts a negligible common ion retardation on the solvolysis, does slow down the perchlorate catalysed reaction by accelerating $k_{-4}$.

Further evidence for two different types of ion-pair is provided by the hydrolysis of [18]O-labelled *p*-chlorobenzhydryl *p*-nitrobenzoate, $ClC_6H_4-CHPh-OC^{18}OC_6H_4NO_2$, in 90 per cent aqueous acetone

[64JA120]. During the course of the reaction the $^{18}O$ label is scrambled between the two carboxyl oxygen atoms and the ester undergoes partial racemisation. Both processes are first order. Under the same reaction conditions exchange between the ester and $^{14}C$-labelled p-nitrobenzoic acid is exceedingly slow showing that dissociation does not proceed as far as the fully separated ions. The addition of sodium azide to the solution suppresses the racemisation but does not interfere with the $^{18}O$ exchange. This again demonstrates that there are two distinct ion-pair inter-mediates. In the intimate ion-pair the two oxygen atoms of the carbox-ylate anion become equivalent and the return process leads to the oxygen scrambling. Racemisation, however, requires that the anion migrate from one side of the benzhydryl cation to the other. Evidently this is difficult within the confines of the intimate ion-pair; it has to wait till the ions are more loosely associated in the solvent-separated ion-pair, but this is readily intercepted by the strongly nucleophilic azide ion.

The exact structures of the ion-pairs are not known; indeed, it is not certain that there are only two discrete energy minima on the path between the covalent RZ and the fully dissociated ions. One can get a picture of what is happening by imagining the process in reverse. The two separated ions will each be surrounded by ordered solvation shells. As they approach, the first interactions will be between solvent molecules in these shells, and as these re-position themselves there may well be several positions of relative stability, eventually arriving at some such structure as $R^{+}\cdots^{\delta-}S-H^{\delta+}\cdots Z^{-}$, with only a single solvent molecule (SH) separat-ing the two approaching ions. One more squeeze and the ions are face-to-face in the same solvent cage, finally collapsing together into covalency.

However, even if we consider only Winstein's two types of ion-pairs there is clearly the possibility of a very wide variety of reaction types (figure 7.3).

Reactions with steps 1, 2 or 3 rate-limiting are unimolecular and we can designate them $S_N1(A)$, $S_N1(B)$, and $S_N1(C)$, respectively; reactions with steps 4 to 7 rate-limiting are bimolecular: $S_N2(A)$ to $S_N2(D)$. Even this does not fully describe all the mechanistic possibilities: an $S_N1(A)$

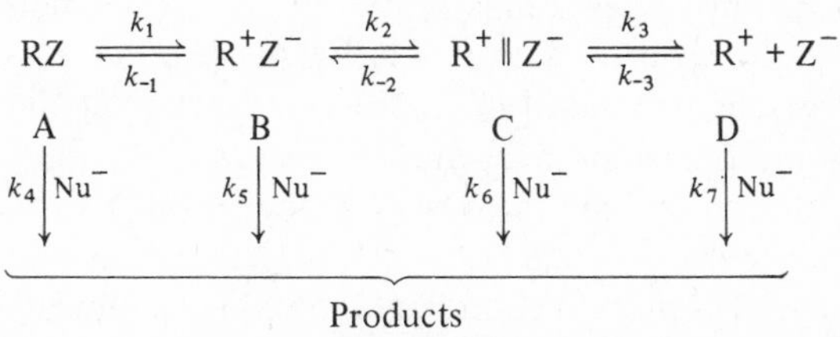

$$RZ \underset{k_{-1}}{\overset{k_1}{\rightleftharpoons}} R^+Z^- \underset{k_{-2}}{\overset{k_2}{\rightleftharpoons}} R^+ \| Z^- \underset{k_{-3}}{\overset{k_3}{\rightleftharpoons}} R^+ + Z^-$$

A      B      C      D

$k_4 \downarrow Nu^-$    $k_5 \downarrow Nu^-$    $k_6 \downarrow Nu^-$    $k_7 \downarrow Nu^-$

Products

Figure 7.3. Ion-pair mechanisms for nucleophilic substitution.

reaction could be terminated by rapid nucleophilic attack on the intimate ion-pair ($k_5 > k_2$) or it could continue along the path of further dissociation ($k_2 > k_5$) with the main product-forming step being 6 or 7.

On this basis Hughes and Ingold's classical $S_N2$ mechanism would become $S_N2(A)$ and their $S_N1$, as originally envisaged with complete dissociation, could include any of the three $S_N1$ mechanisms provided the product was formed through step 7. In fact, though, they included in their $S_N1$ category reactions which showed all possible stereochemical behaviour from total racemisation to total inversion; we can now rationalise this in terms of the difference in product-forming steps. Usually a reaction that does not go further than the intimate ion-pair proceeds with inversion of configuration as in the azide-modified hydrolysis of *p*-chlorobenzhydryl nitrobenzoate described above. (The anisylbutyl system is an exception because the anion does not have to move round to the opposite face of the cation to induce racemisation. Exceptions are also found with particularly stable cations, such as $p\text{-MeOC}_6\text{H}_4\text{CHMe}^+$, whose lifetimes are long enough for the reorganization to occur within the intimate ion-pair [70JA7401].) Any reaction that reaches the fully dissociated ions must produce total racemisation. Partial racemisation can come via step 6 or via a mixture of product-forming steps.

### 7.3    The mechanistic borderline

Winstein's dual ion-pair hypothesis can obviously account for a wide variety of mechanistic behaviour, but because it is so flexible it leaves room for doubts about the precise mechanisms of reactions that fall between the clear cut extremes. In particular there is considerable controversy as to whether borderline behaviour arises from a gradual merging of one mechanism into another or whether it could be due to two mechanisms running in parallel. We shall illustrate the problem by considering two different examples of it.

### 7.3.1    *Solvolyses of alkyl substrates*

There is no doubt that simple primary alkyl substrates react via the classical $S_N2$ mechanism. ('Simple' in this context refers to saturated systems in which there is no possible mesomeric stabilisation of a carbenium ion.) Recently suggestions that even methyl compounds react via ion-pairs, $S_N2(B)$, have been in vogue [73Acc46], but these can be discounted on various grounds [76Acc281]; in particular, it has been shown that the ionisation of primary alkyl halides would require between 60 and 160 kJ mol$^{-1}$ more than the observed activation energies for their solvolyses [73J(P2)1893; 77J(P2)873].

Tertiary substrates, even in the most unfavourable circumstances, react via one of the $S_N1$ mechanisms. For example, the reaction of t-butyl

Table 7.3. *Elimination versus substitution in solvolyses (at 75 °C) of t-butyl substrates: the effect of the leaving group* [63JA1702]

| t-BuZ | | Percentage olefin formation | | |
| Z | Solvent: | $H_2O$ | AcOH | EtOH |
| --- | --- | --- | --- | --- |
| Cl | | 7.6 | 73 | 44 |
| Br | | 6.6 | 70 | 36 |
| I | | 6.0 | | 32 |
| $SMe_2^+$ $ClO_4^-$ | | 6.5 | 12 | 18 |

bromide with radio-labelled bromide ion in acetone, a poor ionising solvent, follows first order kinetics. On the other hand it seems unlikely that simple tertiary compounds normally dissociate as far as the free ions: partial racemisation (third entry in table 7.2) and a dependence of the products on the leaving group (table 7.3) suggest that the reactions usually go through solvent-separated ion-pairs. Even the solvolysis of t-BuSMe$_2^+$ may involve some ion-pairing with the perchlorate counter ion, but it does offer a rough model for the behaviour of the free cation; table 7.3 thus suggests that in water the dissociation is considerable, since all the substrates behave much the same as the sulphonium salt. In other solvents, however, the leaving group plays an important part in the product-forming step, so that dissociation must be limited.

It is not clear whether in these reactions it is step 1 or step 2 of Winstein's scheme (figure 7.3) that is rate-limiting. The return steps $-1$ and $-2$ are definitely slow: there is little common ion retardation, and negligible exchange of radio-chlorine occurs during the acetolysis of t-butyl chloride in the presence of labelled sodium chloride.

Secondary alkyl substrates are less easy to classify, and their solvolyses have therefore been studied extensively with a view to learning more about the $S_N1/S_N2$ borderline.

There are two main schools of thought. The first, exemplified by the work of Shiner and his colleagues [76JA2865; 78JA8133; 79JO2108], accepts the standard Winstein ion-pair model (figure 7.3) with any of the steps 1, 2, 5 or 6 rate-limiting depending on the nature of the secondary alkyl group, of the leaving group and of the solvent (and further complicated because of elimination reactions occurring in parallel with substitution, §8.4). The main evidence comes from the observation of wide variations in the magnitude of secondary kinetic isotope effects in $\alpha$ and $\beta$ deuteriated substrates (§2.4.2).

The major factor contributing to $\alpha$ $k_H/k_D$ is a loosening of the $H-C_\alpha-Z$ bending mode on going to a carbenium-like transition state. When $C_\alpha-Z$

bonding in the transition state is minimal the $\alpha$ kinetic isotope effect is maximal, and at the same time it depends maximally on the nature of Z. This condition would be expected when step 2 is rate-limiting. When step 1 is rate-limiting the $C_\alpha-Z$ bond is less fully broken in the transition state; when steps 5 or 6 are rate-limiting the $C_\alpha-Z$ bond breaking is partly compensated by $HS-C_\alpha$ bond making (where HS is the nucleophilic solvent).

The $\beta$ kinetic isotope effect probably arises mainly from hyperconjugative interaction in an incipient $H-C_\beta-C_\alpha^+$ carbenium ion [79JA5532; 83PPO205] and is thus also a measure of $C_\alpha-Z$ cleavage.

Further evidence comes from the stereochemistry of these solvolyses. They usually occur with complete inversion of configuration, but partial racemisation in very weakly nucleophilic solvents (such as $(CF_3)_2CHOH$) points to reaction via a solvent-separated ion-pair.

McLennan has recently cast doubts on Shiner's interpretation [81J(P2)1316]. He calculates that the maximum $\alpha k_H/k_D$ values to be expected for $S_N1(B)$ reactions (step 2 rate-limiting) are significantly greater than any values that have been observed experimentally, and suggests that in all Shiner's reactions there is some degree of $S_N2$ character, some partial $HS-C_\alpha$ bond making to compensate for the $C_\alpha-Z$ bond breaking.

This is more in accord with the interpretation of secondary alkyl solvolyses put forward by Schleyer and Bentley [76JA7658, 7667; 81JA5466]. They propose that there is a gradual change of mechanism from $S_N1(A)$ to $S_N2(A)$ with no clear dividing line. 2-Adamantyl substrates (7.3) are taken as models for the $S_N1$ end of this mechanistic spectrum. Their solvolyses appear to take place through a transition state in which there is no significant nucleophilic solvation – that is, no interaction between the nucleophilic centre of the solvent and the incipient carbenium ion – though there can, of course, be hydrogen bonded solvation of the anionic component of the ion-pair and some general solvation.

The solvolyses of 2-adamantyl bromide occur some $10^8$ times slower than those of the tertiary 2-methyl-2-adamantyl bromide (7.4). The ratio is more or less independent of the solvent and approaches the value expected for the relative rates of ionisation of secondary and tertiary substrates from measurements of the relative stabilities of carbenium ions in the gas phase and in super-acid solution [70JA2540; 79JA522; 80J(P2)1244]. Likewise the nucleophilicity of the solvent has an insignificant effect on the relative reactivities of 2-adamantyl tosylate and the bridgehead 1-bicyclooctyl tosylate (7.5) in which there can be no trace of any $S_N2$ character in the transition state [81JA5466]. (By contrast, isopropyl tosylate is about 1000 times less reactive than bicyclooctyl

(7.3; R=H)
(7.4; R=Me)          (7.5)          (7.6)

tosylate in $(CF_3)_2CHOH$ and about 80 times more reactive in the much more nucleophilic ethanol.)

Further evidence for the absence of nucleophilic solvation in 2-adamantyl solvolyses comes from the use of the azide probe (§7.1.2) which shows that even the strongly nucleophilic azide ion does not enter into an $S_N2$ reaction with 2-adamantyl substrates, presumably because of steric hindrance (7.6); how much less would the weakly nucleophilic solvent molecule participate in such a transition state!

Schleyer and Bentley therefore use 2-adamantyl sulphonates as a base-line for comparing the behaviour of other secondary alkyl sulphonates, and some primary ones too. They find regular trends in a variety of different mechanistic criteria: the magnitude of the $\alpha$ deuterium isotope effect, reflecting changes in the bonding at the site of substitution; the Grunwald–Winstein $m$ parameter measuring the susceptibility of the substrate to the ionising power of the solvent (§5.2); the effect of the nucleophilicity of the solvent on the rates of solvolysis. All these show a progressive change from methyl substrates, through ethyl, isopropyl, then other bulkier secondary systems, down to the 2-adamantyl baseline. A common origin for these changes is implied by an inverse relation between the $m$ values (which crudely reflect $S_N1$ character) and the sensitivity to solvent nucleophilicity ($S_N2$).

A specially pertinent correlation is one that places all secondary tosylates, ROTos, on a scale between 2-adamantyl and isopropyl via the relation:

$$\log (k/k_0)_R = Q_R \log (k/k_0)_{2\text{-Ad}} + (1 - Q_R) \log (k/k_0)_{i\text{-Pr}}$$

where $k_0$ is the solvolytic rate coefficient in a reference solvent (as in the Grunwald–Winstein equation). The scaling parameter $Q_R$ is effectively independent of solvent, suggesting that each substrate has its characteristic point on the mechanistic spectrum. Sample values of $Q_R$ are:

| R | Me–CH–Me | Et–CH–Me | Et–CH–Et | i-Pr–CH–Me | t-Bu–CH–Me | 2-Ad |
|---|---|---|---|---|---|---|
| $Q_R$ | 0.00 | 0.20 | 0.30 | 0.42 | 0.76 | 1.00 |

These clearly reflect the progressive exclusion of solvent from the transition state, mainly as a consequence of steric repulsion [81JA5466].

Schleyer and Bentley picture this as a gradual merging of solvolysis mechanisms from a classical concerted $S_N2$, through a nucleophilically solvated ion-pair, to the 2-adamantyl ion-pair in which there is no nucleophilic solvation at all:

$$H-S + R-Z \xrightarrow[-Z^-]{\text{Concerted}} H-\overset{+}{S}-R \qquad\qquad S_N2(A)$$

$$H-S + R-Z \longrightarrow [H-\overset{+}{S}\cdots R\,Z^-] \longrightarrow S-R \qquad S_N2\ (\text{Int})$$

$$R-Z \longrightarrow [R^+Z^-] \qquad\qquad\qquad S_N1(A)$$

Figure 7.4 illustrates the changing reaction profiles, the transition states being characterised by a progressively decreasing contact between the nucleophilic solvent and the electrophilic carbon atom of the substrate.

The two models of the $S_N1/S_N2$ borderline are both constructed within the framework of Winstein's ion-pair scheme (figure 7.3) and the crux of the argument is the nature of step 1. In Shiner's model it is truly

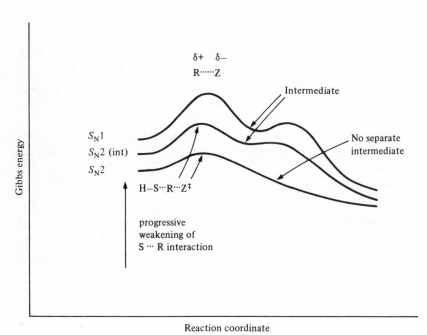

Figure 7.4. The $S_N1/S_N2$ borderline according to the model of Schleyer and Bentley [81JA5466].

unimolecular, with no nucleophilic participation. Except for a few $S_N1(A)$ solvolyses it constitutes a pre-equilibrium, and internal return, the reverse of step 1, is an important element of Shiner's analysis. For Bentley and Schleyer, however, internal return is insignificant. The 2-adamantyl substrates follow the $S_N1(A)$ route with $k_1$ rate-limiting; for other substrates non-limiting behaviour is due to increased participation by the nucleophile, not to a shift to rate-limiting $k_2$ or any other step.

Bunnett [81JA946] has recently been able to make a direct measurement of internal return by $^{18}O$ scrambling experiments:

$$Ph-S^{18}O_2-O-R \underset{k_{-1}}{\overset{k_1}{\rightleftharpoons}} Ph-S^{18}O_3^-\ R^+ \overset{k_{prod}}{\rightarrow} \text{products}$$

He showed that internal return is less important than Shiner supposes, but more so than is suggested by the model of Schleyer and Bentley. In particular, in 2-adamantyl benzenesulphonate solvolyses, $k_{-1}$ is some 2–5 times $k_{prod}$, depending on the solvent. The implication is that step 1 is *not* rate-limiting, as indicated in figure 7.4. It seems that an eventual resolution of the problem may call for elements of both models – alternative rate-limiting steps *and* variable nucleophilic solvation.

### 7.3.2 Reactions of primary benzyl substrates

Many reactions of benzylic substrates are accelerated both by electron-donating substituents in the benzene ring and by electron acceptors [80JA6463]. However, the $\alpha$ deuterium kinetic isotope effect increases monotonically with increasing electron donation, suggesting that there is a progressive weakening of the $C_\alpha-Z$ bond in the transition state (compare the discussion in the preceding section). This behaviour is generally interpreted in terms of a gradual change from a tight $S_N2$ transition state in which an electron-withdrawing aryl group pulls the nucleophile in close to the reaction centre (7.7), to a looser, more $S_N1$-like transition state with substantial carbocation character (7.8), though it has

$$(Nu\cdots CH_2\cdots Z)^- \\ \phantom{xxx}\downarrow \\ \phantom{xxx}Ar$$

(7.7)

$$\overset{\delta-}{Nu}\cdots\cdots\overset{\delta+}{CH_2}\cdots\cdots\overset{\delta-}{Z} \\ \phantom{xxxx}| \\ \phantom{xxx}Ar:$$

(7.8)

also been suggested [79JA3288] that a more or less invariant transition state would be stabilised both by polar electron withdrawal (facilitating attack by an anionic nucleophile) and by mesomeric electron donation (because of conjugation in the transition state with the $sp^2$-like benzylic carbon atom).

The reactions of anilines with benzyl halides in ethanol are bimolecular; in every case the rates depend on both the nucleophile and the

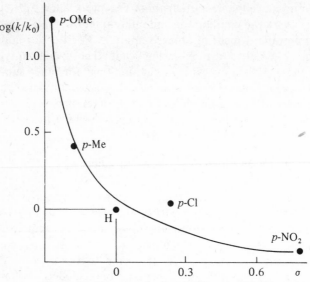

Figure 7.5. Hammett plot for $PhNH_2$ + $ArCH_2Cl$ in EtOH at 50 °C.

Table 7.4. *Hammett reaction constants*, $\rho$, *for the reactions of* $Ar'NH_2$ *with* p-$X-C_6H_4-CH_2Cl$

| X | MeO | Me | H | Cl | NO$_2$ |
|---|---|---|---|---|---|
| $\rho$ | −0.59 | −0.73 | −0.87 | −1.24 | −1.55 |

halide [76JO3364]:

$$Ar'NH_2 + ArCH_2Cl \rightarrow Ar'NHCH_2Ar$$

Changes in the benzyl substrate (Ar) generate a curved Hammett plot (figure 7.5; the curvature remains if $\sigma^+$ is plotted instead of $\sigma$). This again accords with a gradually changing transition state although with an uncharged nucleophile the reaction is not actually accelerated by electron withdrawal in Ar.

Changes in the aniline group (Ar') give linear Hammett plots, but the slope is markedly dependent on the nature of the benzyl group (table 7.4). With a loose transition state, as with p-methoxybenzyl chloride, the aniline nitrogen atom bears little charge and the magnitude of $\rho$ is small. In the tight transition state with p-nitrobenzyl chloride the nucleophile makes its presence more felt, and its nitrogen atom bears a considerable positive charge.

The nucleophile, too, can influence the transition state structure, as shown by the data of table 7.5, which relate to the reactions of various

Table 7.5. *Relative reactivities of* p-*phenoxybenzyl and* p-*methoxybenzyl chlorides in 70 per cent acetone at 20°C* [62CI1287]

| Nucleophile | $k_2$(PhO) $(k_2/10^{-5}\,l\,mol^{-1}\,s^{-1})$ | $k_2$(MeO) | $k_2$(PhO)/$k_2$(MeO) |
|---|---|---|---|
| $PhSO_3^-$ | 20 | 0.16 | 125 |
| $NO_3^-$ | 32 | 0.23 | 139 |
| $F^-$ | 42 | 0.49 | 86 |
| $Cl^-$ | 66 | 0.76 | 87 |
| $Br^-$ | 79 | 5.09 | 15.5 |
| $N_3^-$ | 345 | 71 | 4.9 |

Table 7.6. *Kinetic and activation data for the racemisation of* p-$X$-$C_6H_4$-$CHBrCH_3$ *in acetone at 30°C* [80JO3539]

| X | $k/10^{-3}\,l\,mol^{-1}\,s^{-1}$ | $\Delta H^{\ddagger}/kJ\,mol^{-1}$ | $\Delta S^{\ddagger}/J\,K^{-1}\,mol^{-1}$ |
|---|---|---|---|
| Me | 7.90 | 57 | $-100$ |
| H | 3.65 | 64 | $-80$ |
| $NO_2$ | 16.75 | 70 | $-50$ |

nucleophiles with two *para* substituted benzyl chlorides (p-PhO and p-MeO). In the absence of added nucleophiles the first order rate coefficients for the solvolyses are in the ratio $k_1$(PhO)/$k_1$(MeO) = 135. These reactions are almost certainly $S_N1$: compare [81CC737] which shows that the solvolysis of p-methoxybenzyl bromide in aqueous dioxan is subject to the common ion effect. When the nucleophiles listed in Table 7.5 are added, their reactions with the benzyl chlorides are bimolecular, but the changes in $k_2$(PhO)/$k_2$(MeO) indicate a changing transition state structure. The strongest nucleophile in the series, $N_3^-$, closes in on the substrate giving a tight transition state with relatively little C−Cl cleavage, little charge buildup on the benzyl carbon atom, and little influence from the aryl substituent. Weaker nucleophiles, such as $NO_3^-$, give a much looser transition state with considerable positive charge on the benzyl carbon atom, and the relative reactivities of the two substrates resemble those in the solvolysis.

Recently this tight/loose picture of a variable transition state has been called into question. Activation parameters for the reaction of bromide ion with *secondary* benzyl substrates $ArCHBrCH_3$ were obtained from the kinetics of racemisation (table 7.6). The results show the customary curved Hammett plot, the reaction being accelerated both by the electron-donating methyl and the electron-withdrawing nitro groups. But a tight transition state (p-nitro) would be expected to have more bonding than a loose one, implying a smaller value of $\Delta H^{\ddagger}$ and a more negative $\Delta S^{\ddagger}$: the opposite is found. A dissociation ion-pair mechanism, on the other hand, would have the lowest $\Delta H^{\ddagger}$ for the most stable carbenium ion (*p*-methyl). The $\Delta S^{\ddagger}$ values, however, are not easy to interpret on this model either.

A comparison between the selectivity of various benzyl substrates towards *m*-chloroaniline ($k_N$) and ethanol ($k_E$) for reactions in aqueous ethanol has also been interpreted in terms of an ion-pair model along the lines of the classic Winstein scheme (figure 7.3) [80J(P2)250]. For both *p*-methoxybenzyl and 2-adamantyl substrates the ratio $k_N/k_E$ lies in the range 6–22, depending on the leaving group and solvent mix. This is interpreted as evidence that both types of substrate react via similar mechanisms: $S_N1$, with the main product-forming step being attack on the solvent-separated ion-pair (step 6). However, benzylic substrates with less strongly electron-donating substituents show much greater selectivities, $k_N/k_E$ falling in the range 600–2000, as expected for an $S_N2$ mechanism where nucleophilicity is important. But what is remarkable is that these benzyl substrates are actually more selective than primary alkyl substrates: for example, in 50% aqueous ethanol the value of $k_N/k_E$ for benzyl chloride is 1440; for 1-octyl chloride it is only 640. The conventional variable transition state model would suggest that the benzyl transition state should be looser, more carbenium-like, and less selective than octyl. A tentative PMO analysis of these observations suggests that the more selective benzyl substrates may be reacting via the $S_N2(B)$ mechanism, with step 5 (attack on the intimate ion-pair) rate-limiting (and, of course, also product-controlling).

### 7.3.3    The ultimate borderline

Studies of secondary alkyl solvolyses and of the nucleophilic substitutions of benzylic substrates have taught us much about the mechanisms of reactions in the region of the $S_N1/S_N2$ borderline, but the fundamental nature of that border remains unclear. Few now doubt the inherent variability of the $S_N2$ transition state (though they may dispute a particular interpretation of that model) but does it merge imperceptibly into $S_N1$ as the participation of the nucleophile gradually fades out? There is a school of thought which asserts that, however variable the $S_N2$ transition state, there must ultimately be an abrupt mechanistic change to $S_N1$; any apparent continuity is due to the two mechanisms running in parallel at the border. Gold suggests that this is logically incontrovertible [56J4633]. Matter is quantised; a transition state either does or does not incorporate a molecule of nucleophile; there can be no gradual transition from one molecule to no molecule.

At this ultimate borderline debate is probably as much a matter of semantics as of chemistry. Schleyer, while accepting Gold's point in principle, prefers to imagine a continuous spectrum of mechanisms yet his

Table 7.7. *Second order rate coefficients for*
$RBr + Cl^- \rightarrow RCl + Br^-$ *in*
*dimethylformamide at* $25°C$
$(k_2/10^{-5} \ l \ mol^{-1} \ s^{-1})$ [68J(B)142]

| α-substitution | | β-substitution | |
|---|---|---|---|
| R | $k_2$ | R | $k_2$ |
| Me | 50 000 | | |
| Et | 1 300 | Et | 1 300 |
| i-Pr | 25 | n-Pr | 930 |
| t-Bu | 1.1 | i-Bu | 45 |
| | | Neopentyl | 0.0085 |

arguments are based on the study of what inevitably must be a discontinuous series of discrete compounds. Golds's abrupt boundary would be hard to define in terms of the gradual exponential decay of overlap integrals with distance. Indeed, the more closely we try to scrutinise the borderline the more it seems to hide itself in the mists of uncertainty and statistical mechanics.

## 7.4 Structural and solvent* effects on mechanism

### 7.4.1 *The structure of the substrate*

Methyl and primary alkyl substrates react by the $S_N2$ mechanism, tertiary by $S_N1$; secondary compounds fall near the mechanistic borderline (§7.3.1). The reason for this behaviour is two-fold. The $S_N2$ transition state has a penta-coordinated central carbon atom so the reaction is sterically retarded if there are bulky groups attached to it; the consequences are illustrated in table 7.7 which shows that extra groups at either the α or β position retard the reaction. The results correlate well with calculated steric strain energies [75J(P2)1365]. The reactions of table 7.7 were carried out in a poorly ionising solvent to minimise $S_N1$ competition, but even so the t-butyl compound is certainly reacting via an ion-pair; its true $S_N2(A)$ rate would be very much slower.

The $S_N1$ transition state is sterically less crowded than the ground state and the reactions might be expected to show steric acceleration. In fact the effect is small except in extreme cases and it is often difficult to distinguish between steric and polar effects of the different alkyl groups. Much more important here is the marked stabilisation of a tertiary carbenium ion by inductive and hyperconjugative electron donation from the neighbouring alkyl groups. The magnitude of this stabilization is exemplified by the $10^8$ difference in reactivities of the secondary and

---

* See also chapter 5 for further discussion of solvent effects in nucleophilic substitution reactions.

Table 7.8. *Steric effects in* $S_N$ *reactions of cyclopentyl and cyclohexyl systems* [52JA1894; 52BB427]

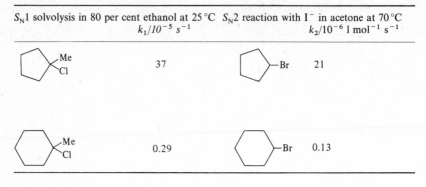

| $S_N1$ solvolysis in 80 per cent ethanol at 25 °C $k_1/10^{-5}$ s$^{-1}$ | $S_N2$ reaction with I$^-$ in acetone at 70 °C $k_2/10^{-6}$ l mol$^{-1}$ s$^{-1}$ |
|---|---|
| 37 | 21 |
| 0.29 | 0.13 |

tertiary 2-adamantyl systems (§7.3.1). The much smaller rate ratios (typically $10^4$) found in t-butyl and isopropyl solvolyses accord with considerable $S_N2$ character (nucleophilic solvation) in the isopropyl reactions.

The effect of steric strain is marked in cyclic systems. Cyclopentyl substrates invariably react faster than cyclohexyl analogues in both $S_N1$ and $S_N2$ reactions (table 7.8). The cyclohexane ring is more or less strain free, with staggered bonds all round, and formation of a trigonal carbon atom, whether in an $S_N2$ transition state or in a carbenium ion, introduces eclipsing interactions (**7.9**). With cyclopentyl derivatives the situation is reversed; a trigonal carbon atom relieves some of the torsional strain present in the ground state.

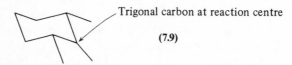

Trigonal carbon at reaction centre

(**7.9**)

Substrates with the leaving group at a bridgehead are normally very unreactive: $S_N2$ reactions are excluded because there is no approach from the rear, and so is complete dissociation to the free carbenium ion, which could not adopt a coplanar configuration and which would be open to solvation from one side only. However, ion-pair formation, though not easy, is possible as table 7.9 illustrates.

Any structural feature that can stabilise a carbenium ion will tend to facilitate an $S_N1$ mechanism. Thus in benzylic systems it seems likely that ionisation is the norm for secondary compounds and can occur in solvolyses even with primary compounds (§7.3.2). Stronger nucleophiles give $S_N2$ reactions with primary benzyl substrates. The rates of reaction

Table 7.9. *Relative rates of solvolysis of bridgehead bromides in 80 per cent ethanol at 25 °C* [67JA582]

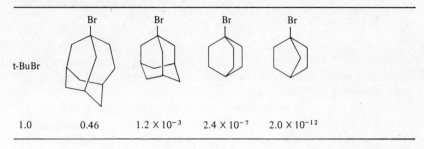

| t-BuBr | | | | |
|---|---|---|---|---|
| 1.0 | 0.46 | $1.2 \times 10^{-3}$ | $2.4 \times 10^{-7}$ | $2.0 \times 10^{-12}$ |

are often similar to those of the corresponding methyl compounds; this suggests there is some resonance stabilisation of the $S_N2$ transition state – just about enough to counteract the steric destabilisation [compare 79CC1131]. Hammett plots for these reactions commonly show shallow minima, electron donors and acceptors both giving slight acceleration. This accords with the idea, discussed above, of a variable $S_N2$ transition state: electron donors stabilise a loose transition state in which a slight positive charge builds up on the benzyl carbon atom, and electron acceptors stabilise a tight transition state with a negatively charged benzylic carbon:

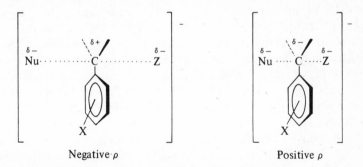

Negative $\rho$        Positive $\rho$

## 7.4.2 The leaving group (nucleofuge)

There has been remarkably little work done on the quantitative interpretation of nucleofugality [77Acc125]. In both $S_N1$ and $S_N2$ transition states the leaving group, $Z$, is taking its bonding electrons out with it, so one might expect some correlation with the stability of $Z^-$ as measured by its basicity. This indeed is found in a qualitative manner, but quantitative correlations appear only with a series of closely related leaving groups such as differently substituted arenesulphonates.

A typical order of reactivity in aqueous solution is as follows; approximate $pK_a$ values of the corresponding acids, HZ, are included for comparison.

$$N_2^+ \gg RSO_3 \simeq ROSO_3 > I > Br > Cl > Me_2S^+ > NO_3 > F > RCO_2 > Me_3N^+$$
$$\quad -5 \text{ to } -2 \qquad -10 \;-9 \;\; -7 \qquad -5 \qquad -1.3 \quad 3.2 \;\; \sim 5 \qquad 10$$

Clearly factors other than basicity are also playing a part; solvent effects in particular are important, and especially so when comparing charged and uncharged leaving groups.

Although the order of nucleofugality is the same in both $S_N1$ and $S_N2$ reactions, the mechanistic balance can be influenced by changing the leaving group. The C–Z bond is more completely broken in an $S_N1$ reaction, so a good nucleofuge will shift a borderline mechanism towards the $S_N1$ side whereas a poor one will have to seek more help from the attacking nucleophile.

Reactions can be catalysed by modifying the leaving group, and this in turn may modify the mechanism. Alkyl fluorides, normally inert towards nucleophiles, are hydrolysed in sulphuric acid solution in which the leaving group is HF rather than $F^-$. Alcohols and ethers (leaving group $pK_{BH^+} \simeq 16$) can similarly react in strongly acidic solution, as in the Zeisel reaction:

$$I^{\curvearrowleft} \qquad Me{-\!\!-}\overset{+}{O}RH \text{ (leaving group } pK_{BH^+} \simeq -2)$$

In aprotic solvents catalysis may occur via hydrogen bonding. t-Butyl bromide in nitromethane reacts at the same rate with a variety of nucleophiles ($Br^-$, $Cl^-$, $NO_2^-$, $H_2O$, EtOH, PhOH) when they are present in low concentration: typical $S_N1$ behaviour. At higher concentrations the hydroxylic nucleophiles exhibit second order kinetics but the rate coefficients do not relate to their nucleophilicities but to their acidities, which follow the opposite order. This suggests that the rate-limiting step is still ionisation, but now assisted by hydrogen bonding: $t\text{-}Bu^{\delta+}\cdots Br^{\delta-}\cdots H{-}OR$ [54J2918, 2930].

Such catalysis, of course, must also be present in protic solvents though it is not immediately detectable. It has the effect of partially levelling the reactivities of the halides since their hydrogen bonding capacity is in inverse order to their nucleofugality; thus the relative rates of reaction of methyl halides with azide ion in dimethylformamide at $25\,^\circ\text{C}$ are: 1:250:2000 for the sequence Cl, Br, I, but in methanol they are 1:60:100 [69CR1].

Lewis acid catalysis is also common, as in the Friedel–Crafts alkylation reactions which use an alkyl chloride with an aluminium chloride catalyst. Bromide and iodide are softer anions than chloride, and their reactions tend to be catalysed by softer Lewis acids. Heavy metal salts, particularly silver and mercury, are often used.

Table 7.10. *Nucleophilicities defined as* n = *log* $(k_{Nu}/k_{H_2O})$ *for reactions with* $CH_3Br$ *in water at* 25 °C [mainly from 53JA141; 54JA4385]

| Nucleophile | n | Nucleophile | n | Nucleophile | n |
|---|---|---|---|---|---|
| $ClO_4^-$ | <0 | $AcO^-$ | 2.7 | $SCN^-$ | 4.4 |
| $H_2O$ | 0.0 | $Cl^-$ | 2.7 | $PhNH_2$ | 4.5 |
| $TosO^-$ | 1 | Pyridine | 3.6 | $I^-$ | 4.7 |
| $NO_3^-$ | 1.0 | $Br^-$ | 3.5 | $CN^-$ | 5.1 |
| $F^-$ | 2.0 | $N_3^-$ | 4.0 | $SH^-$ | 5.1 |
| $SO_4^{2-}$ | 2.5 | $HO^-$ | 4.2 | $SO_3^{2-}$ | 5.1 |

Any catalytic effect on the leaving group will tend to push the mechanism towards $S_N1$. Thus Wagner–Meerwein rearrangements frequently accompany Friedel–Crafts alkylations or the acid catalysed conversion of alcohols into alkyl halides.

### 7.4.3    The nucleophile

(a) *Nucleophilicity.* Both nucleophilicity and basicity are related to electron-donating ability: basicity is a clearly defined thermodynamic property; nucleophilicity, on the other hand, is a kinetic concept and has no universally agreed quantitative definition. Table 7.10 lists values from a scale based on the relative reactivities of nucleophiles towards methyl bromide. Other compilations using other substrates in other protic solvents, are in only semi-quantitative agreement [67CEd89; 68JA319].

Two separate trends can be discerned. First, for nucleophiles that attack via the same atom there is a qualitative correlation between basicity and nucleophilicity: $HO^- > PhO^- > AcO^- > H_2O > ClO_4^-$, for example. Within a more closely related series, such as substituted phenoxides, the correlation becomes quantitative and linear, though steric hindrance, which is less important in protonation than in nucleophilic substitution, may cause deviations [compare 81JO635].

The other trend, much more pronounced, is that large, polarisable nucleophiles are more reactive than smaller, harder ones: $I^- > Br^- > Cl^- > F^-$; $PhSe^- > PhS^- > PhO^-$. There is no doubt that solvation plays a dominant role in this behaviour. In protic solvents anionic nucleophiles are stabilised by hydrogen bonding. This is strongest with the small anions, in which the negative charge is concentrated over a small surface area: the heats of hydration of the fluoride and iodide ions are 490 and 284 kJ mol$^{-1}$, respectively. The nucleophile must shed a considerable part of this solvation shell before it can react. In dipolar aprotic solvents the anions are much less strongly solvated and consequently react much faster than in protic solvents; the acceleration is

Table 7.11. *Second order rate coefficients for reactions of anionic nucleophiles with methyl iodide at 25°C* [69CR1]

| Nucleophile | In methanol $k_2^M/10^{-3}$ l mol$^{-1}$ s$^{-1}$ | In dimethylformamide $k_2^D/10^{-3}$ l mol$^{-1}$ s$^{-1}$ | Acceleration $k_2^D/k_2^M$ |
|---|---|---|---|
| $p$-NO$_2$C$_6$H$_4$O$^-$ | 0.0025 | 10 | 4 000 |
| AcO$^-$ | 0.0025 | 20 000 | 8 000 000 |
| Cl$^-$ | 0.0032 | 2 500 | 800 000 |
| Br$^-$ | 0.079 | 1 300 | 16 000 |
| N$_3^-$ | 0.079 | 3 200 | 40 000 |
| SCN$^-$ | 0.50 | 80 | 160 |
| CN$^-$ | 0.63 | 320 000 | 500 000 |
| $p$-NO$_2$C$_6$H$_4$S$^-$ | 63 | 16 000 | 250 |

greatest for those that were most strongly hydrogen bonded. Table 7.11 gives some examples.

Note the reversal in reactivity of bromide and chloride ions; it extends to the other halides, which in dimethylformamide have fairly similar reactivities but usually in the order $F^- > Cl^- > Br^- > I^-$. The effect with uncharged nucleophiles is much less dramatic; their reactions are seldom accelerated more than ten-fold on going from methanol to dimethylformamide, and may even be slightly retarded.

However, even aprotic solvents exert a very large effect. The recent development of ion cyclotron resonance has allowed $S_N2$ reactions to be measured in the gas phase; rates are typically around $10^{11}$ l mol$^{-1}$ s$^{-1}$, enormously greater than any solution value [80Acc76]. The order of nucleophilicities is considerably different, with CN$^-$, for example, being extremely unreactive [e.g. 77JA4219].

(b) *The α-effect.* There is one important exception to the generalisation that nucleophilicity parallels basicity for systems that react via the same atom. A species possessing two adjacent electronegative elements, HONH$_2$, N$_2$H$_4$, R$_2$S$_2$, etc., is much more strongly nucleophilic than would be expected from its basicity. The hydroperoxide anion (HOO$^-$) is $10^4$ times less basic than hydroxide, but it reacts some 50 times faster with benzyl bromide. This behaviour, known as the α-effect, is due to an interaction between lone-pair orbitals on the adjacent atoms:

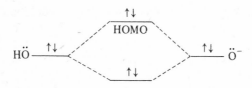

As usual, such an interaction produces two new MOs, one with lower energy and one with higher, leaving an abnormally high-energy HOMO which can react strongly with the low lying LUMO of a relatively soft nucleophile such as an alkyl halide. (Added in proof: A recent report that $HOO^-$ does not show enhanced nucleophilicity in the gas phase casts doubt on the usual explanation of the $\alpha$ effect [83JA2481].)

(c) *Ambident nucleophiles.* Many nucleophiles can react at alternative sites: nitrate, cyanide and enolate anions are examples. Some potentially ambident nucleophiles react (in substitution reactions at carbon) in only one way: the sulphite ion invariably gives sulphonates, $RSO_3^-$, rather than sulphites, $ROSO_2^-$. Others give mixed products, the balance depending on the substrate and the reaction conditions.

There seems to be a general tendency for such nucleophiles to react via their more electronegative site in reactions with considerable $S_N1$ character and via their less electronegative site in $S_N2$-like reactions. Thus primary alkyl halides react with silver nitrite in ether to give mostly the nitroalkane, $RNO_2$, whereas tertiary halides give mostly the nitrite, RONO. Again, sodium cyanide normally gives predominantly the nitrile, RCN, but heavy metal catalysis using silver cyanide encourages formation of the isonitrile, RNC. The rationale for this behaviour may lie in the relative hardness of a carbenium ion compared with the softer alkyl halide. In $S_N1$ reactions the electrostatic interactions tend to be the more important so that attack occurs via the site with the greater negative charge; in $S_N2$ it is orbital overlap that is dominant and attack occurs at the site where the HOMO is more intense, usually the less electronegative atom. This also accords with the pattern of protonation of such ambident species; the proton, a hard electrophile, invariably attacks fastest at the most electronegative atom. Sulphite and nitrite, for example, both protonate at oxygen. (The kinetic product may not always be the more stable thermodynamically; enols are formed by fast protonation of their anions at oxygen, but the keto tautomer is normally more stable.)

However, the hardness/softness balance is only part of the story; solvent effects, imperfectly understood, are also important. Table 7.12

Table 7.12. *Solvent effects on the product distribution in the reaction of sodium 2-naphthoxide with benzyl bromide* [63JA1148]

| Solvent | Per cent O alkylation | Per cent C alkylation | Solvent | Per cent O alkylation | Per cent C alkylation |
|---------|----------------------|----------------------|---------|----------------------|----------------------|
| $Me_2NCHO$ | 97 | 0 | MeOH | 57 | 34 |
| $Me_2SO$ | 95 | 0 | EtOH | 52 | 28 |
| $MeOCH_2CH_2OMe$ | 70 | 22 | $H_2O$ | 10 | 85 |
| tetrahydrofuran | 60 | 36 | $CF_3CH_2OH$ | 7 | 84 |

illustrates the effect of solvent on the site of benzylation of the ambident 2-naphthoxide anion, which can react either at the 1-carbon atom (Friedel–Crafts alkylation) or at the oxygen atom to give the benzyl naphthyl ether. Two opposing effects can be discerned. With aprotic solvents an increase in polarity encourages an $S_N1$ mechanism and favours attack at the harder site, oxygen. Protic solvents should likewise facilitate $S_N1$ reaction, but this effect is now masked because the anionic oxygen atom is strongly deactivated by hydrogen-bonded solvation.

### 7.5 Other $S_N$ mechanisms

*7.5.1 The $S_Ni$ mechanism*

The reactions of alcohols with thionyl chloride to give alkyl chlorides often proceed largely with retention of the stereochemical configuration of the alkyl group. Hughes and Ingold [37J1252] rationalised this in terms of an intramolecular substitution in an intermediate alkyl chlorosulphite, and they designated the mechanism $S_Ni$ (for internal). They originally envisaged an intramolecular $S_N2$ process, but the reaction exhibits features characteristic of ionisation: it is encouraged by good ionising solvents and by substrates that can give stable carbenium ions, and it is often accompanied by Wagner–Meerwein rearrangement. This points to an ion-pair mechanism:

$$ROH + SOCl_2 \xrightarrow{-HCl} ROSOCl \rightleftharpoons R^+ \cdots \underset{Cl^-}{\overset{\bar{O}}{S}}{=}O \longrightarrow RCl + SO_2$$

In the presence of a base such as pyridine, free chloride ions are liberated (with the formation of both $pyH^+$ and $ROSOpy^+$) and the reaction proceeds in a conventional $S_N2$ manner with inversion of configuration.

*7.5.2 Allylic rearrangement: $S_N'$*

Allyl substrates undergo nucleophilic substitutions faster than saturated analogues; the behaviour resembles that of benzyl systems. The reaction is frequently accompanied by rearrangement, designated $S_N'$, as table 7.13 demonstrates.

The first example in the table is presumably a conventional $S_N2$ reaction. The others proceed through ion-pairs:

$$CH_3CH{=}CHCH_2Cl \rightleftharpoons CH_3CH\overset{CH}{\underset{+}{\diagdown}}CH_2 \rightleftharpoons CH_3CH\overset{CH}{\underset{+}{\diagup}}CH_2 \rightleftharpoons CH_3CHClCH{=}CH_2$$

$$\underset{\text{Alternative products}}{\left\lfloor \quad Cl^- \qquad\qquad Cl^- \quad \right\rfloor}$$

Table 7.13. *Allylic rearrangement in nucleophilic substitution*

(A)  $CH_3CH=CHCH_2Cl$ $\xrightarrow{\text{Direct}}$ $CH_3CH=CHCH_2Nu$ (Y)

Rearranged

(B)  $CH_3CHClCH=CH_2$ $\xrightarrow{\text{Direct}}$ $CH_3CHNuCH=CH_2$ (Z)

| | | Substrate | | | |
|---|---|---|---|---|---|
| | | (A) | | (B) | |
| Reaction | Products (per cent): | (Y) | (Z) | (Y) | (Z) |
| 2M EtO$^-$/EtOH, 100 °C | | 100 | 0 | 0 | 100 |
| 0.8M HO$^-$/H$_2$O, 25 °C | | 60 | 40 | 38 | 62 |
| EtOH, 100 °C | | 92 | 8 | 82 | 18 |

Complete dissociation to a free carbenium ion would, of course, give the same mixture of products from both reactants.

Allylic rearrangements provided some of the earliest evidence for the existence of ion-pairs. Racemisation of $MeCH=CHCHZMe$ (Z = p-nitrobenzoate) is four times faster than solvolysis in 90 per cent acetone [60JA2515]. Isomerisation of $CH_2=CHCMe_2Cl$ to $ClCH_2CH=CMe_2$ is likewise faster than acetolysis [51JA1958]. In neither case could the rearrangement arise from complete dissociation because the reactions are not subject to common ion retardation, but both can be rationalised as $S_N2(B)$ processes (figure 7.3).

There remains the possibility that some allylic rearrangements may proceed in a concerted manner, $S_N2'$ [76Acc281; 80T1901; 82JO2517]:

$$Nu^- \quad CH_2=CH-CH_2-Z \qquad\qquad (7.10)$$

Molecular orbital calculations [75JA6615] suggest that in such a process an uncharged nucleophile should attack the allyl system *syn* to the leaving group, with a delocalised 6-electron transition state (**7.10**; compare chapter 16). Such stereochemistry has been observed, but one cannot yet affirm with certainty that this is not another manifestation of the $S_N2(B)$ ion-pair mechanism [79JA2107].

### 7.5.3 Neighbouring group participation

Mustard gas, $S(CH_2CH_2Cl)_2$, the vesicant poison gas of the First World War, is hydrolysed in aqueous solution several thousand times faster than simple primary alkyl chlorides. The alkaline hydrolysis of chiral 2-bromopropanoate, $CH_3\overset{*}{C}HBrCO_2^-$, to the hydroxyacid proceeds with retention of configuration. Hydrolysis of the chloroamine $Et_2NCH_2CHEtCl$ gives the rearranged alcohol, $Et_2NCHEtCH_2OH$.

These three examples typify the behaviour known as neighbouring group participation, the intramolecular association of one group in a molecule with the reactions of another. Each of the above reactions entails a three-membered cyclic intermediate:

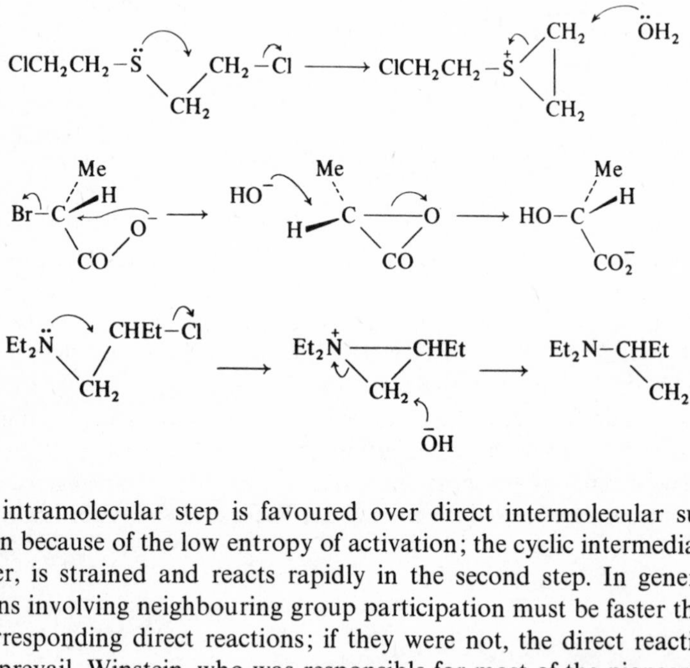

The intramolecular step is favoured over direct intermolecular substitution because of the low entropy of activation; the cyclic intermediate, however, is strained and reacts rapidly in the second step. In general, reactions involving neighbouring group participation must be faster than the corresponding direct reactions; if they were not, the direct reaction would prevail. Winstein, who was responsible for most of the pioneering work in this field, referred to this acceleration as *anchimeric assistance*. It is not always easy to estimate what the rate of the direct reaction would have been, in order to quantify the extent of the acceleration. One of Winstein's methods was to compare the behaviour of *cis* and *trans* disubstituted cyclohexanes. Neighbouring group participation is possible in the *trans* isomer and results in anomalous stereochemical behaviour; the *trans* configuration is retained but an enantiomeric reactant is racemised because the bridged intermediate is in rapid equilibrium with its enantiomer via ring inversion:

In the *cis* isomer the mutual orientation of the two groups precludes the intramolecular step and normal substitution occurs. If one can assume that the *cis* isomer provides a good model for the electronic effects of the neighbouring group, then the rate ratio for the two isomers provides a measure of anchimeric acceleration. Some experimental results for acetolyses are as follows [48JA821; 51JA5458].

| X | OBs | Cl | Br | OAc |
|---|-----|-----|-----|-----|
| $k_t/k_c$ | 1.12 | 3.8 | 810 | 630 |

In fact the brosylate group itself appears to be too weakly nucleophilic to enter into neighbouring group participation, so the close similarity of the *cis* and *trans* rates for the dibrosylates indicates that, despite obvious conformational differences, the *cis* isomer does provide a reasonable model for the unassisted reaction of the *trans*. The halogens show a parallel with the nucleophilicities of the halide ions. Iodine is an even more effective neighbouring group; using Taft's $\sigma^*$ to estimate the unassisted rate, Streitwieser calculates an anchimeric acceleration of 1.5 million [56JA4935]. The acetoxy group is more reactive than would be expected on a simple analogy with nucleophilicity, but it reacts via a much less strained five-membered ring intermediate.

Such behaviour has been confirmed by $^{18}O$ labelling experiments [66JO4234]; if initially the carbonyl oxygen is labelled, half of the label is found in the diol obtained by hydrolysing the diacetate formed in acetolysis.

Neighbouring group participation by an aryl group has already been exemplified in the solvolytic racemisation of *threo*-3-anisylbut-2-yl sulphonates **(7.1)**. The *erythro* isomer is not racemised under similar conditions, though it, too, exhibits the special electrolyte effect. Moreover, there is no interconversion between the *threo* and *erythro* epimers. This accords with the intermediacy of a bridged phenonium ion.

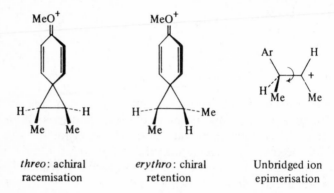

|  |  |  |
|---|---|---|
| *threo*: achiral racemisation | *erythro*: chiral retention | Unbridged ion epimerisation |

Electron-withdrawing substituents on the benzene ring reduce the stability of the phenonium ion and encourage reaction via an unbridged ion-pair; acetolysis of the *threo-p*-nitro compound gives 7 per cent *threo* and 93 per cent *erythro* products. A plot of $\log k$ against $\sigma$ for these solvolyses gives a straight line for electron acceptors and a curve for donors

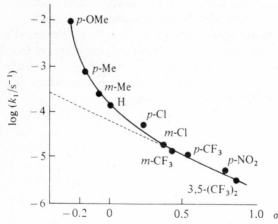

Figure 7.6. Hammett plot for acetolysis of *threo*-3-arylbut-2-yl brosylates at 75 °C.

Table 7.14. *First order rate coefficients* $(k_1/10^{-6}\,s^{-1})$ *for direct and anchimerically assisted solvolyses of* $PhCH_2CH_2OTos$ *at 75 °C* [68JA6546]

| Solvent | Direct | Assisted |
|---------|--------|----------|
| EtOH | 7.0 | 0.042 |
| AcOH | 0.19 | 0.22 |
| HCOOH | 3.9 | 35 |
| CF₃COOH | 0.05 | 1160 |

(figure 7.6); extrapolating the linear part gives an estimate of the unassisted rate for the anchimerically assisted reactions, and thence a measure of the acceleration [71JA5765].

The related 2-arylethyl system, $ArCH_2CH_2OTos$, behaves in a similar manner, though the direct reaction is now presumably $S_N2$. The use of $^{14}C$ labelled $PhCH_2\overset{*}{C}H_2OTos$ enables the rate of rearrangement to be measured, and from this the balance between direct and phenonium ion routes can be determined. The results (table 7.14) provide an insight into the importance of the solvent at the $S_N1/S_N2$ borderline: ethanol, a poor ionising solvent and a relatively strong nucleophile, reacts almost entirely by the direct $S_N2$ route, whereas trifluoroacetic acid, with contrary properties, reacts via the phenonium ion-pair. Confirmation came from the stereochemistry of the *threo*-1,2-dideuteriophenethyl tosylate, which showed no measurable epimerisation on solvolysis in trifluoroacetic acid. The extent of anchimeric assistance in the acetolysis reaction has been independently assessed from the curved Hammett plot; the two methods are in excellent agreement [69JA7508].

A $\pi$ bond can also act as a neighbouring group [72MIb]. 3-$\beta$-Cholesteryl chloride (**7.11**; Z = Cl) undergoes nucleophilic substitution with retention of configuration. Methanolysis of the corresponding tosylate gives the 3-$\beta$-ether if the solution is not buffered, but in the presence of potassium acetate the product is the so-called i-cholesteryl ether (**7.12**; Nu = OMe). Acetolysis similarly gives the i-acetate in a reaction which is about a hundred times faster than the acetolysis of cyclohexyl tosylate, and in acid the i-acetate is rapidly isomerised to the 3-$\beta$-acetate. Winstein proposed that this behaviour was due to the participation of the $\pi$ orbitals of the cholesteryl double bond in the expulsion of the leaving group, with the formation of an intermediate with delocalised positive charge (**7.13**) [48JA838; 56JA4347]. This somewhat resembles the $\pi$ system of the allyl cation: Winstein called it

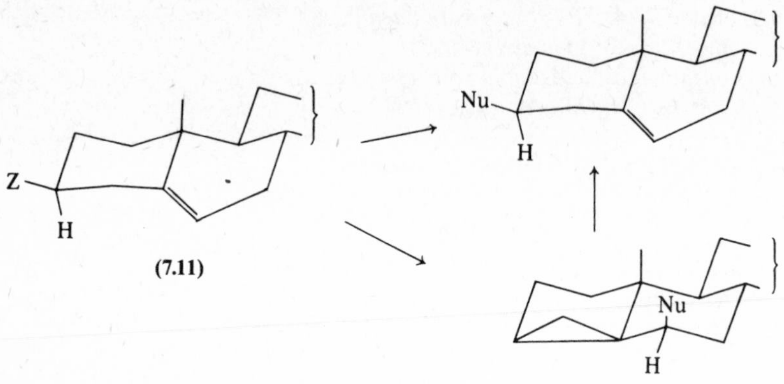

(7.11)

(7.12)

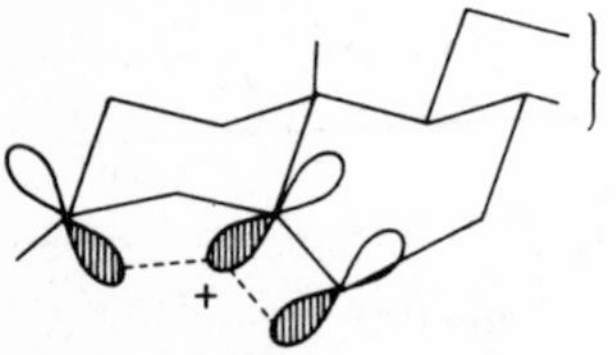

(7.13)

homoallyl, and the interrupted $\pi$ conjugation it contains he called homoconjugation. Like the allyl system it is open to attack at either end, and the formation of the i-product is analogous to the allylic $S_N'$ rearrangement.

Subsequently a much more dramatic example of the same phenomenon was found in the acetolysis of *anti*-7-norbornenyl tosylate (7.14). The reaction is $10^{11}$ times faster than that of the saturated analogue, and again occurs with retention of configuration [56JA592]. The intermediate in this reaction is symmetrical (7.15); the $\pi$ electrons are ideally situated for neighbouring group participation.

The cationic intermediates in these reactions are rather different from normal, tricoordinated carbenium ions. In (7.15) the 7-position carbon is penta-coordinated and the other two participating carbons are tetra-coordinated. In the terminology introduced by Olah [73AG173] these are carbonium ions, though the word is still widely used to describe all

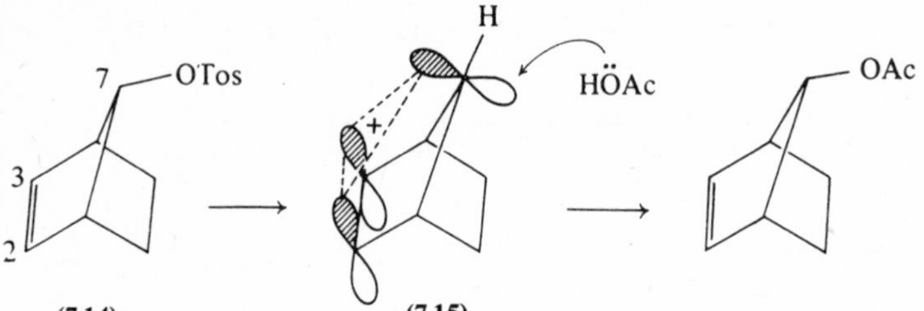

(7.14)

(7.15)

carbocations, and the tetra- and penta-coordinated species are distinguished as *non-classical carbonium ions*.

There was for some time considerable controversy about the reality of such species. Could the results not be due to rapid equilibration between 'classical' carbenium ions?

The evidence is against such an explanation. Introducing a methyl group in the 2-position accelerates the reaction by a factor of about 13; a second methyl at position 3 further increases the rate 11-fold. This strongly indicates that the positive charge is shared between these two positions. If the reaction proceeded through the classical ions the first methyl group would certainly accelerate the reaction by generating a tertiary carbenium centre, but the second methyl group would have little extra effect because both methyl groups could not simultaneously be stabilising the charge [69JA2160].

When a *p*-anisyl group is introduced at the 7-position (**7.16**) the anchimeric acceleration disappears. The reaction is actually much faster than predicted from a Hammett plot of other 7-aryl derivatives, but not faster than that of the saturated *p*-anisyl compound [70JA2549]. Here is confirmation of the point made earlier in the discussion: a reaction with neighbouring group participation must be faster than one without; if it were not the direct reaction would happen instead. In this case the direct reaction via the highly stable classical ion (**7.17**) does indeed take over.

In 1939 Wilson had first suggested the possibility of a non-classical structure for the camphyl cation [39J1188], but for ten years and more his ideas were largely ignored. After Winstein's work, however, the

(**7.16**)  (**7.17**)

concept spread like wildfire, and during the 1960s it became fashionable to draw almost any carbocation as a non-classical structure. Such extravagance is unjustified, as H. C. Brown in particular has demonstrated, and it is now clear that there are fairly stringent limitations on the geometrical requirements for non-classical delocalisation. There remains one highly contentious topic: can the $\sigma$ electrons of suitably oriented C—C or C—H bonds also participate in solvolyses, generating non-classical intermediates? This question is discussed in a different context in §14.3.

### 7.6 Further reading

S. R. Hartshorn, *Aliphatic Nucleophilic Substitution*, Cambridge University Press, London, 1973.

C. A. Bunton, *Nucleophilic Substitution at Saturated Carbon*, Elsevier, London, 1963.

Solvolyses:

T. W. Bentley & P. v. R. Schleyer, *Adv. Phys. Org. Chem.*, 1977, **14**, 1.

J. M. Harris, *Progr. Phys. Org. Chem.*, 1974, **11**, 89.

A. Streitwieser, *Solvolytic Displacement Reactions*, McGraw-Hill, New York, 1962.

Ion-pairs:

S. Winstein, B. Appel, R. Baker & A. Diaz, in *Organic Reaction Mechanisms*, Chemical Society Special Publication No. 19, 1965.

Reactions of secondary substrates:

D. J. Raber & J. M. Harris, *J. Chem. Educ.*, 1972, **49**, 60.

Ambident nucleophiles:

R. Gompper, *Angew. Chem. Internat. Ed.*, 1964, **3**, 560.

Substitutions in aprotic solvents:

A. R. Parker, *Chem. Rev.*, 1969, **69**, 1.

Neighbouring group participation:

J. B. Lambert, H. W. Mark, A. G. Holcomb & E. S. Magyar, *Acc. Chem. Res.*, 1979, **12**, 317.

B. Capon & S. P. McManus, *Neighbouring Group Participation*, Plenum, New York; vol 1, 1976.

B. Capon, *Quart. Rev.*, 1964, **18**, 45.

# 8 Elimination reactions

This chapter covers only one category of elimination: reactions of the type $H-C-C-Z \rightarrow C=C$. These constitute the most important and most thoroughly studied class of eliminations but the total scope is much wider, including the formation of triple bonds or of carbon–oxygen or other hetero-atom double bonds, reactions in which hydrogen is not one of the leaving groups (dehalogenations and fragmentations, for example), and eliminations from non-adjacent atoms.

## 8.1 The $E1$-$E2$-$E1$cB mechanistic spectrum

Much of the early work on eliminations, as with nucleophilic substitutions, is due to Hughes and Ingold; it is summarised in Ingold's book [69MI] and more briefly in the report of one of his lectures [62P265]. They proposed mechanisms analogous to $S_N2$ and $S_N1$: a one-step, concerted, bimolecular reaction and a two-step reaction via a carbenium ion with the first step identical to that in an $S_N1$ reaction:

$$B^- \quad H-C-C-Z \qquad \qquad E2$$

$$H-C-C-Z \; \rightleftharpoons \; Z^- \quad H-C-C^+ \longrightarrow C=C \qquad E1$$

They also envisaged a third mechanism, which Ingold later called $E1$cB (unimolecular elimination from conjugate base), in which a substrate bearing a relatively acidic proton reacts via a carbanionic intermediate:

$$B^- \quad H-C-C-Z \longrightarrow C^- C-Z \longrightarrow C=C \qquad E1cB$$

Clearly one can envisage a whole mechanistic spectrum between the $E1$ and $E1$cB extremes, depending on the relative timing of the $C-H$ and $C-Z$ cleavage; the idealised $E2$ mechanism would fall precisely in the middle with both bonds breaking in unison. This concept of a spread of $E2$ mechanisms is elaborated in §8.1.3. The debate remains, as with $S_N$

reactions, as to whether there can be a continuous progression from bimolecular to unimolecular. There are also the complications introduced by the intervention of ion-pairs; these presumably play much the same role in $E1$ reactions as they do in $S_N1$ (compare table 7.3), but their detailed behaviour has been less studied in this context. Likewise $E1cB$ reactions may occur via carbanion ion-pairs.

### 8.1.1 Characteristics of the E1 and E2 mechanisms

The rate-limiting step for $E1$ and $S_N1$ reactions is the same, and so their kinetic behaviour will be identical. The $E_1$ reaction is first order in substrate, unaffected by added base, and subject to common ion retardation (provided the carbenium ion is sufficiently selective and that ionisation proceeds beyond the intimate ion-pair).

The $E2$ reaction follows second order kinetics, first order in both substrate and base. With solvolytic eliminations there is the usual problem of distinguishing between the pseudo-first order kinetics of a bimolecular reaction and a true first order process.

In an $E1$ reaction there is only a small (secondary) H/D kinetic isotope effect, but a significant isotope effect for the leaving group: $^{34}S/^{32}S$ in $-SMe_2^+$ for example. In $E2$, on the other hand, both the $H-C_\beta$ and $C_\alpha-Z$ bonds are partially broken in the transition state and both show kinetic isotope effects. Their magnitudes can be very variable, depending on the precise structure of the transition state; $k_H/k_D$ usually falls in the range 3 to 8 [e.g. 81JO4242, 4247].

The stereochemical characteristics of the two mechanisms are discussed below (§8.2).

### 8.1.2 Characteristics of the E1cB mechanism

(a) *Kinetics.* Kinetic analysis is not very helpful in attempting to characterise an $E1cB$ mechanism. If the steady-state approximation can be applied to the carbanion in the scheme:

$$B^- + HCCZ \underset{k_{-1}}{\overset{k_1}{\rightleftharpoons}} BH + {}^-CCZ \overset{k_2}{\rightarrow} C{=}C + Z^-$$

we can derive the following kinetic expression.

$$\text{Rate} = \frac{k_1 k_2 [HCCZ][B^-]}{k_{-1}[BH] + k_2}$$

If the ionisation is rate-limiting $(k_{-1}[BH] \ll k_2)$ this reduces to: Rate $= k_1[HCCZ][B^-]$, indistinguishable from $E2$. This extreme is commonly called an irreversible $E1cB$. A rapid equilibrium between reactants and carbanion $(k_{-1}[BH] \gg k_2)$ leads to the more complex relation:

$$\text{Rate} = \frac{k_1 k_2 [HCCZ][B^-]}{k_{-1}[BH]}$$

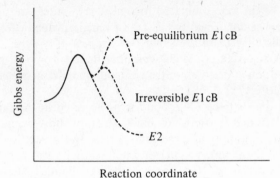

Figure 8.1. E1cB and E2 reaction profiles.

However, when the base is the anion of the solvent, as it usually is, $[BH]$ is constant and the kinetics again reduce to second order. This mechanistic variation is called the pre-equilibrium E1cB. The reaction profiles of figure 8.1 illustrate the difference between these two types of E1cB, and the fine dividing line between the irreversible E1cB mechanism and E2.

There remains a third possibility: a very acidic substrate (for example, with a nitro group activating the $\beta$ proton) can be completely ionised to the anion in an excess of base; there would then be a first order decomposition in the second step, unaffected by changes in $[B^-]$.

(b) *Isotope effects in pre-equilibrium E1cB mechanisms.* The rapid pre-equilibrium can be detected by deuterium exchange; if the reaction is stopped partway, deuterium will be incorporated into the substrate from a labelled solvent. A corollary of this is that such a reaction shows no H/D kinetic isotope effect, though there can be an effect from the leaving group, $Z$. There is also a *solvent* kinetic isotope effect on the initial rate of reaction, at a stage before exchange has become significant, because $k_{-1}$ is smaller in $D_2O$ (transfer of $D^+$ to the carbanion) than in $H_2O$ [81JA2457; see also 83JA265].

(c) *Leaving group effects in irreversible E1cB reactions.* The C–Z bond remains intact during the rate-limiting step of the irreversible E1cB, and the absence of a nucleofugal isotope effect may provide evidence for the mechanism. The dehydrochlorination of DDT, $(p\text{-}ClC_6H_4)_2CH\text{–}CCl_3$, by strong bases is believed to be an example of irreversible E1cB [74J(P2)1373; 76AJ787] but DDD and related compounds, $Ar_2CH\text{–}CHCl_2$, in general react by an E2 mechanism: their dehydrochlorinations are faster than those of their DDT analogues whereas ionisation to the carbanion should be slower in a system with one fewer electron-withdrawing chlorine atom. The *p*-nitro compound, exceptionally, does react at the rate estimated for anion formation, and the $E1cB$ mechanism here is confirmed by $^{35}Cl/^{37}Cl$ kinetic isotope effects. For other DDD substrates these are small but significant, lying in the range 1.0023 to 1.0038 $\pm$ 0.0002, and showing that there is some $C_\alpha-Cl$ cleavage in the transition state; the *p*-nitro compound shows no detectable effect at all [77J(P2)1753, 1758].

For the same reason, that the C–Z bond is not cleaved in the rate-limiting step, the normal pattern of nucleofugality (for example I > Br > Cl > F) may not apply in these eliminations. In the reaction of *trans*-1-chloro-2-fluoroacenaphthene (**8.1**) with potassium t-butoxide, HF is eliminated in preference to HCl because the rate-limiting step is formation of the carbanion, which can be stabilised by delocalisation of the charge into an adjacent chlorine *d* orbital [82JO3237].

(**8.1**)

### 8.1.3 The variable E2 transition state

The idea of a broad spread of *E2* mechanisms, from those with a largely carbanionic $\beta$ carbon atom to those with a largely cationic $\alpha$ carbon atom, was originally promoted by Bunnett [62AG225; 69SPC53]. It has since been modified and extended by several other workers. It is conveniently visualised as a hypothetical three-dimensional potential energy surface mapped onto two-dimensional paper as illustrated in figure 8.2. The diagrams can be thought of in terms of a geographical analogy (compare figure 2.4, §2.3.3): a point in the horizontal plane corresponds to the geometry of the reacting molecules and the vertical height to the potential energy of the system. A reaction is then represented as a pathway from reactants in one valley to products in another, crossing the pass between the two, that is, the transition state. For example, the geometrical variables allowed for in figure 8.2 are the $C_\beta-H$ and $C_\alpha-Z$ distances, going from the equilibrium bond lengths in the reactants to fully separated in the products. The upper pair of diagrams describe a totally symmetrical reaction with a transition state in which the $C_\beta-H$ and $C_\alpha-Z$ bonds are both half broken. The lower pair show the behaviour of a substrate that can stabilise a positive charge on the $\alpha$ carbon atom so that the reaction has significant $E1$ character.

These diagrams can be used to predict the qualitative effect on the transition state of changing some feature of the reaction. If a change

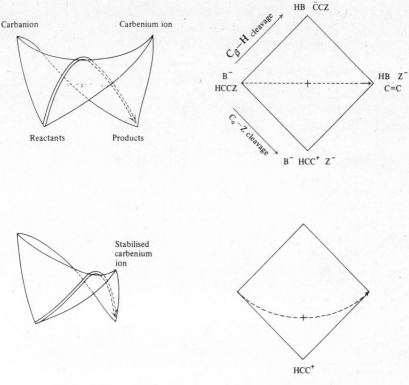

Figure 8.2. The variable E2 transition state. The upper diagrams represent, in perspective and in plan, a totally symmetrical E2 process. The lower diagrams show the effect of lowering the energy of the carbenium ion 'corner' of the system, giving an E2 reaction with E1 character [cf. 70J(B)274; 75JA3102].

lowers the energy of the system in a direction perpendicular to the reaction coordinate, then the transition state moves in the direction of lower energy, as in the E1-like example of figure 8.2. But if the change is parallel to the reaction coordinate then, strange as it seems at first sight, the transition state moves *away* from the direction of lower energy. This follows from the Hammond postulate (§3.4) that the transition state resembles the side of higher energy (figure 8.3). (In the geographical analogy, the summit of a pass is nearer to the higher of the valleys it connects.)

The application of this approach can be illustrated by considering the effects of changing the leaving group (table 8.1) or the attacking base (table 8.2).

Table 8.1 shows that with a decrease in nucleofugality (as shown by the reaction rates) the value of $\rho$ increases, revealing an increasing negative charge on the $\beta$ carbon atom and a more E1cB-like mechanism. A poor

Table 8.1. *Leaving group effects in eliminations from* $ArCH_2CH_2Z$ *using* $EtO^-/EtOH$ *at 30 °C (Data compiled by Bunnett* [62AG225; 69SPC53])

| Leaving group Z | Relative rates (Ar = Ph) | $\rho$ | $k_H/k_D$ (Ar = Ph) |
|---|---|---|---|
| I | 26 600 | 2.07 | |
| Br | 4 100 | 2.14 | 7.1 |
| OTos | 392 | 2.27 | 5.7 |
| Cl | 68 | 2.61 | |
| $SMe_2^+$ | 7.7 | 2.75 | 5.1 |
| F | 1 | 3.12 | |
| $NMe_3^+$ | | 3.77 | 3.0 (at 50 °C) |

Table 8.2. *Effect of base on elimination from* $ArCH_2CH_2NMe_3^+$ *in dimethylformamide at 56 °C* [77JO205]

| Base | $\rho$ | $k_H/k_D$ (Ar = Ph) |
|---|---|---|
| $PhO^-$ | 4.25 | 2.83 |
| $m\text{-}NO_2C_6H_4O^-$ | 3.81 | 2.26 |

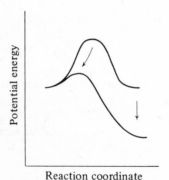

Reaction coordinate

Figure 8.3. The Hammond postulate. As the energy of one side of the reaction coordinate is lowered the structure of the transition state changes towards that of the other side.

nucleofuge corresponds, to a first approximation, to a species for which $Z^-$ has a relatively high energy. The effect of this on the transition state, as shown in figure 8.4, is to increase the extent of $C_\beta$–H cleavage. The H/D kinetic isotope effect confirms this interpretation: the maximum

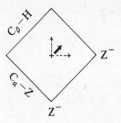

Figure 8.4. Effect on E2 transition state of a poor nucleofuge (high energy $Z^-$). The dotted arrows show a central transition state moving towards $Z^-$ along the reaction coordinate and away from $Z^-$ perpendicular to it. The resultant change in transition state structure is shown by the solid arrow (cf. figure 8.2).

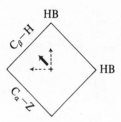

Figure 8.5. Effect on E2 transition state of strong base (low energy HB). The dotted arrows show a central transition state moving away from HB along the reaction coordinate and towards HB perpendicular to it. The resultant change in transition state structure is shown by the solid arrow (cf. figure 8.2).

value at $Z = Br$ corresponds to a transition state in which the proton is about half transferred to the base and the decreasing values accord with progressively more $C_\beta$–H cleavage. Again, the Brønsted $\beta$ value for elimination from cyclohexyl tosylate using a range of substituted thiophenoxides as bases is 0.27; for the bromide it is 0.36, and for the chloride 0.39. The poorer the nucleofuge, the more is the proton transferred from substrate to base in the transition state [66J(B)705].

Table 8.2 shows that the effect of increasing the base strength (lowering the energy of HB) is not, as might be expected, to increase the degree of proton transfer from the substrate. In fact the isotope effect suggests the reverse; the bond is more than half broken in the transition state, so an increase in $k_H/k_D$ corresponds to a decrease in C–H cleavage. The increased carbanion character revealed by the $\rho$ values arises because the stronger base reduces the extent of $C_\alpha$–Z cleavage in the transition state (figure 8.5).

Conversely the dehydrochlorination of DDT, which with a strong base is E1cB (§8.1.2), shifts towards a concerted E2 mechanism when the base is an amine [79JO3344]. With NaOMe as base the value of $\Delta S^\ddagger$ for the elimination is $-12 \, J \, K^{-1} \, mol^{-1}$; with $MeNH_2$ it is $-205$: this large negative value accords with a high degree of association between base and substrate in the transition state.

## 8.2 Stereochemistry

### 8.2.1 *Stereochemistry of* E2 *eliminations*

In the concerted *E*2 mechanism one would expect some stereoelectronic correlation between the H and Z leaving groups. MO calculations and the Principle of Least Nuclear Motion [66JA5525; 69JA7144; 77APO(15)1] both suggest that the preferred orientation for elimination is *anti*-periplanar (compare §6.4.3); if this is not possible a *syn*-periplanar arrangement is better than non-planar (clinal):

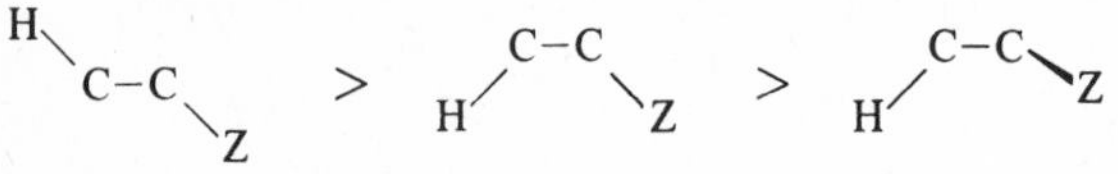

(*a*) *Diastereoisomeric substrates* reveal this preference for *anti* elimination by giving isomeric alkenes [56JA790]:

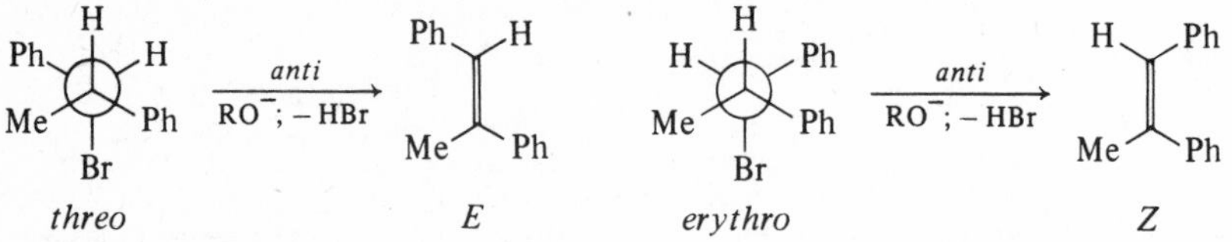
anti
RO⁻; – HBr
threo
E
erythro
Z

Deuterium labelling shows that simpler systems behave in the same fashion [68CC449]:

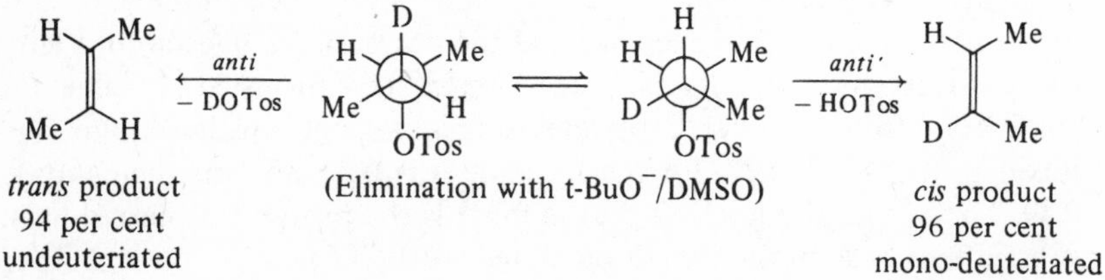
anti
– DOTos
anti'
– HOTos
trans product
94 per cent
undeuteriated
(Elimination with t-BuO⁻/DMSO)
cis product
96 per cent
mono-deuteriated

(*b*) *Cyclohexyl substrates* provide models with restricted and well-defined geometries. Dehydrochlorination of neomenthyl chloride gives a mixture of isomeric 2- and 3-menthenes, (predominantly the latter: §8.3.1). Menthyl chloride gives exclusively 2-menthene; the 3-isomer could only come from a clinal elimination [53J3839]:

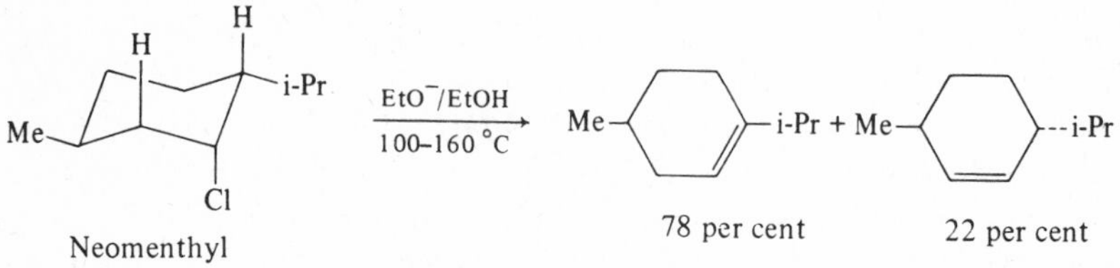
EtO⁻/EtOH
100–160 °C
78 per cent
22 per cent
Neomenthyl

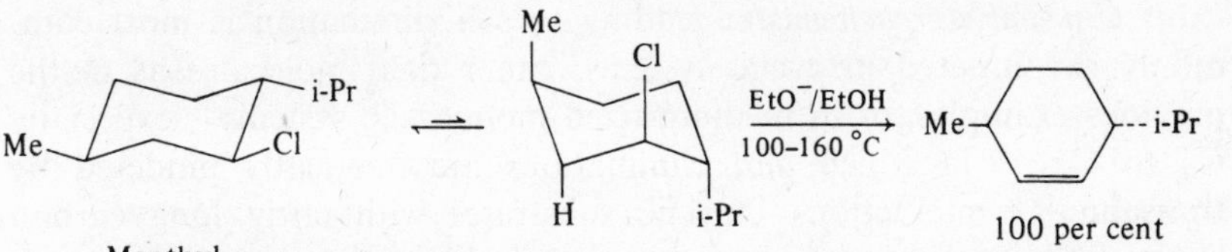

Menthyl

EtO⁻/EtOH 100–160 °C

100 per cent

*β*-Benzenehexachloride, in which no *anti* elimination is possible, reacts some $10^4$ times slower than any other isomer [53J3832]. A very small amount of deuterium exchange occurs during the reaction (0.08 per cent); this may be too small to be significant, but if it implies an *E*1cB mechanism it indicates that the clinal *E*2 is even slower.

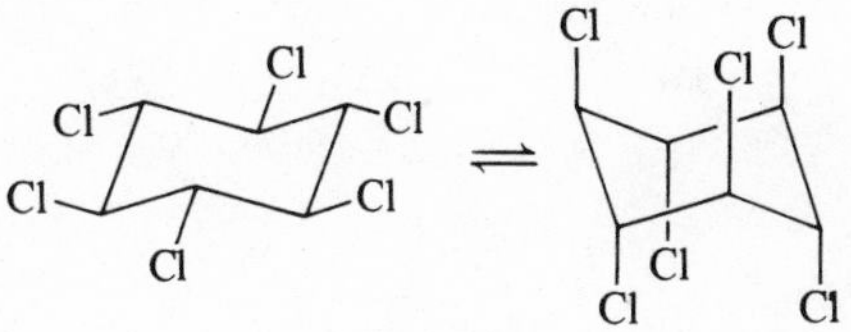

*β*-Benzenehexachloride

(*c*) *Bridged rings* have a rigid framework which allows the relative importance of *syn*-periplanar and *anti*-clinal eliminations to be assessed:

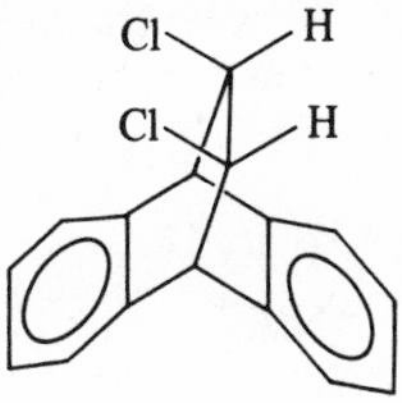

*anti*-Clinal

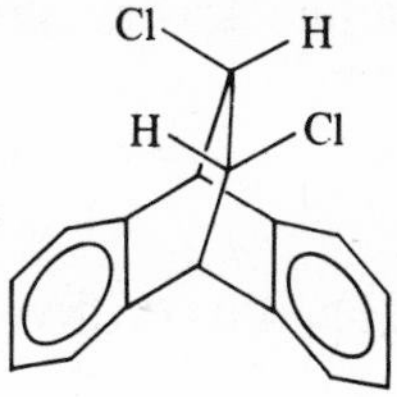

*syn*-Periplanar

| 1 | Relative rates of reaction with HO⁻/EtOH/dioxan, 120 °C | 9 | [52JA2193] |

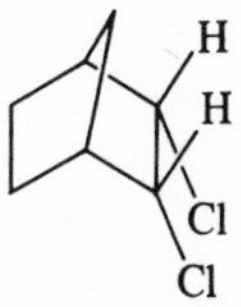

*anti*-Clinal

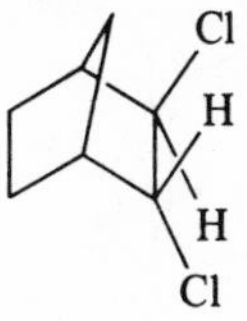

*syn*-Periplanar

| 1 | Relative rates of reaction with $C_5H_{11}O^-/C_5H_{11}OH$, 101 °C | 66 | [57JA3438] |

(*d*) *Unusual stereochemistry* leading to *syn* elimination is most commonly encountered in cyclic systems, either rigid molecules as in the previous examples, or in medium-sized monocyclic systems – expecially $C_9$ to $C_{12}$ – in which *anti* eliminations are frequently hindered by transannular interactions. Acyclic substrates with fairly long carbon chains (at least three carbon atoms on each side of the new double bond) frequently give a high proportion of *syn* elimination in forming the *trans* alkene, but not for its *cis* isomer. The reasons for this are still not fully clear [77JA6699], but it probably arises from steric hindrance to the approach of the base.

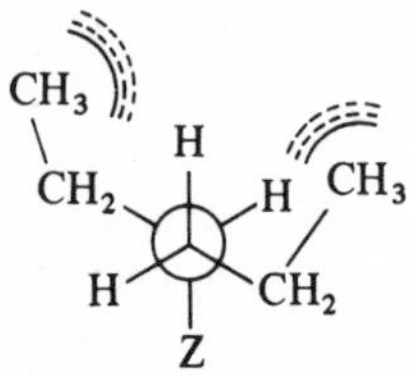

(*e*) *The ylid mechanism* is an alternative mode of reaction for some ammonium and sulphonium salts in which the normal *E*2 mechanism is sterically disfavoured. Pyrolysis of $t\text{-Bu}_2CD\text{--}CH_2NMe_3^+OH^-$ gives $t\text{-Bu}_2C\text{=}CH_2$ together with trimethylamine, 75 per cent of which is monodeuteriated $Me_2NCH_2D$, formed via an ylid intermediate. In this reaction the leaving groups must necessarily be *syn* to each other [63JA1949]:

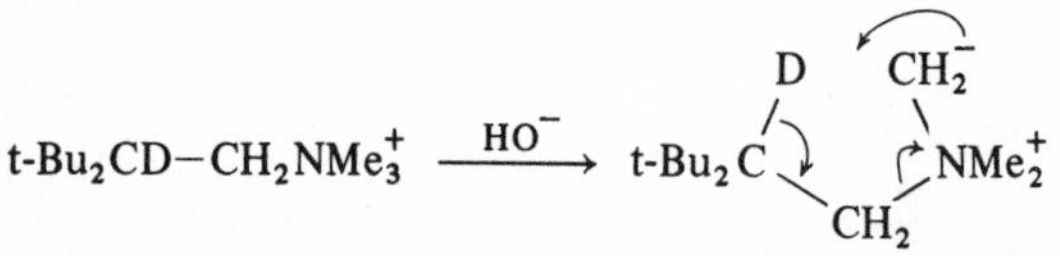

Elimination from the norbornyl system (**8.2**) gives norbornene containing no deuterium, and there is only 6 per cent incorporation of deuterium into the trimethylamine [67JA6701]. This suggests that the *syn-*

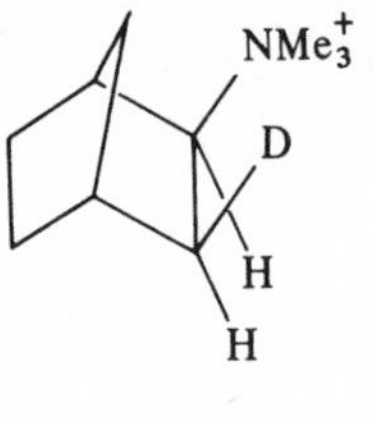

(**8.2**)

periplanar *E*2 competes preferentially with the ylid mechanism unless steric interactions are exceptionally severe.

## 8.2.2 Stereochemistry of E1 and E1cB mechanisms

In the extreme an $E1$ reaction should be completely non-stereoselective because there is no correlation between the two bond-breaking steps. In practice one commonly observes some selectivity, but less than for a related $E2$ reaction. As with $S_N1$ reactions, the intervention of ion-pairs allows the leaving group to participate in the product-forming transition state. Compare the behaviour of menthyl and neomenthyl chlorides under solvolysis conditions with that in $E2$ reactions (previous section) [53J3839]:

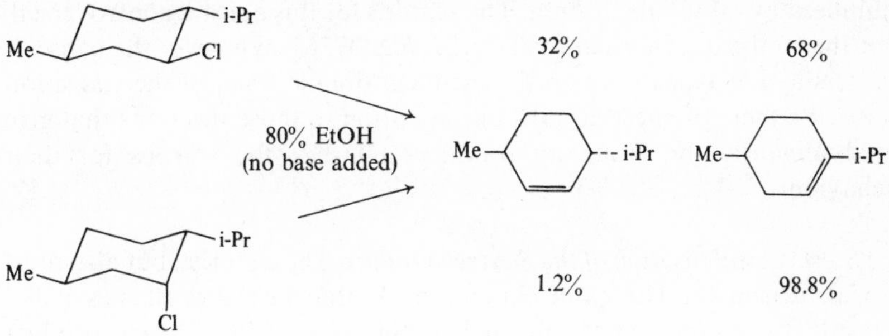

A pre-equilibrium $E1cB$ reaction should also be non-stereoselective, but there is some indication that *syn* elimination is preferred around the border between irreversible $E1cB$ and $E2$ mechanisms [79JA2845]. Evidence in support of this comes from the reactions of 1,2-dihalogenoacenaphthenes (such as **8.1**) [82JO3237]. The *cis* isomers, from which elimination must be *anti*-clinal, appear to react by a concerted $E2$ mechanism: HCl is eliminated much faster than is HF. For the *trans* isomers, however, elimination must be *syn*. With t-BuOK/t-BuOH the mechanism is clearly irreversible $E1cB$ (reactivity: F > Cl ≃ Br, §8.1.2). With EtOK/EtOH the leaving group effects are very small and the mechanism is ambiguous. A possible explanation for the *syn* stereochemistry is that the $E1cB$ transition state can be stabilised by electrostatic attraction between the counter-ion of the attacking base and the leaving group:

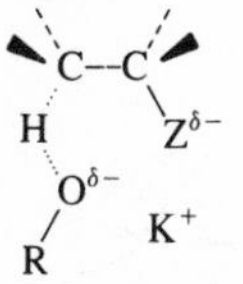

The importance of irreversible $E1cB$ reactions has only recently been recognised, and it may well be that a number of the *syn* eliminations mentioned above occur via this mechanism rather than via $E2$.

## 8.3 Orientation

The dehydrochlorination of 2-chlorobutane gives a mixture of three butenes. Figure 8.6 illustrates this and introduces the necessary terminology.

### 8.3.1 Positional orientation

A predominance of Saytzev elimination is the norm in all *E*1 reactions and in most *E*2. This is not surprising: in both mechanisms there is a considerable degree of double bond character in the product-forming transition state (figure 8.7). A carbon–carbon double bond is stabilised by alkyl substitution. The reasons for this are still controversial, but the fact is uncontested [59T(5)127; 82JO736]. Whatever the cause, it confers a similar stability on the incipient double bond of the transition state. We therefore need to turn our attention to those reactions that give predominantly the Hofmann product and seek the reasons for their behaviour.

(*a*) *Steric destabilisation of the Saytzev product.* The simplest, but also most trivial, reason for Hofmann elimination is that in a few alkenes bulky substituents counteract the normal stabilisation of the Saytzev product [56JA2190]:

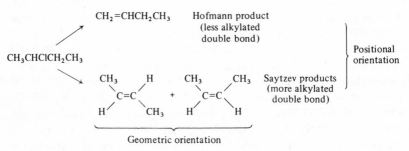

Figure 8.6. Orientation in elimination reactions.

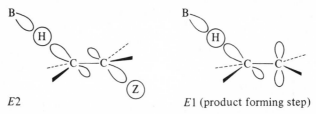

Figure 8.7. Transition state structures with partial double bond character.

t-BuCH$_2$CBr(CH$_3$)$_2$ $\longrightarrow$

$$\underset{\text{86 per cent}}{\overset{\displaystyle t\text{-BuCH}_2}{\underset{\displaystyle CH_3}{\diagdown}}\!\!C\!=\!CH_2}$$

$$\underset{\text{14 per cent}}{\overset{\displaystyle t\text{-Bu}}{\underset{\displaystyle H}{\diagdown}}\!\!C\!=\!C\!\!\overset{\displaystyle CH_3}{\underset{\displaystyle CH_3}{\diagup}}}$$

Table 8.3. *Positional orientation in E2 reactions of*
$CH_3CH_2CH_2CHZCH_3$ *with* $EtO^-/EtOH$ *at 80–130 °C* [56JA2199]

| Z: | Br | I | OTos | SMe$_2^+$ | SO$_2$Me | NMe$_3^+$ |
|---|---|---|---|---|---|---|
| Hofmann (per cent) | 31 | 30 | 48 | 87 | 89 | 98 |
| Saytzev (per cent) | 69 | 70 | 52 | 13 | 11 | 2 |

(b) *Leaving group effects.* Table 8.3 illustrates the influence the leaving group has on positional isomerism in E2 reactions.

There are two competing theories which seek to rationalise this behaviour. Ingold's original proposals emphasised the electronic properties of the leaving group. They were subsequently extended by Bunnett [62AG225; 69SPC53] and others in terms of the variable E2 transition state (figure 8.2). If Z is strongly electron-withdrawing it will tend to stabilise the carbanion and move the transition state into the E1cB part of the mechanistic spectrum. Likewise if Z is a poor nucleofuge the transition state will move in the direction of increased C$_\beta$–H cleavage (compare figure 8.4). Both effects increase the build-up of negative charge on the α carbon atom, a tendency opposed by β alkyl substituents. (The order of stability of carbanions is 1° > 2° > 3°, the opposite of the order for carbenium ions.) Moreover, the shift towards E1cB reduces the double bond character of the transition state and the importance of Saytzev stabilisation diminishes. Consequently, leaving groups such as SMe$_2^+$, SO$_2$Me and NMe$_3^+$, which combine the properties of strong electron withdrawal and relatively low nucleofugality, eliminate preferentially towards the less substituted β carbon atom giving the Hofmann product.

The alternative explanation, due to H. C. Brown [56JA2199], is that Hofmann elimination is largely steric in origin. The three leaving groups in question share another property: they are very bulky. The conformation required for *anti* elimination is therefore destabilised by β alkyl substituents:

$$\underset{\displaystyle R}{\overset{\displaystyle H}{\diagdown}}\!\!\underset{\displaystyle R'}{\overset{}{C_\beta}}\!\!-\!\!C\!\!\diagdown\!\!\overset{}{\underset{\displaystyle Z}{}}$$

Table 8.4. *Effects of $\beta$ substitution on rates of elimination from $RR'C_\beta H-C_\alpha H_2 Z$ with $EtO^-/EtOH$* [60J4054]

| R | R' | Second order rate coefficients ($k_2/10^{-5} \, \text{l} \, \text{mol}^{-1} \, \text{s}^{-1}$) | |
|---|---|---|---|
| | | $Z = SMe_2^+, 64\,°C$ | $Z = NMe_3^+, 104\,°C$ |
| H | H | 79 | 71 |
| H | Me | 29 | 5.2 |
| Me | Me | 10 | 1.7 |
| H | Et | 21 | 2.8 |
| H | i-Pr | 16 | 1.1 |
| H | t-Bu | 0.43 | 0.084 |

Both explanations suggest that the swing to Hofmann is the result of slowing formation of the Saytzev product rather than accelerating formation of the Hofmann product and this is borne out by the facts. Table 8.4 shows that $\beta$ substitution considerably retards reactions with Hofmann-type leaving groups.

In fact it seems that both explanations have some validity. Elimination of HF from fluoroalkanes gives predominantly Hofmann products [65JA3401]. This must be an electronic effect; fluorine is strongly electron-withdrawing and a poor nucleofuge, but it is also small. On the other hand, the dramatic effect of the t-butyl group (last entry in table 8.4) is surely steric in origin.

(c) *The effect of the base.* 2-Bromobutane gives only about 20 per cent of the Hofmann product when treated with potassium ethoxide in ethanol, but with t-butoxide in t-butanol the Hofmann and Saytzev products are formed in about equal amounts. An even more dramatic swing is found in the reactions of $(CH_3)_2CH-CBr(CH_3)_2$ [56JA2193]:

| $RO^-$ (in ROH at $70 \pm 10\,°C$) | $EtO^-$ | $t\text{-}BuO^-$ | $Et_3CO^-$ |
|---|---|---|---|
| Hofmann (per cent) | 21 | 73 | 92 |

There are again two explanations: the stronger bases in poorer ionising solvents encourage reactant-like transition states with somewhat enhanced carbanion character (compare figure 8.5), so $\beta$ alkylation disfavours the Saytzev product electronically. But the stronger bases are also the bulkier, so they will tend to attack the less hindered $\beta$ hydrogen atom. The electronic effect is probably dominant as shown by the existence of linear Gibbs energy relations between orientational preference and the $pK_{BH^+}$ of all but the most bulky bases. Solvation effects are also important, though imperfectly understood [74JA430; 75Acc239; 79JA1176; 79JO4770].

Table 8.5. *Percentage of elimination in solvolyses in 80 per cent EtOH at 25 °C* [50JA1223]

| | | |
|---|---|---|
| $(CH_3)_3C-Cl$ $\longrightarrow$ | $(CH_3)_2C=CH_2$ | 16 per cent |
| $(CH_3CH_2)_3C-Cl$ $\longrightarrow$ | $(CH_3CH_2)_2C=CHCH_3$ | 40 per cent |
| $((CH_3)_2CH)_2C-Cl$ $\atop \phantom{..}CH_3$ $\longrightarrow$ | $(CH_3)_2CH \diagdown C=C(CH_3)_2 \atop CH_3 \diagup$ | 78 per cent |

### 8.3.2    Geometric orientation

In elimination from substrates of the type $RCH_2CHZCH_3$ there is often a fairly close parallel between the *cis/trans* geometric ratio and the Hofmann/Saytzev balance. The *trans* isomer usually predominates, but as the proportion of Hofmann product increases so does the proportion of the *cis* isomer in the remaining Saytzev product. To elaborate on some of the data of table 8.3: the 70 per cent of pent-2-ene obtained from pentyl iodide comprises 54 per cent of the *trans* and only 16 per cent of the *cis* isomer, but the 13 per cent from the sulphonium salt is formed in the ratio 8:5.

A reactant-like or *E*1cB-like transition state has relatively little double bond character and the incipient *cis* alkyl groups interact with each other less strongly than in a more central *E*2 transition state where they are approaching an eclipsed position.

$E1cB$                                $E2$

There are, however, other and very complex factors at work. Thus in reactions of 2-butyl tosylate in t-butanol several bases actually give more of the *cis* than of the *trans* but-2-ene, and in dimethylsulphoxide the normal parallel behaviour of positional and geometric orientation is reversed. Such solvent effects have not been satisfactorily explained.

## 8.4    The balance between elimination and substitution

Let us begin by considering the solvolytic behaviour of a tertiary substrate that ionises to a fully dissociated carbenium ion; the leaving group has no influence on the subsequent divergence between substitution and elimination. The dominant factor is the extent of Saytzev stabilisation of the resulting alkene; table 8.5 is illustrative.

The addition of a base to such a system brings the *E*2 mechanism into

Table 8.6. *Elimination and substitution: reactions with*
*NaOEt/EtOH* [48J2055, 2058]

| Substrate | Second order rate coefficients $(k_2/10^{-5}\,l\,mol^{-1}\,s^{-1})$ | | Alkene (per cent) |
|---|---|---|---|
| | $S_N2$ | $E2$ | |
| $CH_3CH_2Br$ | 172 | 1.6 | 0.9 |
| $CH_3CH_2CH_2Br$ | 55 | 5.3 | 8.9 |
| $(CH_3)_2CHCH_2Br$ | 5.8 | 8.5 | 60 |
| $(CH_3)_2CHBr^a$ | (1.6) | (7.6) | 80 |
| $(CH_3CH_2)_2CHBr^a$ | (1.5) | (13) | 88 |

[a]In the original paper the rates for primary compounds were measured at 55 °C and are as quoted above. Rates for the secondary compounds were measured at 25 °C; in this table the values in parentheses have been increased from the reported values by factors of 28 for substitution and 32 for elimination, which should make them roughly comparable to the true values for the primary substrates.

operation. The order of reactivity in $E2$ is normally $3° > 2° > 1°$ because of Saytzev stabilisation; the steric factors that induce the reverse order in $S_N2$ are not important here because the base is not attacking at the crowded carbon atom. The added base has no effect on the unimolecular processes and (with a tertiary substrate) it cannot participate as a nucleophile in an $S_N2$ reaction, so the net effect is an increase in the proportion of elimination.

With primary substrates the balance of reactivity swings in favour of substitution, but substituents at the $\beta$ carbon atom can play an important role. Table 8.6 illustrates the behaviour of primary and secondary substrates, with Saytzev stabilisation making elimination relatively more easy.

In general the stronger the base the greater the proportion of elimination from a given substrate. A strong base is a hard nucleophile and has greater affinity for the proton than for the softer electrophilic carbon. Moreover, many strong bases are also bulky (t-butoxide, for example) and this, too, discourages substitution. Solvents also play some part in influencing the action of the base, as they do in the differing orientational effects of bases. Solvent effects are also discussed in the following section.

The leaving group also influences the $E2/S_N2$ balance. The cationic $SMe_2^+$ and $NMe_3^+$ groups often encourage elimination to the almost total exclusion of substitution. Such groups generate $E1cB$-like transition states with considerable cleavage of the $C_\beta$—H bond (§8.3.1), and their bulk discourages the formation of an $S_N2$ transition state; the two effects work together.

Eliminations generally have higher enthalpies of activation than substitution because the strong $C_\beta$–H bond has to be cleaved in the *E*2 or product-forming *E*1 transition state. An increase in temperature thus increases the proportion of elimination (compare the footnote to table 8.6).

**8.5     The *E*2C mechanism**

The generalisation that a strong nucleophile that is also a weak base should promote substitution rather than elimination seems reasonable enough, but it is not always true. Halide ions in dipolar aprotic solvents and sulphides in alcohols frequently initiate elimination. For example, $PhS^-$ is a weaker base than $EtO^-$ by a factor of $10^{10}$, but in ethanolic solution it induces elimination from t-butyl chloride about ten times faster.

There is still considerable debate about the detailed mechanism of these reactions. Most models suggest that in the transition state there is some form of interaction between the base/nucleophile and the $\alpha$ carbon atom as well as the $\beta$ hydrogen atom. Because of this the mechanism is described as *E*2C, and the normal *E*2 mechanism, in which the base interacts solely with the $\beta$ hydrogen, is called *E*2H. The *E*2C transition state has been represented as essentially $S_N2$ in character with additional interactions with the $\beta$ hydrogen (**8.3**), but the observed steric effects at the $\alpha$ carbon are less than would be expected for such a structure. More recently a bent *E*2 transition state has been suggested, with the base attracted electrostatically to a positively charged $\alpha$ carbon atom (**8.4**) [75T2999]. In terms of the variable *E*2 transition state a weak base does

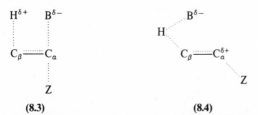

(8.3)                    (8.4)

indeed tend to increase the carbenium ion character and the degree of $C_\alpha$–Z cleavage (compare figure 8.5). Indeed, Bunnett suggests that the proposed involvement of the $\alpha$ carbon is illusory, and that the variable *E*2 transition state may be in itself an adequate mechanistic model [79JO1463]. The leaving group usually has a somewhat larger effect on the rates of *E*2C than of *E*2H reactions – other things being equal. Thus cyclohexyl bromide reacts some 66 times faster than the chloride with $EtO^-$/EtOH, but with $PhS^-$/EtOH the ratio is 140, confirming that the $C_\alpha$–Z bond is rather more fully broken in the *E*2C transition state. This behaviour is also in accord with a qualitative valence bond analysis which suggests that in the E2C transition state there is relatively weak bonding of the α carbon atom both to the base (hence the reduced steric effect) and to the nucleofuge [82JA187].

A characteristic of E2C reactions is a very strong preference for the Saytzev product, usually much more than in E2H. This is synthetically useful, complementing the use of strong bulky bases to encourage Hofmann elimination; CH₃CHBrCH₂CHMe₂ gives 98 per cent of the Saytzev product with tetrabutylammonium bromide in acetone, 96 per cent of the Hofmann product with potassium t-butoxide in t-butanol [75JA2477]. This was originally taken to show a strongly product-like E2C transition state with a well-developed π bond. The evidence is now against this, particularly in that conjugative stabilisation of the incipient double bond does not seem important, as figure 8.8 indicates [76JO3201]. Attempts to rationalise this behaviour are thus far unconvincing.

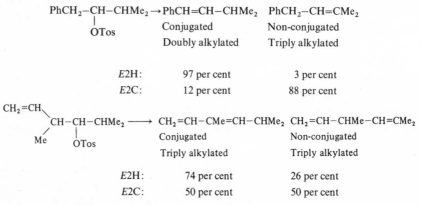

Figure 8.8. Absence of conjugative stabilisation in E2C transition states. E2H reactions were carried out with t-BuOK in t-BuOH, E2C with NBu₄Br in acetone containing a little lutidine.

The effects of substituents on reaction rates are in general agreement with a cationic E2C transition state such as (**8.4**). Alkyl substitution or other electron donation at $C_\alpha$ accelerates the reaction, though not as much as in $S_N1/E1$ solvolyses, in agreement with partial carbenium ion character. β Aryl substituents have only a small effect, in contrast to the typical E2H behaviour illustrated in table 8.1, and this accords with the lack of significant charge build-up on the β carbon atom.

## 8.6 Pyrolytic eliminations

The ylid mechanism (§8.2.1($e$)) is a special case of a more general class of reaction, the pyrolytic eliminations. It is special in that the presence of a base gives it the appearance of belonging to the $E2$ family. Most pyrolytic eliminations occur without added base or solvent, usually in the gas phase. The most important examples are as follows:

Ester pyrolysis
($\sim 400\ ^{\circ}$C)

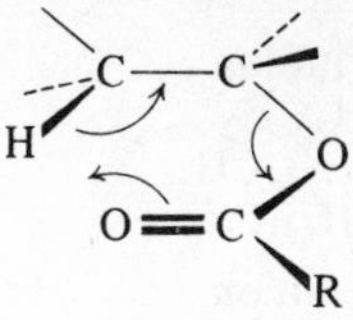

Xanthate ester pyrolysis
the Tschugaev reaction
($\sim 200\ ^{\circ}$C)

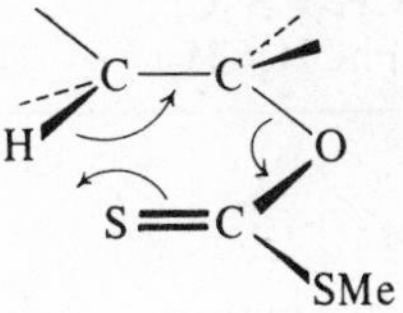

The Cope elimination
($\sim 100\ ^{\circ}$C)
(Sulphoxides behave similarly)

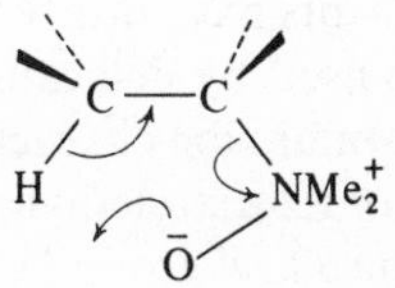

These reactions all exhibit certain features in common: a strong preference for *syn* elimination (particularly pronounced in the Cope elimination), first order kinetics, negative entropies of activation, and lack of response to radical scavengers. These observations accord with the cyclical mechanisms represented above, commonly designated $E_i$. However, $E_i$, like $E2$ is only the central point of a mechanistic spectrum in which there are different degrees of $C_\beta$–H and $C_\alpha$–Z cleavage in the transition states. The extremes would be ion-pair mechanisms, as shown:

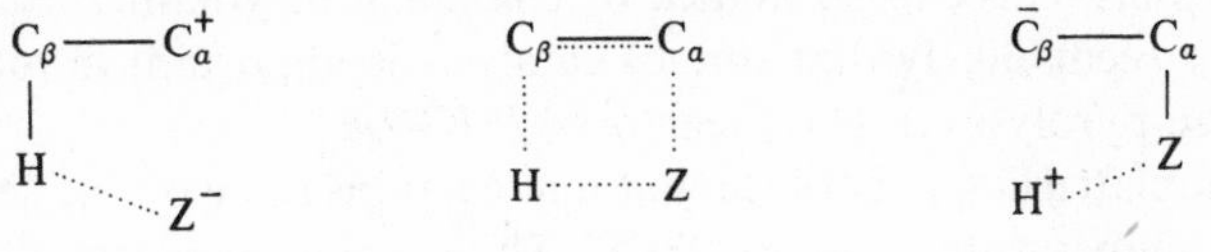

$C^+$ ion-pair    $E_i$    $C^-$ ion pair

Evidence for this mechanistic spread comes, *inter alia*, from a study of substituent effects that reflect the extent of charge build-up on the $\alpha$ and $\beta$ carbon atoms (table 8.7).

Table 8.7. *Substituent effects in pyrolytic eliminations. Relative rates are statistically corrected for the number of β hydrogen atoms*

| Alkyl group | Cope elimination at 100 °C [57JA4720] | Acetate pyrolysis at 400 °C [58J3398; and from data in 76J(P2)280] |
|---|---|---|
| α Substitution | | |
| $CH_3CH_2$ | 1.0 | 1.0 |
| $(CH_3)_2CH$ | 1.3 | 12 |
| $(CH_3)_3C$ | 2.0 | 550 |
| β Substitution | | |
| $CH_3CH_2$ | 1.0 | 1.0 |
| $CH_3CH_2CH_2$ | 0.9 | 1.3 |
| $PhCH_2CH_2$ | 105 | 4.0 |

The Cope elimination clearly lies on the carbanion side of the spectrum; alkyl substituents have little effect but the negative charge at the β carbon atom is stabilised by delocalisation into a benzene ring. Ester pyrolysis exhibits carbenium ion characteristics. Hammett correlations confirm this. In pyrolyses of 1,2-diarylethyl acetates, $ArC_\beta H_2-C_\alpha H(OAc)Ar'$, substituent changes in the α aryl group (Ar = Ph; Ar' variable) correlate with $\sigma^+$ with $\rho = -0.62$, a relatively large value for a high temperature process. Changes in the β aryl group (Ar' = Ph; Ar variable) induce negligible rate changes: $\rho = 0.08$ [61JA3647; 62JA4817]. Again, in benzoate pyrolyses (ROCOAr) the values of $\rho$ for R = Et, i-Pr and t-Bu are, respectively, 0.26, 0.335 and 0.58 [75J(P2)1802]. The positive sign confirms that the ArCOO group has a partial negative character in the transition state, and the magnitudes of $\rho$ show an increasing transition state polarity in the order 1° < 2° < 3°.

There remains some doubt about the precise nature of these pyrolyses: are they true gas phase reactions or are they catalysed by the surface of the reaction vessel? This can be studied by changing the volume/surface area ratio and it seems likely that surface catalysis is important in most high temperature pyrolyses in glass reactors [77JO698].

Another related category of elimination reactions is the pyrolysis of alkyl halides at temperatures around 500 °C. These are particularly difficult to study because of competition from radical mechanisms, both homogeneous and surface-catalysed. Nevertheless they have been fairly thoroughly investigated [69CR33]. Substituent effects indicate a well-developed carbenium ion-pair, more so than in ester pyrolysis. This also accords with the order of halide reactivity I > Br > Cl, and with *ab initio* MO

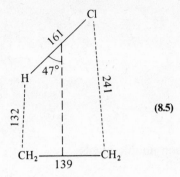

(8.5)

calculations [75JA5975] which predict that the transition state in ethyl chloride pyrolysis has the geometry shown in (**8.5**; interatomic distances shown in pm).

## 8.7    Further reading

A. F. Cockerill & R. G. Harrison, in *The Chemistry of Double-bonded Functional Groups*, ed. S. Patai, Wiley, London, 1977.

W. H. Saunders & A. F. Cockerill, *Mechanisms of Elimination Reactions*, Wiley-Interscience, New York, 1973.

A. F. Cockerill, in *Comprehensive Chemical Kinetics*, eds C. H. Bamford & C. F. H. Tipper, vol. 9, Elsevier, Amsterdam, 1973.

R. A. Bartsch & J. Závada, *Chem. Rev.*, 1980, **80**, 453.

E. Baciocchi, *Acc. Chem. Res.*, 1979, **12**, 430.

C. J. M. Stirling, *Acc. Chem. Res.*, 1979, **12**, 198.

F. G. Bordwell, *Acc. Chem. Res.*, 1972, **5**, 374.

J. F. Bunnett, *Survey Progr. Chem.*, 1969, **5**, 53.

D. V. Banthorpe, in *Studies in Chemical Structure and Reactivity*, ed. J. H. Ridd, Methuen, London, 1966.

J. F. Bunnett, *Angew. Chem. Internat. Ed.*, 1962, **1**, 225.

C. K. Ingold, *Proc. Chem. Soc.*, 1962, 265.

*E*2C mechanism:

D. J. McLennan, *Tetrahedron*, 1975, **31**, 2999.

W. T. Ford, *Acc. Chem. Res.*, 1973, **6**, 410.

Pyrolytic eliminations:

G. G. Smith & F. W. Kelly, *Progr. Phys. Org. Chem.*, 1971, **8**, 75.

A. Maccoll, *Chem. Rev.*, 1969, **69**, 33.

# 9     Addition to carbon–carbon double bonds

## 9.1     Electrophilic addition

### 9.1.1    Addition of bromine

When bromine and ethylene, each meticulously dried, are mixed together in the dark in a vessel whose surfaces are coated with paraffin wax, they do not react together. In the light an addition reaction does take place; this implies a homolytic mechanism which is discussed later in §9.2. Even in the dark, however, the addition occurs if there are traces of moisture present or if the reagents are allowed to make contact with a glass surface [23J(123)3006; 23JA1014]. This suggests a polar mechanism and the inference is confirmed in that the reaction of alkenes with an aqueous solution of bromine is not retarded by radical scavengers.

Ethylene does not react with sodium chloride solution, but if sodium chloride is added to bromine water the reaction gives both $BrCH_2CH_2Br$ and $BrCH_2CH_2Cl$ [25JA2340].

This behaviour is compatible with a two-step mechanism in which the ethylene is subject to electrophilic attack by the bromine:

$$CH_2{=}CH_2 + Br_2 \rightarrow (C_2H_4Br)^+ \, Br^- \rightarrow BrCH_2CH_2Br$$

Substituent effects are also in accord with this: electron-donating groups attached to the double bond accelerate the reaction; electron-withdrawing groups retard it. The first two columns of table 9.1 illustrate the effect of alkyl substitution.

The isomeric but-2-enes react with bromine to give stereospecifically the *anti* adducts; that is, the *trans* alkene gives exclusively the *meso* dibromide and the *cis* alkene gives the racemic dibromide [69JA1469]:

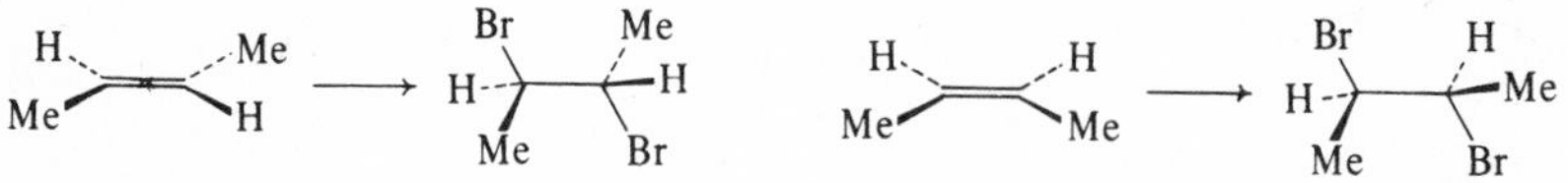

This stereochemical behaviour had long been accepted as the norm for additions to alkenes, even though the experimental evidence was limited;

*196*

Table 9.1. *Second order rate coefficients* $(k_2/l\,mol^{-1}\,s^{-1})$ *for reactions of alkenes with bromine in methanol at* $25\,°C$ [79JO1173, 2758]

| Alkene | $k_2$ | Alkene | $k_2$ | Alkene | $k_2$ |
|---|---|---|---|---|---|
| $CH_2=CH_2$ | 4.7 | $Me_2C=CH_2$ | $3.7 \times 10^4$ | $PhCH=CH_2$ | $1.5 \times 10^3$ |
| $MeCH=CH_2$ | $4.0 \times 10^2$ | $Me_2C=CHMe$ | $1.3 \times 10^6$ | $PhCH\overset{c}{=}CHMe$ | $7.9 \times 10^2$ |
| $MeCH\overset{c}{=}CHMe$ | $2.4 \times 10^4$ | $Me_2C=CMe_2$ | $1.2 \times 10^7$ | $PhCH\overset{t}{=}CHMe$ | $3.2 \times 10^3$ |
| $MeCH\overset{t}{=}CHMe$ | $1.3 \times 10^4$ | | | $PhCMe=CH_2$ | $1.4 \times 10^5$ |

as early as 1912 Frankland was able to refer to 'The Principle of Trans Addition' in his presidential address to the Chemical Society [12J(101)654]. It was rationalised in 1937 by the proposal that the cationic intermediate in the addition was a cyclic bromonium ion (9.1); ring opening in the second step, accompanied by Walden inversion, would lead to the observed *anti* addition [37JA947]:

(9.1)

Similar bromonium ions have also been invoked as intermediates in anchimerically assisted nucleophilic substitutions (§7.5.3). Their existence has been demonstrated in a series of elegant $^1H$ and $^{13}C$ n.m.r. studies by Olah [67JA4744; 68JA947, 2587; 74JA3565]. However, he generated them in liquid sulphur dioxide solution by reactions of the type: $BrCH_2CH_2F + SbF_5 \rightarrow (C_2H_4Br)^+SbF_6^-$, and so this cannot be taken as conclusive evidence that they are also formed as intermediates in the very different conditions of the alkene–bromine addition reactions.

The effects of alkyl substitution around the double bond provide further evidence for a bridged transition state. All three dimethylethylenes react at similar rates (table 9.1); if the charge were localised on one carbon atom one would have expected that isobutene, which could give the tertiary carbenium ion $Me_2C^+CH_2Br$, would react enormously faster than the but-2-enes.

For many years it was believed that *anti* addition was more general than in fact it is. Thus, although the but-2-enes react stereospecifically, the $\beta$-methylstyrenes, $PhCH=CHMe$, give mixtures of the *erythro* and *threo* dibromides; *anti* addition still predominates, but up to 28 per cent of the *syn* adduct is formed [69JA1469].

This suggests that the nature of the cationic intermediate – and of the related transition state – varies with the structure of the alkene. The opposite extreme from the cyclic bromonium ion would be an open carbenium ion, and one can envisage structures, either symmetrical or unsymmetrical, with the charge delocalised between carbon and bromine:

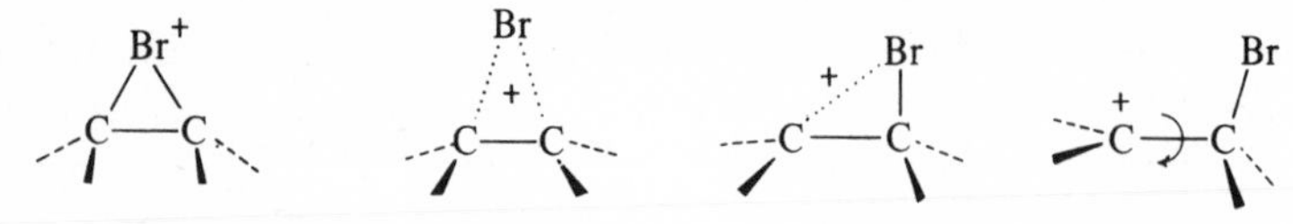

Electron-donating substituents on the double bond increase the carbenium ion character of the intermediate; electron withdrawal places more of the charge on the bromine. With $\beta$-methylstyrenes some of the charge can be delocalised into the benzene $\pi$ system giving an unsymmetrical bridged intermediate **(9.3)**. The effect of methyl substituents on the styrene double bond (column 3 of table 9.1) contrasts with their effect in the ethylene series: $\alpha$-methylstyrene is much more reactive than either of the $\beta$ isomers, implying that in the transition state there is a considerable positive charge on the $\alpha$ carbon atom.

If the open carbenium ion **(9.4)** were an intermediate then two geometrically isomeric alkenes would each give an identical mixture of diastereoisomeric dibromides. Partial loss of stereospecificity, as with the $\beta$-methylstyrenes, can arise via the relatively slow rotation about the C–C bond of a loosely bridged intermediate (which must pass through a carbenium *transition state*):

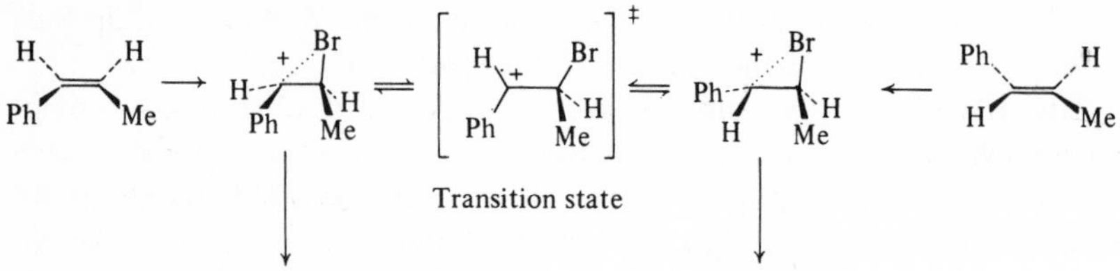

Even this picture is an oversimplification. In nucleophilic solvents (acetic acid and methanol have been most widely used) the dibromide is always accompanied by some monobromide/mono-solvent adduct, formed by solvolysis of the cationic intermediate:

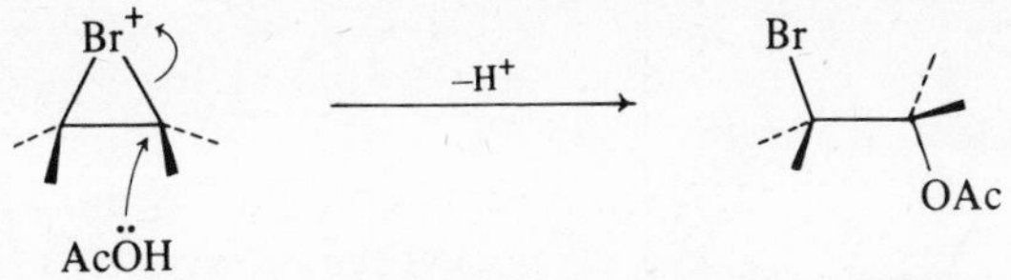

However, even with the $\beta$-methylstyrenes this adduct is formed almost exclusively by *anti* addition [69JA1469]. It seems likely, therefore, that the bromonium ion is formed initially as an intimate ion-pair with the bromide counter-ion on the same side of the hydrocarbon framework as the bromonium bridge. Sterically and stereoelectronically this is an unfavourable arrangement for immediate collapse to the *syn* adduct; instead it must reorient itself (via a solvent-separated ion-pair?) so that the bromide ion can approach the rear of the $C-Br^+$ bond. Rotation of the $C-C$ bond may or may not occur during this process. Such reorientation is unnecessary for solvent attack; there will always be suitably positioned solvent molecules.

The relative importance of loosely bridged intermediates and open carbenium ions is often hard to assess. An interesting series of studies on the addition of bromine to substituted stilbenes shows that the balance can be swayed fairly easily [72JO1770; 73JO493; 74JO2441]. A Hammett plot of $\log(k/k_0)$ for reactions with $XC_6H_4CH=CHPh$ indicates a mechanistic change (figure 9.1).

This behaviour was interpreted in terms of two alternative carbenium ion intermediates:

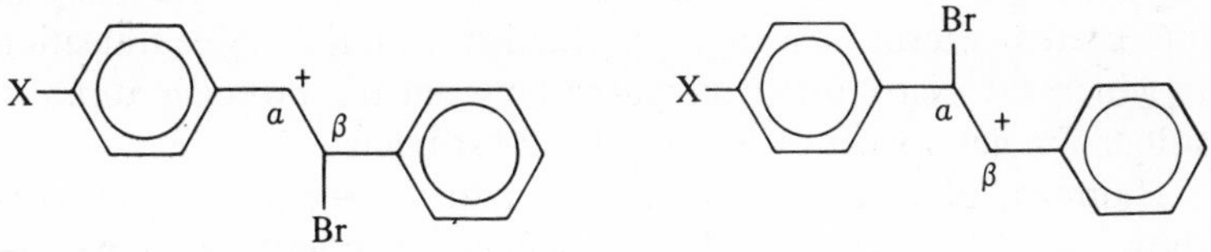

For strongly electron-donating X groups the $\alpha$ carbenium ion is stabilised by through conjugation and the rates of reaction correlate with $\sigma^+$ with $\rho = -5.0$. If X is strongly electron-attracting then the unsubstituted benzene ring offers better stabilisation and the reaction goes through the $\beta$ carbenium ion; this is less sensitive to substituent changes and the value of $\rho$ falls to $-1.5$. For intermediate substituents both routes can run in parallel.

When both benzene rings bear electron-withdrawing substituents the rate of bromine addition is faster than that calculated by extrapolation from the Hammett plots. This was taken to mean that in these reactions a third mechanism, via a bridged bromonium ion, provided an alternative to the destabilised carbenium ions; all three routes might be playing a

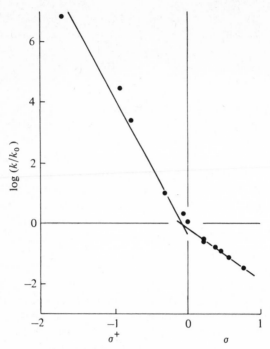

Figure 9.1. Hammett plot for addition of bromine to monosubstituted stilbenes in methanol at 25 °C [72JO1770].

part in the overall reaction. However, the calculations were all based on the hypothesis of a symmetrical bromonium ion; the data could also be interpreted in terms of a single mechanism with a variable transition state in which the distribution of charge between the bromine atom and the stilbene moiety depends on the ring substituents.

Hammett plots of bromine addition to styrene and $\beta$-methylstyrenes show similar but less clearly defined changes of slope. For $\alpha$-methylstyrenes a Yukawa–Tsuno plot (§3.1.3, equation 3.4) is appropriate because through-conjugation between the benzene ring and the potential carbenium centre on $C_\alpha$ is restricted by steric twisting:

A plot of $\log(k/k_0)$ against $\{\sigma + 0.65(\sigma^+ - \sigma)\}$ is accurately linear with $\rho = -4.26$. The relatively large magnitude of $\rho$ points to a substantial positive charge on $C_\alpha$ in the transition state (in accord with the substituent effect on reaction rates shown in table 9.1), and the linearity of the plot indicates that there is little change in transition state structure with changes in the aryl substituents [78JA7645].

At low concentrations of bromine the reactions with alkenes follow second order kinetics, as expected. The description $Ad_E2$ is sometimes used to describe this behaviour. At higher concentrations, however, a third order term becomes significant ($Ad_E3$):

$$\text{Rate} = k_2[\text{alkene}][Br_2] + k_3[\text{alkene}][Br_2]^2$$

A synchronous termolecular process seems highly improbable, so the likely explanation is a rapid pre-equilibrium between the alkene and one molecule of bromine, followed by a rate-limiting attack of a second bromine molecule:

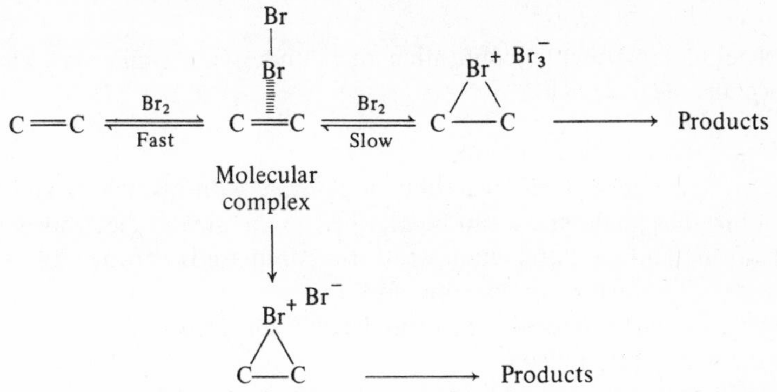

Iodine–alkene molecular complexes are well known and are formed exothermically. For bromine the evidence for the existence of such complexes comes mainly from the observation of transient absorption in the electronic spectrum when bromine and alkene solutions are mixed [67SA(23A)2279]. The frequency of this absorption depends partly on the ionisation energy of the alkene but also on the separation of the carbon and bromine atoms in the molecular complex. It has therefore been used to measure the steric interaction between the bromine and alkene as they approach one another. A similar analysis has been carried out for oxymercuration reactions (for example, $C=C + Hg(OAc)_2 \rightarrow$ AcO$-$C$-$C$-$HgOAc). These are generally believed to follow a similar mechanistic path to that of bromine addition, going through a bridged mercury cation, but they seem anomalous in that they are retarded by alkyl substituents on the double bond in contrast to bromine addition as exemplified in table 9.1. However, when the steric factors are allowed for, the two reactions are found to respond identically to the electronic effects

Table 9.2. *Product distribution in reactions of chlorine with β-methylstyrenes (per cent; the balance of 100 per cent comprises substitution products)* [65JA5172; 80JO1401]

|  | Dichloride | | Chloroacetate | |
|  | anti | syn | anti | syn |
| --- | --- | --- | --- | --- |
| *trans* alkene | | | | |
| in CCl$_4$ at 0–5 °C | 38 | 46 | | |
| in AcOH at 25 °C | 24.5 | 39 | 21.5 | 8.5 |
| *cis* alkene | | | | |
| in CCl$_4$ at 0–5 °C | 29 | 62 | | |
| in AcOH at 25 °C | 14 | 47 | 16 | 14 |

(*Note:* the *anti* product from the *trans* alkene is the same as the *syn* product from the *cis* alkene, and vice versa)

of alkyl substituents, confirming their underlying similarity [81JA2783; see also 82JA7599].

### 9.1.2 Addition of chlorine

In general the reactions of chlorine with alkenes resemble those of bromine; differences can be ascribed to the greater electronegativity of chlorine and to its reluctance to form compounds through an expanded octet. The kinetics of chlorine addition are almost always second order showing that the formation of molecular complexes, if it occurs at all, is kinetically insignificant.

Stereospecific *anti* addition occurs with the but-2-enes and other simple alkenes, so these reactions presumably involve a bridged chloronium ion. The effects of alkyl substitution agree with this: as with bromine addition, isobutene reacts at about the same rate as the but-2-enes. However, addition of chlorine to β-methylstyrene is much less stereoselective than that of bromine (table 9.2).

The pattern is not simple, but the preponderance of *syn* addition of chlorine and the low stereoselectivity in the solvent reaction suggests that the intermediate is essentially an open carbenium ion-pair:

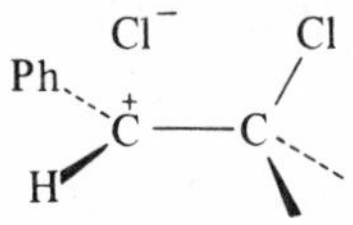

Rotation about the C–C bond, reorientation of the counter ion and collapse to products are probably all occurring at comparable rates, so that a detailed analysis is all but impossible.

### 9.1.3    Addition of hydrogen chloride and hydrogen bromide

These reactions, apparently simple, are in fact very complex and the intimate details of their mechanism remain somewhat obscure. Substituent effects indicate that the rate-limiting step is protonation of the double bond; thus the order of reactivity of alkenes is: $R_2C=CHR \simeq R_2C=CH_2 \gg RCH=CHR \simeq RCH=CH_2$, and the rates of reactions of substituted styrenes correlate with $\sigma^+$. Clearly, therefore, the reaction involves an intermediate with carbenium ion character, but the mechanism is certainly more complex than the following:

$$C=C \xrightarrow{\text{HHal}} CH-C^+ + Hal^- \rightarrow CH-CHal$$

Reaction through a free carbenium ion would be unstereoselective but this is not found. Reaction through a carbenium ion-pair might be expected to facilitate *syn* addition, and this does happen in, for example, the addition of DBr to acenaphthylene [63JA2245]:

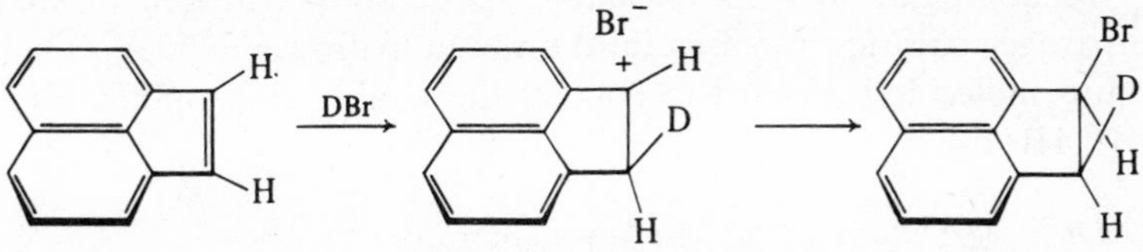

However, this is not the normal behaviour. In the reaction of hydrogen chloride with cyclohexene in acetic acid, for example, *anti* addition is considerably favoured over *syn*.

A reasonable explanation is that the *anti* adduct is formed in a termolecular $Ad_E3$ process involving attack of a second halide ion on a molecular complex:

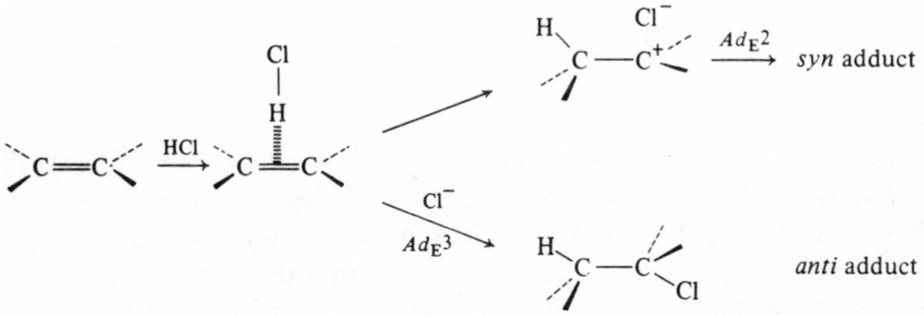

In accord with this it is found that the alkyl acetate by-product is always formed by predominantly *anti* addition (acetic acid attack on the molecular complex), and that with increasing concentrations of hydrogen chloride or, particularly, of added chloride salts, the proportion of acetate to chloride falls and the fraction of chloride produced by *anti* addition rises [70JA2816].

1,2-Dimethylcyclohexene on protonation gives a tertiary carbenium ion; at low concentrations of hydrogen chloride it undergoes preferentially *syn* addition, but the same pattern of reactivity is followed, *anti* addition becoming dominant with increasing hydrogen chloride concentration [71JA2445].

Hydrogen bromide behaves similarly but gives a much greater proportion of the *anti* adduct than does hydrogen chloride under similar conditions. Moreover, an increase in the concentration of hydrogen bromide is much more effective in suppressing acetate formation than is one of hydrogen chloride [74JA4534]. This must relate to the increased dissociation of hydrogen bromide and to the greater nucleophilicity of the bromide ion; the mechanistic balance is tipped strongly in favour of $Ad_E3$.

In agreement with this the kinetics of hydrogen bromide addition are frequently of the form:

$$\text{Rate} = k_3[\text{alkene}][\text{HBr}]^2 + k_3'[\text{alkene}][\text{HBr}][\text{Br}^-]$$

In non-polar solvents the additions of both hydrogen bromide and hydrogen chloride are often third order in hydrogen halide; probably the third molecule is needed to convert the weakly electrophilic HHal into $H^+ HHal_2^-$.

### 9.1.4 Orientation: the Markownikov rule

Additions of unsymmetrical electrophiles to unsymmetrical alkenes can give alternative products; in practice most such reactions give only one of them [e.g. 33JA2531; 39JO428]:

$$CH_3CH=CH_2 + HBr$$
$$CH_3CH_2CH_2Br$$
$$CH_3CHBrCH_3$$

It is now over a century since Markownikov established that the normal pattern in additions of HX was for the H to add to the less substituted end of the double bond and the X to add to the more substituted [1870LA(153)256]. This is commonly rationalised in terms of the relative stabilities of the alternative intermediate carbenium ions: $CH_3CH^+CH_3 > CH_3CH_2CH_2^+$. In fact, of course, the explanation must be in terms of the relative energies of the product-forming transition states, but it is likely that their charge distributions will not be greatly different from those in the carbenium ions.

The Markownikov rule originally applied only to the additions of HX to hydrocarbon substrates but its principles can be extended to other reactions. For example there is a very old report [1892LA(267)300] that $CH_2=CHNMe_3^+$ reacts with HI under vigorous conditions to give the 'anti-Markownikov' adduct $ICH_2CH_2NMe_3^+$: Markownikov orientation would go via the cation $\overset{+}{C}H_3CH-\overset{+}{N}Me_3$ with positive charges on adjacent atoms.

Vinyl chloride, $CH_2{=}CHCl$, gives exclusively the Markownikov adduct; with HCl the product is the *gem*-dichloride $CH_3CHCl_2$. In this case one could not readily have predicted which of the alternative carbenium ion intermediates would be the more stable. The 2-chloroethyl cation, $^+CH_2CH_2Cl$, is less stable than $C_2H_5^+$ because of the electron-withdrawing polar effect of chlorine. In the 1-chloroethyl cation the positive charge is nearer to the chlorine atom and the polar destabilisation is greater, but this is counteracted by the mesomeric donation of lone-pair electrons from chlorine into the carbenium centre:

$$[CH_3-\overset{+}{C}H \rightarrow \overset{..}{C}l \leftrightarrow CH_3-CH{=}\overset{+}{C}l]$$

The experiment shows that the latter effect is enough to lower the energy of the 1-chloro cation below that of the 2-chloro isomer. However, additions to vinyl chloride are slower than those to ethylene under similar reaction conditions, so the net effect of the chlorine is still electron-withdrawing: the mesomeric donation does not outweigh the polar withdrawal, and the 1-chloroethyl cation is less stable than $C_2H_5^+$. This behaviour is entirely analogous to that of chlorobenzene in electrophilic substitution reactions (§10.1.3).

Orientation in additions that go via bridged halonium ions is less predictable. For example, the percentages of Markownikov adduct in the reactions of propene with HOCl, HOBr, and ClBr are 91, 79, and 54, respectively, [50AS39; 58J36]. There is clearly a fairly fine balance of effects in the product-forming transition state, in which the nucleophile is attacking the halonium ion:

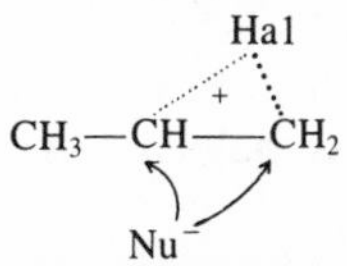

Steric interactions inhibit attack at the central carbon atom and favour the anti-Markownikov product. On the other hand the same carbon bears the greater positive charge. The magnitude of the electronic effect depends both on the electrophilic halogen and on the nucleophile. The difference between HOCl and HOBr reflects the greater reluctance of chlorine to accept part of the positive charge, thus increasing the importance of the electronic effect and favouring the Markownikov adduct. The reactions of HOBr and ClBr indicate that hydroxide, a hard nucleophile (§6.4.2) is attracted more strongly by the positive charge on the central carbon atom than is the somewhat softer chloride.

## 9.2    Radical addition

Radical addition to a carbon–carbon double bond can occur by a typical chain reaction:

Initiation:    $X-Y \rightarrow X^{\bullet}\ Y^{\bullet}$

Propagation: $\begin{cases} X^{\bullet} + C{=}C \rightarrow X-C-C^{\bullet} \\ X-C-C^{\bullet} + X-Y \rightarrow X-C-C-Y + X^{\bullet} \end{cases}$

The occurrence of a radical mechanism can normally be diagnosed by the typical conditions needed for initiation – photolysis or the introduction of radical initiators such as peroxides – and by the retarding effect of radical scavengers. Thus the radical polymerisation of ethylene to give low-density polyethylene is initiated by traces of oxygen:

$$\dot{O}{-}\dot{O} \quad CH_2{=}CH_2 \rightarrow \dot{O}-O-CH_2CH_2^{\bullet} \quad \text{etc.}$$

But higher concentrations of oxygen suppress polymerisation by acting as a chain stopper.

Some of the early studies on the addition of hydrogen bromide to alkenes reported the formation of mixed Markownikov and anti-Markownikov adducts. In 1933 Kharasch showed that the latter was formed via a radical mechanism initiated by traces of peroxides [33JA2468]. When benzoyl peroxide is added to the reaction mixture the anti-Markownikov addition becomes dominant; for example, with styrene:

$$(PhCOO)_2 \rightarrow PhCO_2^{\bullet} \xrightarrow{-CO_2} Ph^{\bullet} \xrightarrow{HBr} PhH + Br^{\bullet}$$

$$\begin{cases} Br^{\bullet} + PhCH{=}CH_2 \rightarrow Ph\dot{C}H-CH_2Br \\ Ph\dot{C}H-CH_2Br + HBr \rightarrow PhCH_2-CH_2Br + Br^{\bullet} \end{cases}$$

The controlling factor is the stabilisation of the transition state *en route* to $Ph\dot{C}HCH_2Br$, in which the unpaired electron is delocalised into the benzene $\pi$ system.

In general, however, the orientation in radical addition reactions is less easily interpreted than in electrophilic additions [78APO51; 80T701; 82AG401]. Monosubstituted alkenes, $XCH{=}CH_2$, are always attacked preferentially at the unsubstituted end, but except for substituents with $\pi$ bonds that can conjugate with the new radical centre (as in the addition of HBr to styrene described above) the nature of X has little effect on the rate of reaction: $CH_2{=}CHCH_3$ and $CH_2{=}CHCF_3$ are almost equally reactive towards $CH_3^{\bullet}$ radicals. In these reactions the main factor is steric: $CH_3CH_2-\dot{C}HCH_3$ is less compressed than $\dot{C}H_2-CH(CH_3)_2$. Likewise bulky radicals show a strong preference for the less hindered end of the double bond.

The significance of polar effects in radical reactions is often under-estimated. The $CH_3^{\cdot}$ radical attacks ethylene about ten times faster than tetrafluoroethylene; for the $CF_3^{\cdot}$ radical the ratio is reversed. In the broadest definition of the terms [83PAC1281] the $CH_3^{\cdot}$ radical is a nucleophile – a reagent that is attracted to a region of low electron density – and the $CF_3^{\cdot}$ radical is an electrophile. The difference is manifest in a switch of orientation in their reactions with $CHF{=}CF_2$: the $CH_3^{\cdot}$ radical preferentially attacks the $CF_2$ carbon, whereas the $CF_3^{\cdot}$ radical goes for the CHF.

Most radical additions occur unstereoselectively; thus the photo-catalysed addition of $BrCCl_3$ to both *cis*- and *trans*-but-2-enes gives the same mixture of diastereoisomers [55JA4638]. With hydrogen bromide, however, the *anti* adduct is often formed with high stereoselectivity [57JA2653]; presumably the reaction involves a bridged radical with bromine *d* orbital participation:

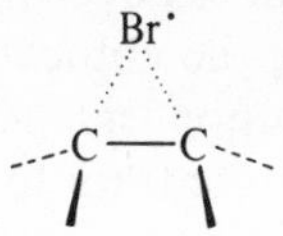

## 9.3 Nucleophilic addition

Ethylene is subject to attack only by exceptionally reactive nucleophiles: t-butyllithium gives $\text{t-BuCH}_2\text{CH}_2\text{Li}$, primary carbanions being less unstable than tertiary. A similar reaction occurs in the anionic polymerisation of ethylene and other simple alkenes, catalysed by alkyl-aluminium compounds and titanium salts.

Polyhalogenoalkenes are more susceptible to nucleophilic attack. Ethanol adds to $CF_2{=}CCl_2$ in the presence of sodium ethoxide to give $EtOCF_2CHCl_2$; the orientation is presumably governed by delocalisation of negative charge into chlorine *d* orbitals:

$$[\text{EtOCF}_2{-}\text{CCl}_2^- \longleftrightarrow \text{EtOCF}_2{-}\text{CCl}{=}\text{Cl}^-]$$

A single strongly electron-withdrawing substituent is enough to facilitate nucleophilic addition; $CH_2{=}CH{-}CF_3$, for example, does not undergo normal electrophilic addition even with strong acids [70JA7596] but it does react with ethanol/ethoxide via the nucleophilic mechanism.

The Michael reaction (using the term in its broadest sense) is the commonest and most important example of nucleophilic addition to a carbon–carbon double bond, activated in this case by conjugation with a carbonyl or other π functional group:

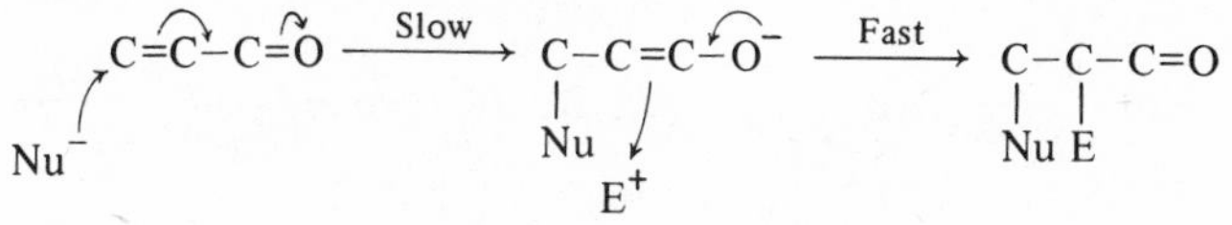

Acid catalysis is also possible, with nucleophilic attack remaining the rate-limiting step:

$$C=C-C=O \underset{\text{Fast}}{\overset{H^+}{\rightleftharpoons}} C=C-C=\overset{+}{O}H \overset{\text{Slow}}{\longrightarrow} \underset{\underset{E^+}{Nu}}{\overset{}{C-C=C-\overset{\cdot\cdot}{O}H}} \overset{-H^+}{\longrightarrow} \underset{Nu\ E}{C-C-C=O}$$

The nucleophilic mechanism can generally be diagnosed by the retarding effect of alkyl groups attached to the double bond; for example, $HOSO_2^-$ reacts some 500 times slower with $CH_2=CMe-CN$ than with $CH_2=CH-CN$.

The addition of bromine to Michael-type substrates can be either electrophilic or nucleophilic, depending on the precise structure of the alkene and on the reaction conditions. In acetic acid solution the addition to acrylic acid, fairly slow at first, is strongly autocatalysed. This is not simple acid catalysis: the addition of perchloric acid does not significantly accelerate the reaction but hydrogen bromide does; the autocatalysis is induced by hydrogen bromide formed in substitution side-reactions. Moreover, the reaction is retarded by adding water contrary to normal bromine additions.

The reaction probably involves nucleophilic addition, initiated by attack of $HBr_3$. Substituent effects are in accord with this [45J129]:

|  | $CH_2=CHCOOH$ | $MeCH=CHCOOH$ | $Me_2C=CHCOOH$ | in HOAc + HBr |
|---|---|---|---|---|
| $k_2/10^{-4} \mathrm{l\,mol^{-1}\,s^{-1}}$ | 70 | 8.5 | 30 | at 24 °C |

Crotonic acid reacts more slowly than acrylic, as expected for a nucleophilic mechanism, but a second methyl group accelerates the reaction again, showing that the mechanism has switched to electrophilic. This is confirmed by the absence of catalysis by hydrogen bromide in the last reaction.

## 9.4 Hydroboration

Diborane reacts slowly with ethylene in the gas phase to give $(C_2H_5)_3B$. The complex kinetics have been interpreted in terms of pre-equilibria amongst $BH_3$, $B_2H_6$ and $B_3H_9$, with the trimer being the species that actually attacks the ethylene [54JA835]. In ether solvents the reaction is greatly accelerated and has become the basis of the important hydroboration syntheses developed by H. C. Brown. Oxidative scission of the alkyl borane with alkaline hydrogen peroxide generates the corresponding alcohol:

$$R_3B \overset{HOO^-}{\longrightarrow} \underset{\underset{R}{|}}{R_2\bar{B}-O-\overset{\frown}{O}H} \rightarrow R_2B-OR \rightsquigarrow B(OR)_3 \overset{HO^-}{\longrightarrow} 3\ ROH$$

Addition in these reactions is stereospecifically *syn*; hydroborations of 1-methyl- and 1,2-dimethyl-cyclopentene and cyclohexene all give less than 3 per cent of the *anti* alcohols. (Note that in the rearrangement step during cleavage of the alkyl borane the stereochemistry of the migrating alkyl group is retained.) This suggests a concerted, cyclic transition state with the C−H and C−B bonds being formed together. Protium/tritium and $^{10}$B/$^{11}$B isotope effects confirm this [72JA6083]. So does the easy elimination of $R_2BH$ from trialkylboranes: a two-step process would call for an activation energy at least equal to the C−B bond energy (about 350 kJ mol$^{-1}$) but this is far too high to accord with the actual conditions; if the elimination is concerted then, by the principle of microscopic reversibility, so is the addition.

Hydroboration is also a highly regioselective reaction. Terminal alkenes of the type $RCH=CH_2$ give 93 to 94 per cent of the primary alcohol; the dialkylated compounds $R_2C=CH_2$ give more than 99 per cent. Steric factors are important: in the cyclic transition state the borane residue is preferentially aligned with the less bulky end of the carbon–carbon double bond. Electronic effects also pay a part. Hydroboration of styrenes, $ArCH=CH_2$, always gives predominantly the primary alcohol, but the degree of regioselectivity depends on the aryl substituent; $\log(2°/1°)$ correlates with $\sigma^+$ suggesting a transition state such as (**9.5**) [60JA4708; 66JA5851].

$$\text{Ar}\overset{\delta+}{\underset{\vdots}{\text{—CH}}}\overset{\delta-}{\cdots\cdots}\text{CH}_2$$
$$\underset{\text{H}}{\overset{\delta-}{\vdots}}\cdots\cdots\underset{\text{B}}{\overset{\delta+}{\vdots}} \qquad\qquad (9.5)$$

The effects of alkene substituents on reaction rates are relatively small in additions with diborane itself where steric and electronic effects may be fairly finely balanced; thus the relative reactivities of n-Pr−CH=CH$_2$ and EtMeC=CH$_2$ are 1.2:1. The kinetic isotope effects also point to a balance between steric and electronic factors. The $^1$H/$^3$H effect falls off with increasing substitution on the alkene suggesting that steric interactions reduce the extent of C−H bond making and B−H bond breaking in the transition state. Conversely the $^{10}$B/$^{11}$B effect increases slightly, indicative of increased B−C bond making with increasing alkene nucleophilicity [72JA6083].

With bulkier boranes steric effects become more significant. In reactions with '9-BBN' (9-borabicyclo[3.3.1]nonane: a cyclooctane ring bridged across the 1,5 positions by >BH) styrenes are hydroborated almost exclusively at the less hindered primary position regardless of substituents in the benzene ring, although the presence of electronic factors is revealed by a Hammett correlation with $\sigma^+$ with the relatively

Table 9.3. *Relative reactivities of alkenes and chloroalkenes towards 9-BBN (arrows indicate site of attack by B)* [82JA4907]

| | | | | | |
|---|---|---|---|---|---|
| A | 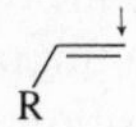 | 1.0 | E | 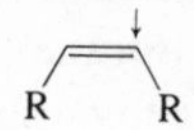 | 1/150 |
| B | 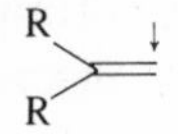 | 2.0 | F | 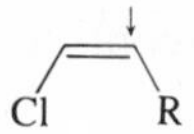 | 1/11 000 |
| C | 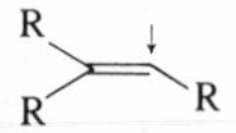 | 1/88 | G | 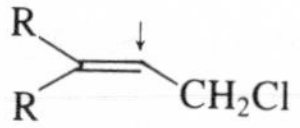 | 1/160 |
| D | 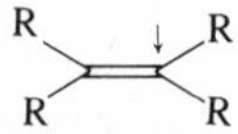 | 1/16 000 | H | 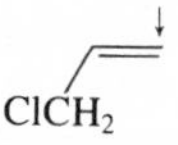 | 1/25 |

R represents an unbranched alkyl group: differences between R groups do not greatly alter reactivity.

low $\rho$ value of $-0.5$ [80JO5306]. The interplay of steric and electronic effects is also seen in the relative reactivities of alkenes and chloroalkenes towards 9-BBN shown in table 9.3. The left hand column of table 9.3 shows that an $\alpha$ alkyl substituent slightly accelerates attack at the $\beta$ position (electronic) but greatly retards $\alpha$ attack (steric). A chlorine atom attached directly to the double bond deactivates the $\beta$ position (compare E and F in table 9.3) as a result of polar electron withdrawal partly compensated by mesomeric donation (9.6). Its deactivating effect on the $\alpha$ position is great: $Me_2C=CHCl$ is attacked preferentially at the sterically crowded tertiary position. A comparison between entries C and G shows a slight deactivation of the $\alpha$ position by the polar electron withdrawal of the allylic chlorine atom, but the effect on the $\beta$ position (entries A and H) is much greater: presumably the polar effect is reinforced by electron-*withdrawing* hyperconjugation with the C–Cl bond (9.7).

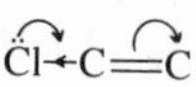

(9.6)

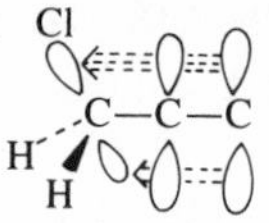

(9.7)

Although the transition states in these reactions certainly resemble (9.5) their precise natures remain unclear. A $[2\sigma + 2\pi]$ concerted process involving only the $\sigma$ electrons of the B–H bond and the $\pi$ electrons of the double bond defies the rules for pericyclic processes (chapter 16).

However, the hydroboration step is very exothermic $(BH_3 + C_2H_4 \rightarrow C_2H_5BH_2; \Delta H^{\ominus} \simeq -140 \text{ kJ})$. This probably means that the transition state is very reactant-like; in such a case the antibonding interactions that destabilise a forbidden pericyclic transition state are relatively undeveloped and the normal rules may not apply. Alternatively the new carbon–boron bond may be formed via the vacant boron $p$ orbital that is orthogonal to the B–H bond (9.8). In this case the reaction,

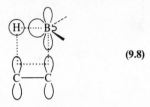

(9.8)

though concerted, is not strictly pericyclic. We should also consider the important role played by the solvent. It seems likely that the reactive species is the ether–$BH_3$ complex and the bonding in the transition state will certainly be modified by involvement of electrons from the oxygen atom of the ether.

## 9.5 Further reading

Electrophilic addition:

G. H. Schmid & D. G. Garratt, in *The Chemistry of Double-bonded Functional Groups*, ed. S. Patai, Wiley, London, 1977.

R. Bolton, in *Comprehensive Chemical Kinetics*, eds C. H. Bamford & C. F. H. Tipper, vol. 9, Elsevier, Amsterdam, 1973.

P. B. D. de la Mare, *Electrophilic Halogenation*, Cambridge University Press, London, 1976.

P. B. D. de la Mare & R. Bolton, *Electrophilic Additions to Unsaturated Systems*, 2nd edn., Elsevier, Amsterdam, 1982.

R. C. Fahey, *Topics Stereochem.*, 1968, **3**, 237.

F. Freeman, *Chem. Rev.*, 1975, **75**, 439.

Radical addition:

J. M. Tedder, *Tetrahedron*, 1980, **36**, 701; *Angew. Chem. Internat. Ed.*, 1982, **21**, 401.

J. M. Tedder & J. C. Walton, *Adv. Phys. Org. Chem.*, 1978, **16**, 51.

B. A. Bohm & P. I. Abell, *Chem. Rev.*, 1962, **62**, 599.

J. I. G. Cadogan & M. J. Perkins, in *The Chemistry of the Alkenes*, ed. S. Patai, Wiley-Interscience, New York, 1964.

B. Giese, *Angew. Chem. Internat. Ed.*, 1983, **22**, 753 (C–C bond formation).

Nucleophilic addition, the Michael reaction:

S. Patai & Z. Rapoport, in *The Chemistry of the Alkenes*, ed. S. Patai, Wiley-Interscience, New York, 1964.

E. D. Bergmann, D. Ginsburg, & R. Pappo, *Org. React.*, 1959, **10**, 179.

Pericyclic addition reactions are discussed in chapter 15.

Additions to carbon–carbon triple bonds (not discussed in this chapter) are reviewed in:

G. H. Schmid (electrophilic addition), J. I. Dickstein and S. I. Miller (nucleophilic addition), in *The Chemistry of the Carbon–Carbon Triple Bond*, ed. S. Patai, Wiley-Interscience, Chichester, 1978.

G. Melloni, G. Modena, & U. Tonellato, *Acc. Chem. Res.*, 1981, **14**, 227 (comparison with alkenes).

# 10    Aromatic electrophilic substitution

In 1960 it seemed that the study of aromatic electrophilic substitution, and in particular of its archetype, nitration, was all but complete. The vast majority of reactions appeared to follow a single well-established mechanism, and it remained but to tidy up a few relatively minor uncertainties. Since then a resurgence of activity has cast doubts on some established ideas, confirmed others, and revealed a rich mechanistic variety.

## 10.1    Nitration
### 10.1.1    The nature of the electrophile
Nitration of aromatic substrates can be accomplished under a wide variety of conditions. Nitric acid itself is the most commonly used reagent, either in aqueous solution, or mixed with sulphuric or perchloric acids, or in various organic solvents such as acetic acid or nitromethane. Other reagents include nitronium salts, commonly $NO_2^+BF_4^-$ in sulpholan, and acetyl nitrate, $AcONO_2$, the mixed anhydride formed by reaction between nitric acid and acetic anhydride.

It seems probable that in all these reactions the active electrophile is the nitronium ion, $NO_2^+$ [50J2400], though in the reactions in acetic anhydride its role is rather different from that in the other reactions (§10.1.6). In an excess of sulphuric acid nitric acid is converted quantitatively into nitronium hydrogen sulphate:

$$HNO_3 + 2H_2SO_4 \rightleftharpoons NO_2^+ + H_3O^+ + 2HSO_4^-$$

Evidence for this comes from Raman spectroscopy ($NO_2^+$ at $1400\,cm^{-1}$), from conductimetric experiments, from electrolysis which shows that the nitrogen-containing species migrates to the cathode, and most convincingly from cryoscopy which demonstrates that each nitric acid molecule dissolves in sulphuric acid with the generation of four entities. (Actually the cryoscopic factor is 3.77 because of the equilibrium

$$H_3O^+ + HSO_4^- \rightleftharpoons H_2O + H_2SO_4$$

The difference from 4.0 is as predicted from the known base strength of water.)

In aqueous sulphuric acid the concentration of nitronium ion falls and so does the rate of nitration; with the more reactive substrates nitration still takes place even when there is no detectable nitronium ion left. Nevertheless the evidence is that even in the absence of any added sulphuric acid the reaction still normally involves nitronium; $^{18}$O isotope exchange between nitric acid and water occurs at a rate identical to that of nitration of highly reactive substrates, suggesting that both processes involve a common slow step [52J4917; 58J2420; see also 79J(P2)133]:

$$\text{HNO}_3 \xrightarrow[\text{Slow}]{\text{H}^+} \text{H}_2\text{O} + \text{NO}_2^+ \underset{\text{H}_2\text{O*}}{\overset{\text{ArH}}{\rightleftharpoons}} \begin{array}{l} \text{ArNO}_2 \\ \text{HNO}_3^* \end{array}$$

In agreement with this the reactions under such conditions are zeroth order in the concentration of the aromatic substrate. Less reactive substrates follow first order kinetics, showing that rate control has switched to the actual substitution step. A similar effect can be observed during the nitration of toluene with an aqueous solution of nitric and sulphuric acids; the kinetics change from zeroth order at low concentrations of sulphuric acid to first order as the concentration is increased [74CC293]. (The reaction rates also depend on the concentration of nitric acid, but this is commonly used in large excess.)

Acidity function correlations also provide support for the involvement of nitronium ion; nitration rates closely follow the $H_R$ function, applicable to systems of the type

$$\text{ROH} + \text{H}^+ \rightleftharpoons \text{R}^+ + \text{H}_2\text{O}$$

The formation of the nitronium ion is analogous. By contrast, if the reactive electrophile were simply protonated nitric acid, $(\text{H}_2\text{ONO}_2)^+$, a better correlation would have been expected with $H_0$ or $H_A$.

Reactions with nitric acid in nitromethane or acetic acid also show a changeover from zeroth order kinetics to first order, depending on the reactivity of the substrate and the conditions of the reaction. In the zeroth order reaction the rate-limiting step is again the formation of nitronium ion:

$$2\text{HNO}_3 \underset{\text{Fast}}{\overset{(1)}{\rightleftharpoons}} \text{NO}_3^- + \overset{+}{\text{H}}_2\text{ONO}_2 \underset{\text{Slow}}{\overset{(2)}{\rightleftharpoons}} \text{H}_2\text{O} + \text{NO}_2^+ \underset{(3)}{\overset{\text{ArH}}{\rightleftharpoons}} \text{ArNO}_2$$

Such reactions are retarded by addition of nitrate ion which reduces the concentration of $(\text{H}_2\text{ONO}_2)^+$ and therefore the rate of nitronium formation. Water also retards the reaction, but less markedly. Significantly nitrate ion does not change the kinetic behaviour of the reaction but as water is added the kinetics switch from zeroth to first order. First order kinetics arise when the nitronium ion is destroyed more slowly by reaction with the substrate than by reversion to the nitric acidium ion (step 3 < step $-2$). Water can change the balance between these two steps, but nitrate ion only interferes with step 1.

It should be added that in these organic solvents the kinetic dependence on nitric acid concentration is of a high order and varies with solvent; several molecules of nitric acid are clearly involved in solvation.

### 10.1.2 The Wheland intermediate

The well-known two-step substitution process was established most convincingly by the kinetic isotope studies of Melander [e.g., 49NA(163)599].

$$\text{NO}_2^+ + \underset{}{\bigcirc} \underset{\substack{k_{-1} \\ \text{Slow}}}{\overset{k_1}{\rightleftharpoons}} \underset{+}{\overset{\text{H}\quad\text{NO}_2}{\bigcirc}} \underset{\substack{k_2 \\ \text{Fast}}}{\overset{-\text{H}^+}{\longrightarrow}} \overset{\text{NO}_2}{\bigcirc}$$

A concerted substitution or a two-step process with $k_2 < k_{-1}$ would involve a hydrogen transfer in the rate-limiting step and would therefore exhibit a primary kinetic isotope effect with deuteriated or tritiated substrates. Isotope effects are indeed observed, but usually these are very small ($k_H/k_T$ is rarely greater than 1.3) and unexceptional for secondary effects. However, the introduction of very bulky groups on either side of the position of substitution can reduce $k_2$ by impeding the nitro group as it comes into coplanarity with the ring. In 2,4,6-tri-t-butyltoluene this is enough to make step 2 rate-limiting, and $k_H/k_D$ is 3.7 [68JA2105].

The cationic intermediate, in which the positive charge of the nitronium ion is transferred to the benzene $\pi$ system, is commonly called the Wheland intermediate, though in fact Wheland originally proposed it as the structure of the transition state [42JA900], unaware that Pfeiffer and Wizinger had already correctly identified it as an intermediate [28LA(461)132]. Ions of this type are formed on protonation of aromatic hydrocarbons with, for example, HCl/AlCl$_3$ or HF/BF$_3$. The resulting salts, ArH$_2^+$ MHal$_4^-$, are coloured and electrically conducting. With DCl or DF and Lewis acid catalysis arenes undergo protium–deuterium exchange via such salts.

### 10.1.3 Orientation and reactivity

The transition state for the formation of the Wheland intermediate has the positive charge shared between the attacking electrophile and the benzene ring (as discussed in more detail in the next

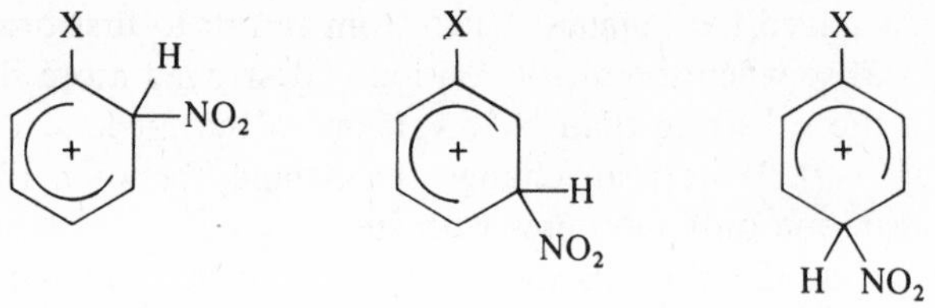

Figure 10.1. Wheland intermediates for *ortho*, *meta* and *para* nitration of $C_6H_5X$.

section). For attack on a monosubstituted benzene there are three isomeric intermediates (excluding the *ipso* intermediate, §10.1.8) (figure 10.1) and three transition states whose energies depend on the interactions between the substituent and the positive charge on the ring. It is simpler to discuss reactivity in terms of the structure of the intermediates, making the reasonable assumption that the pattern of charge distribution around the ring will be similar both in the intermediate and in the corresponding transition state.

The positive charge is concentrated predominantly at the positions *ortho* and *para* to the point of attack, as shown by NBMO calculations (§6.2.2) or by mesomeric symbolism:

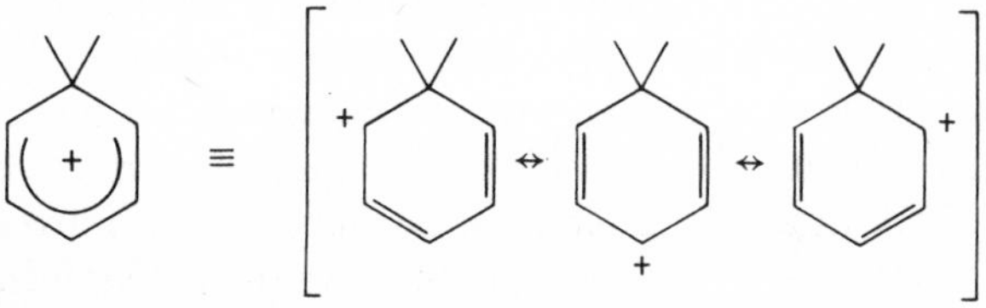

It follows that electron-donating substituents, particularly at the *ortho* and *para* positions, will stabilise the transition state and accelerate the reaction whereas electron withdrawal will have the opposite effect. This results in the well-known pattern of reactivity: electron donors give fast reaction, mainly at the *ortho* and *para* positions; electron acceptors give slow reaction, mainly *meta*. This is exemplified by the nitrations of toluene and of acetophenone, for which partial rate factors are as shown:

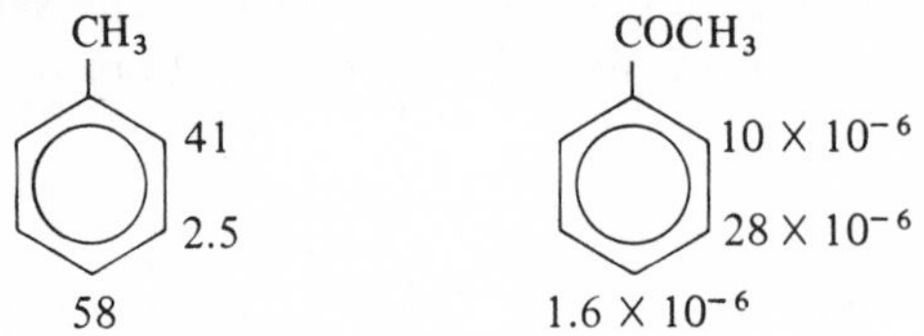

A partial rate factor is the rate of reaction at the position in question relative to that of one position in benzene. For example, toluene is nitrated 24 times as fast as benzene ($HNO_3/HOAc/H_2O$; 45 °C), and 56.5 per cent of the product is *ortho*-nitrotoluene, whence the *ortho* partial rate factor is:

$$f_0 = 24 \times \frac{56.5}{100} \times \frac{6}{2} = 41$$

Reactivity of two   Statistical
*ortho* positions of   correction
toluene relative to
all six positions of
benzene

Reactivity at the *meta* and *para* positions correlates with Hammett's $\sigma^+$ (chapter 3) so one may write

$$\log f_p = \rho\sigma_p^+ \quad \text{and} \quad \log f_m = \rho\sigma_m$$

The apparently anomalous behaviour of the halogens, which deactivate the ring but encourage reaction at the *ortho* and *para* positions, is in fact as predicted. They are the only common substituents with positive values of $\sigma_p^+$; the mesomeric transfer of positive charge from the ring to a *para* (or *ortho*) halogen is enough to stabilise these transition states relative to the *meta* one, but not enough to counteract the polar electron withdrawal. (For fluorine $\sigma_p^+$ is actually slightly negative but the *para* position of fluorobenzene is nevertheless marginally less reactive than a position in benzene; the difference is within the normal limits of error of the Hammett equation.)

### 10.1.4 Selectivity and the nature of the transition state

The energy of the Wheland intermediate (**10.1**) is closely similar to that of the protonated hydrocarbon (**10.2**). The attacking electrophile is separated from the positively charged $\pi$ system by a tetrahedral carbon atom so that its influence is reduced and the difference between H and $NO_2$ is of secondary importance.

(**10.1**; E = $NO_2$)

(**10.2**; E = H)

Substituents on the rest of the ring, however, particularly *ortho* or *para* to the position of attack, can make a big difference to the energy of the system as discussed in the previous section.

If the transition state for substitution closely resembles the Wheland

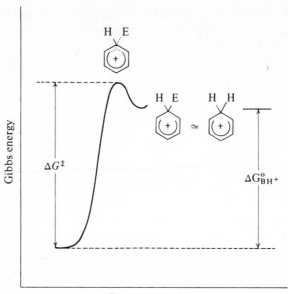

Figure 10.2. Reaction profile for a substitution with a transition state closely resembling the Wheland intermediate. Substituent effects on the reaction rate closely resemble those on basicity (table 10.1).

intermediate then the activation energy, $\Delta G^{\ddagger}$, will be similar to the energy of protonation of the hydrocarbon, $\Delta G^{\ominus}_{BH^+}$ (figure 10.2) and the relative rates of substitution should parallel the corresponding substrate basicities. This is indeed found for some substitutions; table 10.1 illustrates the pattern for bromination reactions.

Nitration, however, is much less sensitive to substituent changes in the substrate. Toluene, which is brominated 600 times faster than benzene, is nitrated only some 20 times as fast. This indicates that the transition state for nitration is not so far along the reaction coordinate and there is less of the positive charge delocalised around the ring (figure 10.3).

Intramolecular selectivity – the relative reactivities of different positions around a substituted benzene ring – should also be related to the nature of the transition state. The greater the charge borne by the ring, the greater will be the difference between the *meta* and *para* positions. (Steric interactions can interfere with the reactivity of the *ortho* position; see §10.3.) Figure 10.4 illustrates this, and Hammett analysis again provides a convenient way of quantifying the correlation.

$$\log f_p = \rho \sigma_p^+ \quad \text{and} \quad \log(f_p/f_m) = \rho(\sigma_p^+ - \sigma_m)$$

Intermolecular selectivity can be measured by $\log f_p$ and intramolecular selectivity by $\log(f_p/f_m)$. A plot of one against the other for a series of

Table 10.1. *A comparison between the basicities of polymethylbenzenes (logarithms of relative stabilities of $ArH_2^+BF_4^-$) and their relative rates of bromination with $Br_2/85$ per cent AcOH ($\log k/k_0$) [see 71Acc240]*

| Methyl substituents | Relative basicities | $\log(k/k_0)$ |
|---|---|---|
| None | 0 | 0 |
| Me | 2.90 | 2.78 |
| $1,2\text{-Me}_2$ | 3.90 | 3.72 |
| $1,3\text{-Me}_2$ | 6.00 | 5.71 |
| $1,4\text{-Me}_2$ | 3.51 | 3.40 |
| $1,2,3\text{-Me}_3$ | 6.30 | 6.22 |
| $1,2,4\text{-Me}_3$ | 6.30 | 6.18 |
| $1,3,5\text{-Me}_3$ | 8.80 | 8.28 |
| $1,2,3,4\text{-Me}_4$ | 7.30 | 7.04 |
| $1,2,3,5\text{-Me}_4$ | 8.30 | 8.62 |
| $1,2,4,5\text{-Me}_4$ | 7.00 | 6.45 |
| $\text{Me}_5$ | 8.30 | 8.91 |

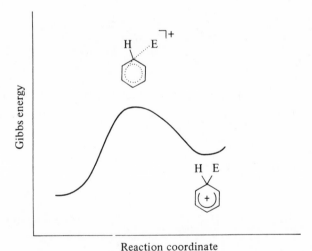

Figure 10.3. Reaction profile for a substitution with an early transition state as in nitration.

different electrophilic substitution reactions – nitration, bromination and so on – with a particular substrate should give a straight line of slope $\sigma_p^+/(\sigma_p^+ - \sigma_m)$. For a wide variety of reactions this indeed found, as the first four entries of table 10.2 illustrate. Selectivity, in these examples, does seem to provide a good insight into the nature of the transition state.

Table 10.2. *Intermolecular and intramolecular selectivity in electrophilic substitution reactions of toluene* [63APO35]

| Reaction | $f_p$ | $f_m$ | $f_p/f_m$ | $\log f_p/\log (f_p/f_m)^a$ |
|---|---|---|---|---|
| Ethylation (EtBr/GaBr$_3$/C$_2$H$_4$Cl$_2$; 25 °C) | 6.02 | 1.56 | 3.87 | 1.33 |
| Nitration (HNO$_3$/HOAc; 45 °C) | 57.6 | 2.52 | 22.9 | 1.30 |
| Acetylation (AcCl/AlCl$_3$/C$_2$H$_4$Cl$_2$; 25 °C) | 750 | 4.80 | 156 | 1.31 |
| Bromination (Br$_2$/AcOH/H$_2$O; 25 °C) | 2425 | 5.45 | 445 | 1.28 |
| Nitration (NO$_2^+$ BF$_4^-$/sulpholan; 25 °C) | 3.19 | 0.14 | 22.7 | 0.37[b] |

[a]The value of $\sigma_p^+/(\sigma_p^+ - \sigma_m)$ for Me is 1.29.
[b]See following section.

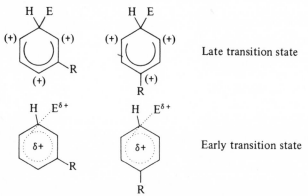

Figure 10.4. The relation between intermolecular and intramolecular selectivity. In a late transition state, with most of the positive charge on the positions *ortho* and *para* to the point of attack, a *para* substituent can exert a large effect (intermolecular selectivity) and there is a large difference between the effects of *meta* and *para* substituents (intramolecular selectivity). In an early transition state less of the charge is borne by the ring, the effect of a *para* substituent is smaller, and so is the difference between *meta* and *para*.

## 10.1.5 Encounter control

In 1961 Olah described some nitration experiments for which the selectivity relation broke down [61JA4571]. Using nitronium salts in sulpholan he found that the reactivity of toluene relative to that of benzene fell to 1.7 or less; even mesitylene reacted only 2.7 times as fast as benzene whereas its relative rate of bromination is almost 200 million. However, the orientation is not markedly different from that in other nitrations (last entry in table 10.2). How can one rationalise the almost

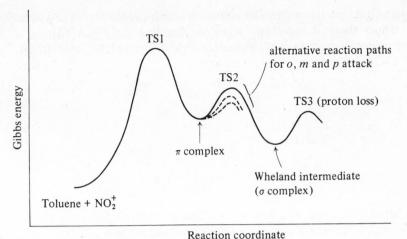

Figure 10.5. The theory of rate-limiting π complex formation.

total loss of intermolecular selectivity while intramolecular selectivity remains unaltered?

Olah's interpretation [see 71Acc240] was that there must be two intermediates with their corresponding transition states (figure 10.5).

The second intermediate would be the Wheland intermediate or σ complex, so called because the electrophile is attached to the substrate by a conventional σ bond, and its preceding transition state TS2 resembles that in 'normal' nitrations so that it leads to a normal product distribution. The first intermediate, Olah suggested, was a π complex in which the electrophilic nitronium ion was associated with the benzene π system without being specifically bonded to a particular carbon atom.

There is no doubt about the existence of complexes of this type [70CR295]. Aromatic and olefinic hydrocarbons form loosely associated adducts with a wide variety of electrophilic species: quinones, picric acid, halogens, silver ions and many others. Some of these adducts have been isolated; others have been detected spectroscopically or by solubility or thermochemical properties of the mixed reagents. Thus the solubilities of the hydrogen halides in arene or alkene solutions in aprotic solvents reveal the presence of colourless 1:1 adducts with heats of formation of the order of $10 \text{ kJ mol}^{-1}$. These solutions are non-conducting, and their properties are in marked contrast to those of the σ complexes, $\text{ArH}_2^+ \text{ MHal}_4^-$ (**10.1**), formed in the presence of $\text{MHal}_3$ Lewis acids.

These weak complexes have been called variously charge-transfer complexes, donor–acceptor complexes, π complexes, and molecular complexes. The terms have been used almost interchangeably, but one can draw a distinction between those complexes in which association is primarily due to intermolecular induction and dispersion forces and those in which there is significant covalent bonding [*cf.* 79JA783]; it is convenient to describe these as molecular and π complexes, respectively, [70CR295], as in the following examples:

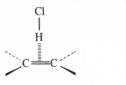 

Alkene–hydrogen chloride    Alkene–proton π complex
molecular complex      (bridged carbonium ion)

Thus Olah envisaged a benzene–nitronium ion π complex derived from interaction between a benzene HOMO and a nitronium LUMO:

The novelty of his thesis was not in the suggestion that such species should exist, but that their formation was rate-limiting. The transition state for this process, TS1, might well resemble a loose molecular complex, so he looked for a correlation between his rates of nitration and the stabilities of molecular complexes analogous to that found between bromination rates and σ complex stabilities and illustrated in table 10.1. Certainly the relatively small substituent effects on nitration rates are comparable in magnitude to the relative stabilities of molecular complexes with hydrogen chloride (table 10.3), but the quantitative correlation is not particularly good. (Note that the data of table 10.1 are presented logarithmically and cover a reactivity range of the order of $10^9$; in table 10.3 the range is less than 3:1.)

There remains some controversy about Olah's results, but most commentators now believe that they arise as a consequence of relatively slow mixing during the nitration [see 71Acc248; 72AG874]. Many of these nitration reactions are extremely rapid, and conventional kinetic measurements are very difficult. The normal practice is to measure relative rates in a competition between two substrates; for example, the relative reactivities of toluene and benzene are determined by measuring

Table 10.3. *A comparison between the stabilities of molecular complexes of polymethylbenzenes with hydrogen chloride, and rates of nitration with nitronium fluoroborate in sulpholan at 25 °C* [see 72AG874]

| Methyl substituents | Relative stabilities of HCl complexes | Relative rates of nitration |
|---|---|---|
| None | 1.00 | 1.00 |
| Me | 1.51 | 1.67 |
| 1,2-Me$_2$ | 1.81 | 1.75 |
| 1,3-Me$_2$ | 2.06 | 1.65 |
| 1,4-Me$_2$ | 1.65 | 1.96 |
| 1,3,5-Me$_3$ | 2.60 | 2.71 |

the amounts of nitrotoluene and nitrobenzene formed when a mixture of the two is nitrated with a deficiency of the electrophile. If the nitration reaction is exceptionally fast the limiting factor can become the rate at which the reagent and substrates are able to come into contact.

Two pieces of evidence in particular support the proposition that this is what is happening in Olah's experiments. Ridd showed that the nitration of dibenzyl (PhCH$_2$CH$_2$Ph) with nitronium tetrafluoroborate in sulpholan gives an anomalously high proportion of dinitro derivatives [70J(B)797]; this probably arises from reaction occurring during mixing in regions of the inhomogeneous mixture where the local concentration of nitronium ions is higher than the average.

During an extensive series of studies of nitration, Moodie and Schofield found that whereas toluene is nitrated in sulphuric acid some 20 times faster than benzene, the introduction of further electron-donating substituents does not increase the reactivity in proportion; the xylenes and mesitylene all react at about twice the rate of toluene in marked contrast to their behaviour in bromination (table 10.1). This again suggests that the point has been reached when the rate is limited by the rate of encounter between nitronium ions and substrate molecules rather than by the rate of the chemical reaction [68J(B)800]. In this second example, however, the rate-limiting process is diffusion on the microscopic scale rather than macroscopic mixing. Values of the enthalpy of activation for these diffusion-controlled reactions agree with those calculated for a two-stage process involving the dissociation of nitric acid to nitronium ion followed by a viscosity-determined diffusion of nitronium ions to the substrate [72J(P2)127].

Similar behaviour is observed in phosphoric acid, but because it is more viscous than sulphuric the rate of diffusion is lower and even for benzene this can become the limiting factor [81J(P2)848]; conversely aqueous nitric acid is less viscous than sulphuric and the encounter limit is raised [81J(P2)94].

The two approaches, $\pi$ complex and encounter control, are not totally incompatible. The fact that positional selectivity is retained in encounter controlled reactions shows that there must be another intermediate, I, prior to the Wheland intermediate:

$$\text{ArH} + \text{NO}_2^+ \underset{}{\overset{\text{Encounter control}}{\rightleftharpoons}} \text{I} \rightleftharpoons \begin{array}{l} \text{Positionally} \\ \text{isomeric} \\ \text{Wheland} \\ \text{intermediates} \end{array}$$

This initial intermediate is commonly called an encounter-pair and represented as $\text{ArH} \cdot \text{NO}_2^+$. There is no satisfactory evidence about the nature of this species. It could be that the two components experience no significant mutual attraction, being merely held together by a 'wall' of surrounding solvent molecules. Alternatively it has been suggested that there may be a complete electron transfer so that I is a radical/radical–cation pair $\text{ArH}^{\bullet +} \cdot \text{NO}_2^{\bullet}$ [77JA5516]. Between these two extremes one can envisage a loose association in some form of molecular or $\pi$ complex. This is the attitude that Olah now appears to adopt: describing nitrations using $\text{AgNO}_3/\text{BF}_3$ in acetonitrile, a homogeneous reaction mixture in which mixing appears not to be a problem, he refers to 'formation of the $\pi$ complex (or encounter-controlled step)' [81JO3533].

### 10.1.6 Nitration in acetic anhydride

Nitration using nitric acid in acetic anhydride has for long been somewhat enigmatic, but the essential features at last seem to be established [78APO1]. The reaction shows the characteristic pattern of diffusion control: the rate is proportional to the concentration of the aromatic substrate but is relatively insensitive to its nature, with mesitylene being only slightly more reactive than *m*-xylene. Intramolecular selectivity resembles that in other nitration reactions, suggesting that the electrophile is $\text{NO}_2^+$. However, the formation of the electrophile is never rate-limiting; even using 1M mesitylene the kinetics remain first order in the mesitylene concentration. (Apparent zeroth order kinetics have been reported, but in fact they arise from a medium effect [74J(P2)600].) This leads to a seemingly paradoxical situation. If the reaction involved a sequence similar to that for diffusion controlled nitration in sulphuric acid, such as:

$$\text{Precursor} \underset{k_{-1}}{\overset{k_1}{\rightleftharpoons}} \text{NO}_2^+ \xrightarrow{k_d[\text{ArH}]} \text{Encounter-pair} \rightsquigarrow \text{Products}$$

where $k_d$ is the rate coefficient for a diffusion-controlled process, and if, however fast step 2 may be, step 1 is never rate-limiting, then $k_{-1}$ must be even greater than $k_d[ArH]$ and the lifetime of the nitronium ion must be exceedingly short. Indeed, calculations show that it is so short (probably less than $10^{-10}$ s) that the maximum possible diffusion limit, $k_d[ArH][NO_2^+]$, could not approach the observed rate of reaction!

Ridd's interpretation is that the species which initially diffuse together must be present in relatively high concentrations and that the nitronium ion is subsequently produced *within* this pre-formed encounter-pair. He suggests the mechanism:

$$ArH + AcONO_2 \;\overset{(1)}{\rightleftharpoons}\; ArH \cdot AcONO_2$$

$$ArH \cdot AcONO_2 + Ac_2\overset{+}{O}H \;\overset{(2)}{\rightleftharpoons}\; ArH \cdot AcO\overset{+}{N}OOH + Ac_2O$$

$$ArH \cdot AcO\overset{+}{N}OOH \;\overset{(3)}{\rightleftharpoons}\; ArH \cdot NO_2^+ \cdot AcOH \;\overset{(4)}{\underset{-AcOH}{\longrightarrow}}\; \overset{+}{Ar}\!\!\underset{NO_2}{\overset{H}{<}} \;\overset{-H^+}{\longrightarrow}\; ArNO_2$$

The first-formed encounter-pair is derived from neutral acetyl nitrate, which is abundant. This is then protonated and the resulting cation tautomerises to give the nitronium-containing species $ArH \cdot NO_2^+ \cdot AcOH$, which collapses to the Wheland intermediate with the loss of a molecule of acetic acid. For the less reactive substrates step 4 is rate-limiting, but for mesitylene and similar highly reactive arenes it is probably step 3.

The protonating agent in step 2 is shown as $Ac_2OH^+$ and not $HNO_3$. The very rapid reversion of the electrophile to its precursor requires that the reverse of step 2 should be fast; $NO_3^-$ is present in very low concentration, too low to be significant in abstracting the proton from $(AcONOOH)^+$, so by the principle of microscopic reversibility $HNO_3$ is not significant on the other side of the equation. The initial pre-equilibria between the nitric acid and acetic anhydride, leading both to $AcONO_2$ and to $Ac_2OH^+$, are in accord with the observed high order (typically about 3) in the stoicheiometric concentration of nitric acid.

### 10.1.7    Nitration via nitrosation

Many nitrations are retarded by the addition of nitrites, the anticatalytic effect arising mainly from the deprotonation of $(H_2ONO_2)^+$ by nitrite, which is a stronger base than nitrate. With very reactive substrates, however, the effect is reversed and nitrites accelerate the reaction. Moreover, many organic compounds are oxidised by nitric acid and the resulting formation of nitrous acid leads to autocatalysis in these reactions.

There appear to be two quite distinct mechanisms for this process. The first [50J2628] involves initial nitrosation (*p*-nitrosophenol has been isolated during the nitration of phenol) followed by oxidation of the nitroso compound. Kinetic analysis suggests that the main electrophiles in the nitrosation are the nitrosonium ion, $NO^+$, and the less active but more abundant dinitrogen tetroxide, $N_2O_4$, which is the mixed anhydride of nitric and nitrous acids. The balance between direct nitration and nitrosation–oxidation is complex. In dilute aqueous solution, where the concentration of nitronium is low, the reaction follows the nitrosation route almost exclusively. In strong acid the high reactivity of the nitronium ion makes itself felt and nitrosation may become so insignificant that the anticatalytic effect of nitrite becomes apparent. In nitromethane or sulpholan, nitration with nitric acid can reach the encounter-controlled limit at about 300 times the rate for benzene, but some substrates react faster than this in autocatalysed processes. This must indicate that nitrosation has taken over from nitration, the nitrosating electrophiles being present in greater concentration than nitronium ions [69J(B)1].

More recently Ridd has found that in 85 per cent sulphuric acid the nitration of dimethylaniline (or rather, of the anilinium cation) is catalysed by nitrous acid by a mechanism that cannot be nitrosation–oxidation because nitration is faster than nitrosation under the same conditions [79J(P2)618]. The formation of tetramethylbenzidine, (*p*-$Me_2N-C_6H_4-)_2$, as a major by-product is evidence for the presence of the radical-cation $Me_2N-Ph^{\cdot+}$, and the e.s.r. spectrum of the corresponding radical is observed during the nitration of *NN*-dimethyl-*p*-toluidine (§10.1.8) [81J(P2)518]. It is suggested that the nitration goes via the following sequence:

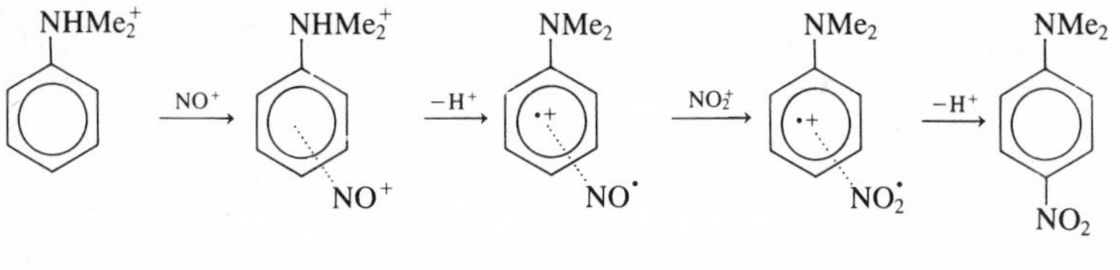

### 10.1.8 Ipso *nitration and the addition-elimination mechanism*

Despite the vast amount of data which has been accumulated over the years on the effect of substituents on *ortho, meta* and *para* reactivity, the possibility of electrophilic attack on the substituted carbon atom, the *ipso* position, has until recently been given scant attention. To the extent that it was considered at all it was assumed to be a trivial side-equilibrium as shown below.

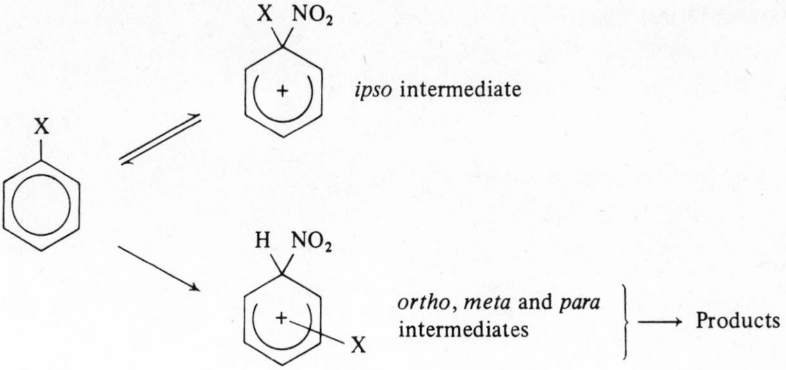

X   NO₂

*ipso* intermediate

X

H   NO₂

*ortho*, *meta* and *para* intermediates } ⟶ Products

However, the *ipso* intermediate can lie directly on product-forming reaction paths. If the group X is a suitable leaving group it can be displaced as X⁺. This is a well-known reaction when X is an electropositive group, such as Me₃Si, though this type of displacement is commonest with the proton as the electrophile. But even alkyl groups can be displaced in this fashion; the nitration of *p*-cymene (*p*-isopropyltoluene) leads to some 10 per cent of *p*-nitrotoluene among the products [74JA4335]. Bromine and iodine are also readily displaced; chlorine is not, the nitro group itself being a better electrofuge [76J(P2)1089].

A second mode of reaction is a 1,2-shift of the electrophile leading via a second intermediate to the *ortho* product:

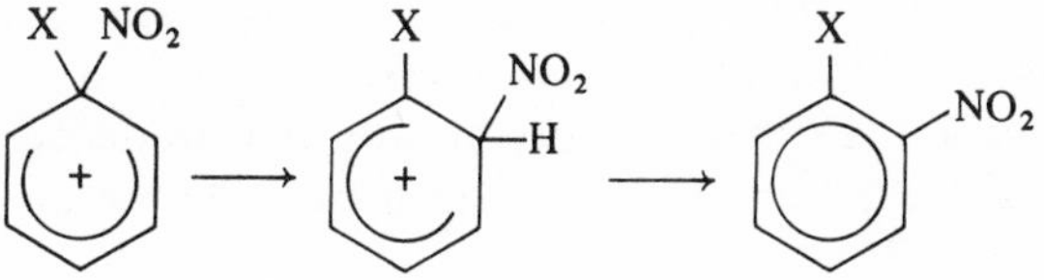

X   NO₂          X   NO₂          X
                     —H              NO₂

It is not always easy to assess the proportion of *ortho* product formed in this way rather than by direct attack, but it can be considerable. The nitration of durene (1,2,4,5-tetramethylbenzene) in sulphuric acid is encounter-controlled. Product analysis is complicated by dinitration, but when the mononitration stage is isolated it is found that the amount of 3-nitrodurene formed varies markedly with the concentration of the aqueous sulphuric acid. Above 85 per cent it is the only significant product; below 60 per cent it constitutes only about 12 per cent of the product mix: the other 88 per cent comes from various reactions of the *ipso* Wheland intermediate with the relatively nucleophilic solvent. The implication is that only 12 per cent of the 3-nitrodurene formed in concentrated acid actually comes from direct attack on the 3 position: the rest is formed via rearrangement after initial *ipso* attack, as illustrated below [81J(P2)1358]:

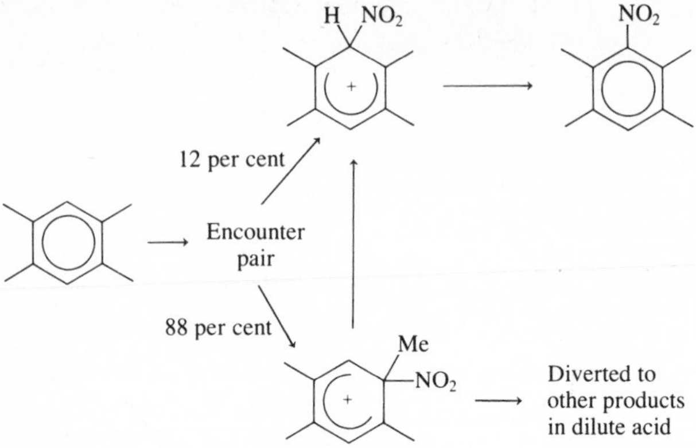

There is also the possibility of the migration of the substituent rather than the electrophile. With a disubstituted substrate this will lead to rearrangement, but this is not a common side-reaction in nitrations.

Finally, the *ipso* intermediate can suffer nucleophilic attack to give an adduct containing a cyclohexadiene ring. This is more likely than with the normal Wheland intermediates because it cannot revert to an aromatic system simply by shedding a proton. Reactions of this type have been widely studied for acetyl nitrate nitrations. Thus the reaction of *p*-cymene with acetyl nitrate at 0 °C gives a mixture of products including *p*-nitrotoluene (see above), the two normal nitration products, and two stereoisomeric cyclohexadiene adducts formed by acetate attack on the *ipso* cation [74JA4335]:

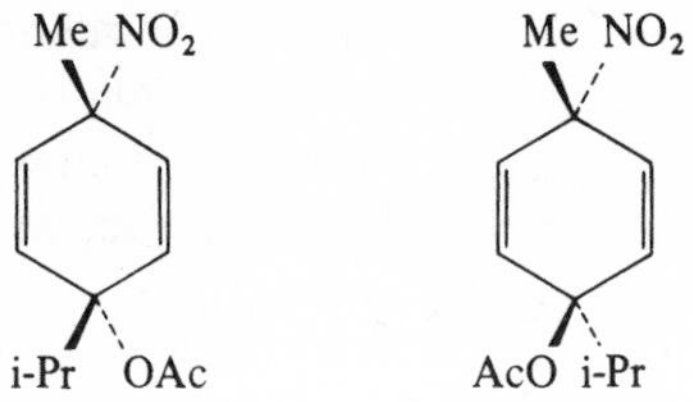

In sulphuric acid these adducts undergo elimination with rearrangement to give 4-isopropyl-2-nitrotoluene, but in acetic acid they lose HNO$_2$ to give the 3-acetoxy compound; this type of side-product is common in acetyl nitrate nitrations.

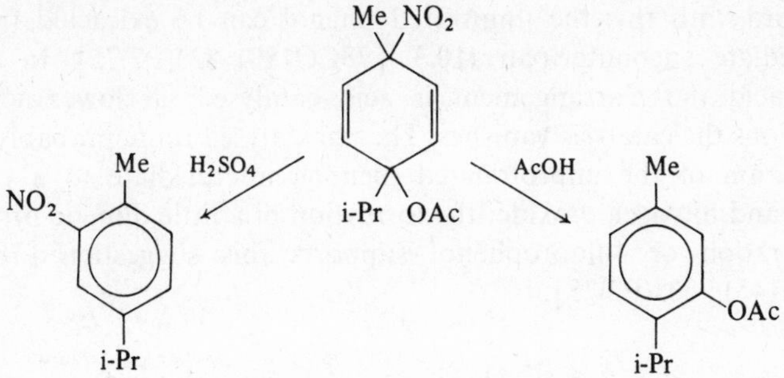

Strongly electron-donating substituents facilitate attack at the *para* position even when this already bears another substituent – that is, the attack is *ipso* with respect to the second substituent. For example, nitration of 4-Me–$C_6H_4$–$NMe_2$ in sulphuric acid, which appears to take place via the nitrosation/radical-cation route described above (§10.1.7), can be followed by n.m.r. spectroscopy. This reveals that the major product (2-$NO_2$-4-Me–$C_6H_3$–$NMe_2$) is formed in a two-stage process with initial attack at the 4 position and subsequent 1,3-rearrangement of the nitro group [81J(P2)518]:

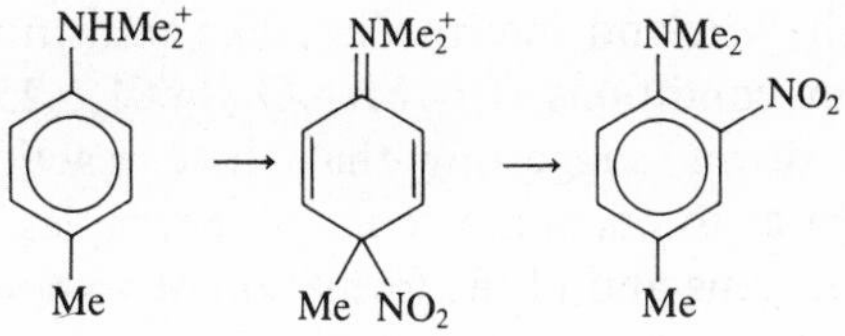

If the positions *ortho* to the amino group are blocked the cationic *ipso* intermediate can be precipitated from solution as the $PF_6^-$ salt [80CC926].

The nature of the 1,3-rearrangement has been elucidated by studies of the nitration of phenols and phenyl ethers. Nitration of *p*-methyl- or *p*-chloro-anisole in aqueous sulphuric acid gives not only the normal *o*-nitro derivative but also the corresponding *o*-nitrophenol (**10.4**; X = Me, Cl). [18]O Labelling shows that the phenolic group comes from solvent suggesting an *ipso*-adduct mechanism with subsequent rearrangement of the nitronium group:

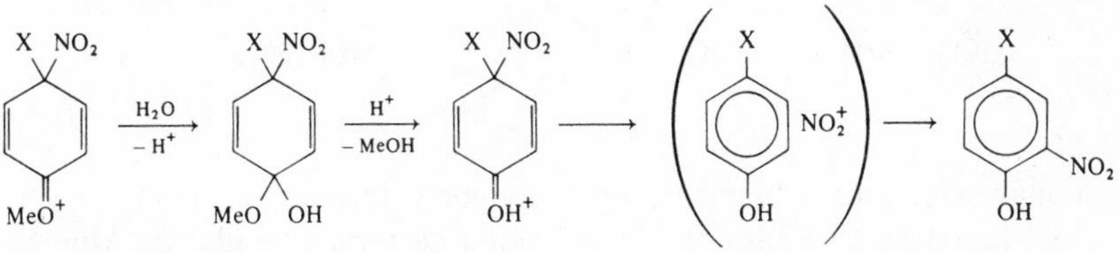

In accord with this the unnitrated phenol can be extracted from the intermediate encounter-pair (**10.3**) [78CC180; 83J(P2)75]. In concentrated acid the rearrangement is acid catalysed; at lower acid concentrations the catalysis vanishes. The uncatalysed route probably entails dissociation of the unprotonated dienone intermediate to a phenoxy radical and nitrogen dioxide: the formation of a little dimeric product in the nitration of 4-fluorophenol supports this suggestion [78JA973; 79J(P2)1451; 83J(P2)75]:

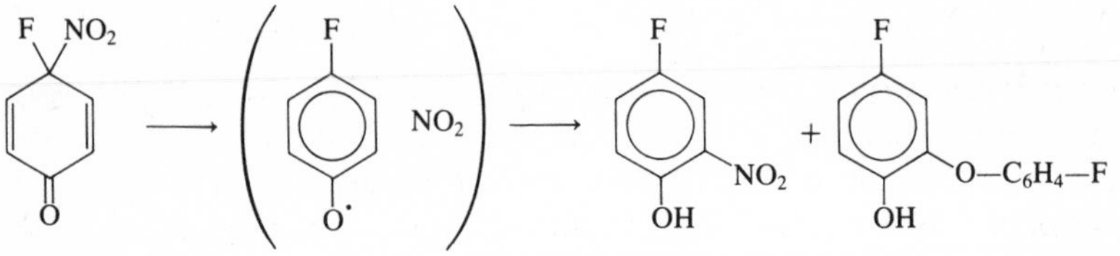

## 10.2    Other electrophilic substitutions

### 10.2.1   Halogenation

Chlorination and bromination exhibit many features in common with nitration. The example has already been given of bromination reactions that pass through Wheland intermediate-like transition states (table 10.1); under different conditions ($Br_2/MeNO_2/FeCl_3$; 25 °C) the selectivity relation breaks down, suggesting that these reactions, like nitrations, can become subject to encounter control. Examples are also known of *ipso* attack by halogens and of the formation of arene–halogen adducts either as side-products or as intermediates in the addition–elimination mechanism of substitution [74Acc361].

However, there remain doubts about the nature of the electrophile in some of these reactions. The more reactive substrates are chlorinated by chlorine in acetic acid solution; the electrophile in these reactions is simply molecular chlorine. The reactions follow second order kinetics (rate = $k_2[\text{ArH}][\text{Cl}_2]$), and apart from normal kinetic electrolyte effects the rates are not influenced by acids, bases, or chloride or acetate ions. This seems to exclude any such species as $Cl^+$, AcOCl or $(\text{AcOHCl})^+$, and accords with a simple two-step mechanism:

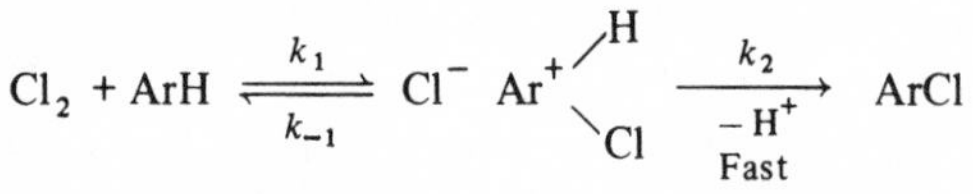

$$\text{Cl}_2 + \text{ArH} \underset{k_{-1}}{\overset{k_1}{\rightleftharpoons}} \text{Cl}^- \; \text{Ar}^+\!\!\overset{H}{\underset{Cl}{\diagup\!\!\diagdown}} \xrightarrow[\text{Fast}]{\underset{-H^+}{k_2}} \text{ArCl}$$

Similar behaviour is found in other organic solvents.

Bromination also follows a fairly similar pattern, save that the kinetics, while remaining first order in the aromatic substrate, are often more complex with respect to bromine. This probably reflects the easier reversibility of the first step. Applying the steady-state approximation to the Wheland intermediate leads to the rate equation:

$$\text{Rate} = \frac{k_1 k_2 [\text{ArH}][\text{Br}_2]}{k_{-1}[\text{Br}^-] + k_2}$$

For chlorination the term $k_{-1}[\text{Cl}^-]$ in the corresponding expression is negligible and the simple second order relation follows. For bromination the two terms in the denominator may be comparable so that common ion retardation can be observed (compare §7.1.2). Moreover the concentration of bromide ion is inversely related to that of molecular bromine through the equilibrium $\text{Br}^- + \text{Br}_2 \rightleftharpoons \text{Br}_3^-$, so the reaction can involve orders higher than unity in $[\text{Br}_2]$. Under these circumstances an H/D kinetic isotope effect appears. This varies with $[\text{Br}^-]$, indicating that $k_{-1}[\text{Br}^-] > k_2$ and step 2 has become rate-limiting.

More problematical are the halogenations by so-called 'positive bromine' and 'positive chlorine'. Reactions with bromine-free hypobromous acid are acid catalysed:

$$\text{Rate} = k_3 [\text{ArH}][\text{HOBr}][\text{H}^+]$$

This was taken to indicate that the electrophile in these reactions was either $\text{BrOH}_2^+$ or $\text{Br}^+$. In strongly acidic media this is undoubtedly true, but recent estimates of the concentrations of these species present in the reaction mixtures suggest that in less acidic solutions the observed rates of reaction are anything up to a million times faster than the rate of encounter between either form of 'positive bromine' and substrate. A linear correlation between the rates of bromination of various substrates and their rates of nitration suggests that there is no fundamental change in the nature of the attacking electrophile. (If some reactions involved one electrophile and some another there would be a discontinuity in the plot.) An explanation has been proposed which includes two alternative pre-equilibria with a common rate-limiting step [73J(P2)1321]:

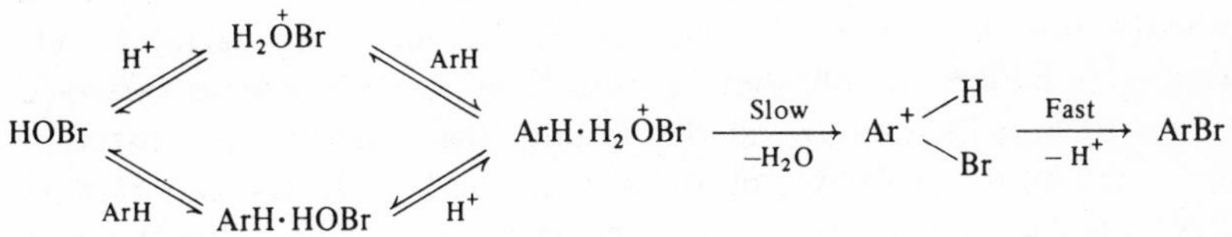

The upper route is effectively the originally accepted mechanism with the intervention of a molecular or $\pi$ complex between the arene and $\text{BrOH}_2^+$; this is encouraged by strongly acid solutions and relatively less reactive substrates. In weakly acidic media and with more reactive substrates the lower route is preferred, involving initial complexation with unprotonated hypobromous acid (which is present in high concentration) and subsequent protonation of the complex (compare nitration with acetyl nitrate, §10.1.6). In either case the overall reactivity and selectivity follows a regular pattern because the rate and product-controlling step is the collapse of the protonated complex to the Wheland intermediate. The existence of pre-equilibria involving proton transfer is confirmed by an $H_2O/D_2O$ solvent isotope effect.

A recent study of halogenation reactions (and also of mercuration) [81JA7240] has provided an interesting insight into the nature of the transition state *en route* to the Wheland intermediate (§10.1.4). On mixing the arene and the electrophile a transient u.v. absorption is observed, decaying as the substitution proceeds. This is associated with a charge-transfer transition from a weakly associated ground state, for example $ArH \cdot Br_2$, to an excited ion-pair, $ArH^{\cdot +} \cdot Br_2^{\cdot -}$. The key observation is a linear relation:

$$\Delta G^{\ddagger} - \Delta G_0^{\ddagger} = E - E_0$$

where $\Delta G^{\ddagger}$ is the Gibbs energy of activation for the substitution, $E$ the charge-transfer transition energy, and the subscript 0 refers to benzene as an arbitrary reference substrate. The unit slope implies that the chemical activation process is closely similar to the spectroscopic excitation, and suggests that the transition state may resemble an ion-pair, similar to some of the structures proposed for encounter-pairs in nitration (§10.1.5) or nitrosation/nitration (§10.1.7). After due allowance is made for solvation energies, it is found that the rates of addition of bromine to alkenes follow the same correlation with charge-transfer transition energies, suggesting that the nature of the initial attack is the same in both systems [82JA7599].

### 10.2.2 Sulphonation

Most studies of sulphonation have been carried out using aqueous sulphuric acid or oleum as the sulphonating agent. Kinetic investigations are complicated because there is such a wide variety of species present in sulphuric acid solutions, their relative concentrations depending on the $H_2O:SO_3$ ratio; they include $H_2SO_4$, $H_3SO_4^+$, $H_5SO_5^+$, $HSO_4^-$, $H_3O^+$, $H_5O_2^+$, $SO_3$, $S_3O_9$ and $H_2S_2O_7$. Further problems arise because sulphonation, unlike most electrophilic aromatic substitutions, is a reversible process.

In aqueous sulphuric acid below 80 per cent w/w of acid the rates of sulphonation correlate linearly with the activity of $H_3SO_4^+$; above about 85 per cent there is a similar correlation with the activity of $H_2S_2O_7$. It seems therefore, that these are the two main electrophiles. They can be regarded as $SO_3$ solvated by $H_3O^+$ and by $H_2SO_4$, respectively; over the range 85 to 100 per cent sulphuric acid the concentration of $H_3O^+$ falls and that of undissociated $H_2SO_4$ rises, the crossover occurring at about 91 per cent. $H_2S_2O_7$ is the stronger electrophile and so its reaction becomes dominant at a somewhat lower concentration than this [68R24]. Thus the mechanism for these reactions is:

$$ArH + SO_3 \cdot X \underset{}{\overset{\text{Slow}}{\rightleftharpoons}} Ar^+ \underset{SO_3^-}{\overset{H}{\diagdown}} \xrightarrow[\pm H^+]{\text{Fast}} Ar{-}SO_2OH$$

$$+ X \qquad\qquad (X = H_3O^+, H_2SO_4)$$

Below about 95 per cent sulphuric acid the substrate H/D kinetic isotope effect is insignificant but it increases at higher acid concentrations. This suggests that the second step, proton transfer, is becoming rate-limiting and in agreement with this the rate falls below that which would be predicted by extrapolating the correlation with $H_2S_2O_7$ activity. At lower acid concentrations the proton can be abstracted by the $HSO_4^-$ anion – a strong base in sulphuric acid – but its concentration falls rapidly with increasing acid concentration leaving sulphuric acid itself as the only significant base present [67R865].

In oleum the rate of substitution increases rapidly. The electrophile is now likely to be the unsolvated $SO_3$ molecule. The rate of reaction depends both on $[SO_3]$ (strictly, the vapour pressure of $SO_3$ over the oleum) and on the acidity function $H_0$. This was at first taken to mean that the electrophile was $SO_3H^+$; this cannot be so because in deuteriated oleum the rate of sulphonation is reduced whereas the concentration of $SO_3D^+$ is higher than that of $SO_3H^+$ in protiated oleum of the same strength. Probably the mechanism is as follows:

$$ArH + SO_3 \overset{\text{Fast}}{\rightleftharpoons} Ar^+ \underset{SO_3^-}{\overset{H}{\diagdown}} \xrightarrow[\text{Slow}]{H^+} Ar^+ \underset{SO_2OH}{\overset{H}{\diagdown}} \xrightarrow[-H_3SO_4^+]{(?) H_2SO_4} ArSO_2OH$$

### 10.2.3 Friedel–Crafts reactions

(a) *Alkylations.* The kinetic study of Friedel–Crafts alkylation reactions is beset with difficulties: the reaction rates are highly susceptible to traces of moisture; they are frequently too high for accurate measurement; the reaction mixtures tend to separate into inhomogeneous layers; the initial reaction products isomerise easily and polyalkylation is hard to avoid; the kinetic behaviour varies with solvent.

Despite these problems the essential features of the reactions now seem established. A series of benzylations using substituted benzyl chlorides with aluminium chloride as catalyst in nitrobenzene solution follow third order kinetics:

$$\text{Rate} = k_3[\text{ArH}][\text{RCl}][\text{AlCl}_3]$$

though the dependence on $[\text{ArH}]$ is less accurately first order than for the other two species [53JA6285]. The mechanism is probably:

$$\text{RCl} + \text{AlCl}_3 \underset{k_{-1}}{\overset{k_1}{\rightleftharpoons}} (\text{RCl}\cdot\text{AlCl}_3) \xrightarrow{\overset{k_2[\text{ArH}]}{\longrightarrow}} \text{Ar}^+\underset{\text{R}}{\overset{\text{H}}{\diagup}} \text{AlCl}_4^- \xrightarrow{\text{Fast}} \text{ArR} + \text{HCl} + \text{AlCl}_3$$

with step 2 rate-limiting. The kinetic expression for this is:

$$\text{Rate} = \frac{k_1 k_2 [\text{RCl}][\text{AlCl}_3][\text{ArH}]}{k_{-1} + k_2[\text{ArH}]}$$

and if $k_{-1} \gg k_2[\text{ArH}]$ this reduces to the observed third order relation with $k_3 = K_1 k_2$. The precise nature of the intermediate represented as $(\text{RCl}\cdot\text{AlCl}_3)$ has been somewhat controversial. In principle it could range from a loose molecular complex to a dissociated carbenium ion system:

| $\text{RCl}\cdot\text{AlCl}_3$ | $\overset{+}{\text{R}}-\overset{-}{\text{Cl}}-\text{AlCl}_4$ | $\text{R}^+\text{AlCl}_4^-$ | $\text{R}^+ + \text{AlCl}_4^-$ |
|---|---|---|---|
| Molecular complex | Coordinated adduct | Ion-pair | Dissociated ions |

In fact, of course, its structure must depend on the nature of R and on the solvent, and it will also differ if different Lewis acid catalysts are used: the reaction is also a *nucleophilic* substitution as far as the halide is concerned and we need to consider the $S_N1/S_N2$ mechanistic spectrum discussed in chapter 7. It is unlikely that many reactions proceed as far as the free carbenium ion; this would be expected to react extremely rapidly with the arene, so that step 1 would be rate-limiting and $[\text{ArH}]$ would not appear in the rate equation. On the other hand the electrophile is probably more polarised than the loose molecular complex. Friedel–Crafts alkylations exhibit very low intermolecular and intramolecular selectivity (table 10.2), indicating that relatively little charge is transferred to the benzene ring in a rather reactant-like transition state – behaviour characteristic of highly reactive electrophiles. The alkyl groups are susceptible to Wagner–Meerwein rearrangement during the reaction which is also indicative of a highly polarised C–Hal bond.

Some 1:1 complexes between Lewis acids and alkyl halides are isolable at low temperatures. They undergo slow halogen exchange as revealed by radio-labelling or by treating alkyl bromides with metal chlorides. The exchange rate is $3° > 2° > 1°$, which is consistent with ion-pair formation but does not exclude the coordinated adduct, which could undergo exchange via a 1,3-shift:

$$
\begin{array}{c}
\text{Cl} \\
\mid \\
\overset{R}{\underset{\underset{\displaystyle\quad:\text{Cl}}{\Big|}}{\overset{\displaystyle\frown}{B}r^{+}}}-\text{M}^{-}-\text{Cl}
\end{array}
\longrightarrow
\begin{array}{c}
\text{Cl} \\
\mid \\
\text{Br}-\text{M}^{-}-\text{Cl} \\
\mid \\
\text{Cl}^{+} \\
/ \\
R
\end{array}
$$

Substituent effects do not help to resolve the problem. It has been suggested that reaction via carbenium ions could be detected if faster reactions occurred with halides which gave less stable (and so more reactive) carbenium ions. However, the observed rate coefficient $k_3$ is the product of the equilibrium constant for complex formation, $K_1$, and the rate coefficient for electrophilic attack, $k_2$. Electron donation in the R group stabilises the complex, whatever its structure, thus increasing $K_1$, but at the same time reducing $k_2$. The former effect seems to be dominant; the relative reactivities towards toluene of different alkyl bromides with gallium bromide catalysts are: Me 1, Et 14, i-Pr $2 \times 10^4$, t-Bu $8 \times 10^5$ [56JA6249; 59JA3315].

The most likely mechanistic behaviour would seem to be a gradual transition from a coordinated complex with methyl halides to an ion-pair with t-butyl, but this still awaits experimental confirmation.

(*b*) *Acylations.* Kinetic studies of Friedel–Crafts acylations frequently reveal rate relations that are first order in substrate and acid chloride but of variable order in catalyst. Infrared spectra of acetyl chloride/aluminium chloride mixtures reveal the presence of free acetylium ions, $CH_3CO^+$, and of 1:1 and 1:2 complexes; X-ray studies have confirmed the existence of the acetylium ion and show that the 1:1 complex is coordinated via oxygen [74AG1]:

$$
CH_3-C\overset{\displaystyle \overset{+}{O}-\bar{A}lCl_4}{\underset{\displaystyle Cl}{\diagdown}}
$$

The 1:2 complex may involve further complexation at chlorine.

The substitution reaction therefore probably involves the rate-limiting attack of any or all of these electrophiles (depending on the nature of the substrate, acyl chloride and solvent) on the aromatic substrate, hence the variable rate dependence on the concentration of the catalyst. This is also affected by complex formation between catalyst and molecules of the ketone product. In accord with this, low temperature n.m.r. spectra of Friedel–Crafts acetylation mixtures show that a 1:1:1 complex of $ArH \cdot CH_3COCl \cdot AlCl_3$, probably with the structure of a Wheland intermediate, is formed in concentrations that are proportional to the rate of reaction [83JO302].

### 10.2.4 Diazo coupling reactions

The weakly electrophilic diazonium ion, $ArN_2^+$, couples only with activated aromatic substrates such as amines or phenols. The reactions are retarded by acid, in which amines are protonated to give the unreactive ammonium ions. The fact that phenols behave in a similar manner indicates that the substrate is actually the phenoxide anion, and this conclusion is confirmed by the unreactivity of anisole towards most diazonium salts: in general phenols and phenyl ethers have similar reactivities in electrophilic substitution reactions. In alkaline solution the reactions are also retarded because the diazonium ion is converted into the diazotic acid, $ArN=NOH$.

After these specific effects of hydronium and hydroxide ions are allowed for it is found that some diazo coupling reactions, but not all, are subject to general base catalysis. This arises if removal of the proton from the Wheland intermediate is kinetically significant. For the reaction sequence:

$$ArN_2^+ + Ar'H \underset{k_{-1}}{\overset{k_1}{\rightleftharpoons}} \overset{ArN=N}{\underset{H}{\diagdown}}\!\!\!Ar'^+ \xrightarrow{k_2[B]} ArN=NAr' + BH^+$$

the rate equation is:

$$k_{obs} = k_1 k_2[B]/(k_{-1} + k_2[B])$$

If the first step is rate-limiting ($k_{-1} \ll k_2[B]$) the terms in $[B]$ fall out and $k_{obs} = k_1$. At the other extreme, when $k_{-1} \gg k_2[B]$, the rate is proportional to $[B]$. When the two terms are comparable the rate depends on $[B]$ in a non-linear way. All three types of behaviour have been observed in diazo coupling reactions, but the non-linear catalysis is the most interesting case for mechanistic studies because under such conditions small changes in the reacting system can have large effects on the kinetics. The phenomenon of base catalysis in electrophilic aromatic substitutions is not confined to diazo coupling reactions but they are the most convenient to study because they can be carried out in dilute aqueous solution at moderate pH.

Those couplings which are subject to base catalysis also exhibit H/D and H/T kinetic isotope effects when the hydrogen atom at the point of attack is isotopically replaced; this is to be expected for reactions in which hydrogen transfer is kinetically important. Note in particular that $k_2$ is sensitive to the isotopic composition of the substrate whereas $k_{-1}$ is effectively independent of it. Consequently an increase in the partition factor, $k_{-1}/k_2[B]$, moving the kinetic control towares the second step, increases both the sensitivity of the reaction to base catalysis and the magnitude of the kinetic isotope effect.

This is illustrated by the reactions of *p*-chlorobenzenediazonium with the anions of the naphtholsulphonic acids (**10.5–10.7**; arrows indicate the positions at which coupling occurs). The reaction with 1-naphthol-4-sulphonic acid (**10.5**) is not subject to base catalysis, nor does it show a kinetic isotope effect [58H2274]; clearly this is a reaction in which the first step is rate-limiting. In 2-naphthol-6,8-disulphonic acid (**10.6**) the site

(**10.5**)          (**10.6**)          (**10.7**)

of substitution is sterically crowded; this reduces $k_2$ by impeding the approach of the base and by hindering the ArN=N group as it moves into coplanarity with the naphthalene rings. With this substrate the reaction is strongly catalysed by pyridine and it shows a large kinetic isotope effect. As the concentration of pyridine is increased its effectiveness as a catalyst decreases, and so does the magnitude of the isotope effect, as expected from the appearance of [B] in the denominator of the partition factor. 2,6-Lutidine (dimethylpyridine) is a less good catalyst than is pyridine because it leads to an even more crowded transition state for proton abstraction, reducing $k_2$ still further; the kinetic isotope effect is larger than with pyridine, and the reaction rate varies in a more nearly linear fashion with the concentration of the catalyst [74JA7222; 55H1597, 1617, 1623].

1-Naphthol-3-sulphonic acid (**10.7**) is intermediate between the other two substrates, both in its sensitivity to catalysis and in the magnitude of the kinetic isotope effect. It reveals an interesting differential intramolecular isotope effect. The 4 position, being more hindered than the 2 position, is more susceptible to base catalysis and exhibits a larger kinetic isotope effect; the 2:4 product ratio is therefore greater for the 2,4-dideuteriated compound than for the unlabelled one [58H2274].

## 10.3   *Ortho:para* ratios

Orientation effects in electrophilic substitution reactions of mono-substituted benzenes are, in general, readily explained in terms of the effect of the substituent on the transition state, modelled by the Wheland intermediate (§10.1.3). For reactions *meta* and *para* to the

Table 10.4. *Partial rate factors for nitration of PhR using acetyl nitrate at 0 °C* [60J4885]

|        | R = Me | Et   | i-Pr | t-Bu |
|--------|--------|------|------|------|
| $f_0$  | 49.7   | 31.4 | 14.8 | 4.5  |
| $f_m$  | 1.3    | 2.3  | 2.4  | 3.0  |
| $f_p$  | 60.0   | 69.5 | 71.6 | 75.5 |

substituent the effects can be quantified in terms of Hammett correlations with $\sigma^+$.

Normally one expects *ortho* substituent effects to resemble *para*, since in the Wheland intermediate the positive charge is concentrated at the positions *ortho* and *para* to the point of attack. The most obvious cause for discrepancy is steric, and this is manifest in the partial rate factors for nitrations of alkyl benzenes (table 10.4).

One may argue the significance of the relatively small changes in $f_m$ and $f_p$, but there is no doubt that the dramatic fall in $f_o$ value arises from steric interactions.

Bulky electrophiles, too, are less reactive towards the more hindered *ortho* position; the ratios of $f_o$ to $f_p$ for Friedel–Crafts alkylations of toluene using alkyl bromides with gallium bromide are: MeBr 0.81, EtBr 0.47, i-PrBr 0.30, t-BuBr 0.00 [56JA6255].

Specific interactions between a solvent and the substrate or the electrophile can similarly increase the steric crowding in the *ortho* transition state. In the Friedel–Crafts benzoylation of toluene using PhCOCl/AlCl$_3$ in 1,2-dichloroethane the ratio of $f_o$ to $f_p$ is 0.052; in nitrobenzene, which complexes with the catalyst to produce a bulkier electrophile, the ratio falls to 0.039 [57JO719; 59JA3308]. In the nitration of acetanilide using nitric acid in aqueous sulphuric acid the $f_o$:$f_p$ ratio changes from 0.03 in 98 per cent sulphuric acid to 0.4 in 66 per cent. This huge relative difference can be attributed to hydrogen bonding between the amide group and the sulphuric acid, an interaction that affects the *ortho* position much more than the *para* [77J(P2)1693].

However, there are factors other than steric ones at work. Substrates bearing $\pi$ attractive functional groups (nitro, carbonyl, nitrile, sulphonyl, etc.) give, as expected, predominantly the *meta* product, but the *ortho* position is invariably considerably more reactive than the *para*: compare the partial rate factors for the nitration of acetophenone where $f_o/f_p = 6$ (§10.1.3).

Several explanations for this have been put forward, but the most likely seems to be that the mesomeric interaction between the substituent and ring is almost totally lost in the *para* intermediate (and therefore to a lesser extent in the transition state too) as illustrated by the following representations in which canonical forms bearing two positive charges on adjacent atoms have been neglected because they will be of very high energy:

ortho                                    para

The effect of halogen substituents is exemplified in table 10.5. The *meta* positions are strongly deactivated and the *para* positions are slightly deactivated; the differences here between the halogens are relatively small. The *ortho* position of fluorobenzene is very much less reactive than the *para* position; the differences are less marked for the other halogens, the *o:p* ratio rising in the order Cl < Br < I. Part of the explanation for this behaviour undoubtedly lies in the strong polar effect of fluorine, destabilising the *ortho* transition state. There may also be a contribution to the $f_o$ values from reaction via *ipso* attack and subsequent rearrangement; iodine is certainly the least deactivating halogen for *ipso* attack and fluorine the most. There is not enough information available to assess the significance of this, but it seems likely that many quantitative assessments of *ortho* reactivity – and in a wide variety of reactions, not only with halide substrates – are complicated by an unknown contribution from *ipso* reactions. Moreover, the rearrangement of an *ipso* intermediate to its *ortho* isomer is in competition with direct nucleophilic attack leading to

Table 10.5. *Partial rate factors for nitration of PhHal using $HNO_3/67.5$ per cent $H_2SO_4$* [70J(B)347]

| Hal = | F | Cl | Br | I |
|---|---|---|---|---|
| $f_o$ | 0.045 | 0.067 | 0.077 | 0.17 |
| $f_m$ | 0.0021 | 0.0018 | 0.0016 | 0.0049 |
| $f_p$ | 0.58 | 0.25 | 0.20 | 0.40 |

cyclohexadienes and other products derived from them (§10.1.8) and the balance between these two processes depends on the reaction medium. Concentrated sulphuric acid is so weakly nucleophilic that it traps none of the *ipso* intermediate, but with increasing dilution there can be a corresponding decrease in the amount of *ortho* product formed. Some substrates can be subject both to this partitioning of the *ipso* intermediate and to specific solvent – substrate interactions (see above); the resulting variation in isomer proportions with changes in the solvent can be complex [75J(P2)648].

## 10.4    Reactivity of polycyclic aromatic substrates

Electrophilic substitutions have been used widely as test beds for various quantitative theories of reactivity. Polycyclic aromatic hydrocarbons are particularly susceptible to molecular orbital treatments and there is an adequate range of experimental data to compare with theoretical predictions. There are several good surveys of the various methods [e.g. 61MI; 66APO73]. Examples of the two main types are given here by way of illustration.

(a) *The isolated molecule approach* bases its estimate of reactivity on the properties of the $\pi$ electron system of the ground state hydrocarbon. It is essentially a PMO method, assessing the way in which the attacking electrophile can perturb the $\pi$ electron distribution of the substrate in the initial stages of the reaction.

A typical example of this approach is due to Fukui, who defines a term he calls 'superdelocalisability' for each position $r$ in the substrate [57JCP(27)1247]:

$$S_r = \sum_{i=1}^{\text{HOMO}} \frac{c_i^2(r)}{E_i(r)}$$

The sum is taken over all the occupied $\pi$ molecular orbitals; $c$ is the coefficient of each MO at position $r$ and $E$ is the energy of the orbital below the non-bonding level. The idea is that $S_r$ measures the ease with which the $\pi$ system can be perturbed as an electrophile approaches position $r$, and so it indicates the ability of that position to form a new bond to the electrophile. The $\pi$ electron density at each position is measured by $c^2$, and the higher energy orbitals (small $E$) are the more easily polarised (compare the second order perturbation equation [6.5], §6.1.3, and the discussion in §6.4.1(b)).

(b) *Localisation methods*, on the other hand, use the Wheland intermediate as a model for the transition state. An estimate is made of the difference between the $\pi$ electron energies of the aromatic substrate and of the Wheland intermediate, and this is assumed to be proportional to the activation energy for the substitution. The simplest of these methods, due to Dewar, uses NBMO calculations (§6.2.2) and therefore demands no more than 'back-of-an-envelope' arithmetic [56J3581].

If two odd alternant hydrocarbon fragments are bonded together to form an even alternant system the change in $\pi$ energy is approximately $2c_a c_b \beta$ where $c_a$ and $c_b$ are the coefficients of the NBMOs of the two fragments at the point of attachment and $\beta$ is the overlap integral. In aromatic substitutions one can regard the substrate as an even alternant system comprising the odd alternant Wheland intermediate bonded to the $p$ orbital of a one-carbon fragment; thus with benzene:

One-carbon fragment $b$

Wheland intermediate
fragment $a$

$\pi$ system of
Wheland intermediate

$\pi$ system of
benzene

The energy difference between the two is therefore $2(c_{a1} + c_{a2})c_b\beta$, where $c_{a1}$ and $c_{a2}$ are the coefficients of the NBMO at the two positions on the Wheland intermediate adjacent to the position being attacked; this is an estimate of the energy required to take two of the $\pi$ electrons from the aromatic system and localise them in a $\sigma$ bond to the electrophile with the consequential redistribution of the remaining $\pi$ electrons around the Wheland intermediate. For a single $p$ orbital $c_b$ is unity and for all aromatic C–C bonds $\beta$ can be taken as constant; hence Dewar defined his 'reactivity number' as:

$$N_r = 2(c_{a1} + c_{a2})$$

For example, to compare the reactivities of benzene and of the 1 and 2 positions of napthalene the calculations are:

$1/\sqrt{3}$  $1/\sqrt{3}$

$1/\sqrt{11}$  $2/\sqrt{11}$

$2/\sqrt{8}$

$1/\sqrt{8}$

$N_r = 2(1/\sqrt{3} + 1/\sqrt{3})$
$= 2.31$

$N_r = 2(1/\sqrt{11} + 2/\sqrt{11})$
$= 1.81$

$N_r = 2(2/\sqrt{8} + 1/\sqrt{8})$
$= 2.12$

This predicts, in agreement with experiment, that the order of reactivity is 1-Naph > 2-Naph > Ph.

Table 10.6. *Reactivity factors and experimental reactivities* [$S_r$ values from 57JCP(27)1247; 54BJ423. $f$ values from 71MI$a$]

| Substrate | Position | $S_r$ | $N_r$ | $\log f$ | Key to figure 10.6 |
|---|---|---|---|---|---|
| Anthanthrene | 6 | 1.504 | 1.03 | 5.27 | $a$ |
| Benzene | 1 | 0.833 | 2.31 | 0.00 | $b$ |
| Benzo[$a$]pyrene | 6 | 1.408 | 1.15 | 5.10 | $c$ |
| Chrysene | 6 | 1.044 | 1.67 | 3.54 | $d$ |
| Coronene | 1 | 0.991 | 1.80 | 3.06 | $e$ |
| Naphthalene | 1 | 0.994 | 1.81 | 2.67 | $f$ |
| | 2 | 0.873 | 2.12 | 1.70 | $g$ |
| Perylene | 3 | 1.195 | 1.33 | 4.89 | $h$ |
| Phenanthrene | 1 | 0.977 | 1.86 | 2.56 | $j$ |
| | 2 | 0.860 | 2.18 | 1.96 | $k$ |
| | 3 | 0.893 | 2.04 | 2.48 | $l$ |
| | 4 | 0.939 | 1.96 | 1.90 | $m$ |
| | 9 | 0.997 | 1.80 | 2.69 | $n$ |
| Pyrene | 1 | 1.115 | 1.51 | 4.23 | $p$ |
| Triphenylene | 1 | 0.928 | 2.12 | 2.78 | $q$ |
| | 2 | 0.833 | 2.31 | 2.78 | $r$ |

Within limits both these methods of assessing reactivity can be remarkably successful. Table 10.6 lists some values of $S_r$ and $N_r$, together with partial rate factors for acetyl nitrate nitrations. Figure 10.6 is a plot of the same data.

The most obvious limitation of both approaches is that no account is taken of steric effects; for example, the least reactive position of phenanthrene is the highly hindered position 4, though both $S_r$ and $N_r$ values predict that it should be more reactive than positions 2 or 3. The data plotted as ◐ in figure 10.6 are all for reactions with approximately the same steric requirements; they involve a single *peri* interaction as in attack at the 1-position of naphthalene. By a judicious choice of scales for the $S_r$ and $N_r$ axes the points for these reactions can be made to coincide within the accuracy of the graph – a remarkable agreement between the two methods and, as can be seen, an excellent correlation with the experimental reactivities.

The agreement is less good for substrates with other steric requirements, though there is still a reasonable semi-quantitative correlation. The major deviations occur with exceptionally reactive substrates ($a$ and $c$ in the table and figure). It is virtually certain that these do not follow the same mechanism as the other nitrations; that will become limited by encounter control and the nitrosation route will have set in.

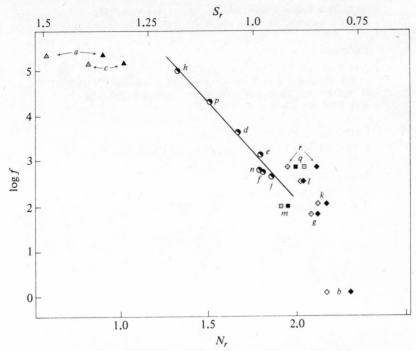

Figure 10.6. Correlations between superdelocalisabilities ($S_r$), Dewar reactivity numbers ($N_r$), and reactivities of polycyclic arenes towards acetyl nitrate. The letters correspond to the key in table 10.6. Key: ◓ reaction sites subject to one *peri* interaction; correlation with both $S_r$ and $N_r$. The straight line is plotted from these points. For other types of substrate solid points represent correlations with $N_r$ and open points with $S_r$: △ ▲ reaction sites subject to two *peri* interactions. ◇ ◆ reaction sites subject to no *peri* interactions. □ ■ reaction sites hindered as in the 4-position of phenanthrene.

## 10.5    Further reading

R. O. C. Norman & R. Taylor, *Electrophilic Substitution in Benzenenoid Compounds*, Elsevier, Amsterdam, 1965.

K. Schofield, *Aromatic Nitration*, Cambridge University Press, Cambridge, 1980.

P. B. D. de la Mare & J. H. Ridd, *Aromatic Substitution: Nitration and Halogenation*, Butterworth, London, 1959.

R. Taylor, in *Comprehensive Chemical Kinetics*, eds C. H. Bamford & C. F. H. Tipper, vol. 13, Elsevier, Amsterdam, 1972.

J. G. Hoggett, R. B. Moodie, J. R. Penton & K. Schofield, *Nitration and Aromatic Reactivity*, Cambridge University Press, Cambridge, 1971.

P. B. D. de la Mare, *Electrophilic Halogenation*, Cambridge University Press, Cambridge, 1976.

S. R. Hartshorn & K. Schofield, *Progr. Phys. Org. Chem.*, 1973, **8**, 278 (nitration).

L. M. Stock, *Progr. Phys. Org. Chem.*, 1976, **12**, 21 (nitration).

H. Cerfontain, *Mechanistic Aspects in Aromatic Sulfonation and Desulfonation*, Wiley-Interscience, New York, 1968.

G. A. Olah, *Friedel–Crafts Chemistry*, Wiley, New York, 1973.
D. L. H. Williams, *Adv. Phys. Org. Chem.*, 1983, **19**, 381 (nitrosation).

### Encounter control:

J. H. Ridd, *Adv. Phys. Org. Chem.*, 1978, **16**, 1.
P. Rys, P. Skrabal & H. Zollinger, *Angew. Chem. Internat. Ed.*, 1972, **11**, 874.
G. A. Olah, *Acc. Chem. Res.*, 1970, **4**, 240 ($\pi$ complex approach).
J. H. Ridd, *Acc. Chem. Res.*, 1970, **4**, 248.

### Orientation, reactivity and selectivity:

L. M. Stock & H. C. Brown, *Adv. Phys. Org. Chem.*, 1963, **1**, 35.

### *Ipso* reactions:

R. B. Moodie & K. Schofield, *Acc. Chem. Res.*, 1976, **9**, 287.
P. B. D. de la Mare, *Acc. Chem. Res.*, 1974, **7**, 361.

# 11 Addition to the carbonyl group and related reactions

Addition to the carbon–oxygen double bond (carbon–nitrogen behaves similarly) is a less complicated mechanistic problem than addition to the carbon–carbon double bond (chapter 9). The vast majority of reactions involve nucleophilic attack on the carbonyl carbon atom and electrophilic attack, usually protonation, on the oxygen:

$$\text{>C=O + H-Z} \longrightarrow \text{>C} \overset{\text{OH}}{\underset{\text{Z}}{<}}$$

There is not the wide variation of mechanistic types which is found with alkene additions, orientation is predetermined, and stereochemistry is irrelevant. Acid and base catalysis, however, can be complicated: bases may activate the nucleophile; acids can activate the carbonyl group but may consume the nucleophile. Many of the initial carbonyl adducts undergo further reaction, and examples of this behaviour are discussed in later sections. First we shall consider some typical examples of the simple addition process.

## 11.1 Simple addition

One of the earliest of any organic reaction mechanisms to be investigated [Lapworth: 03J(83)995; 04J(85)1206] was the formation of cyanohydrins, the addition of HCN. The reaction rate is proportional to the concentrations of aldehyde or ketone and of cyanide ion. It is subject to specific base catalysis in that pH controls the dissociation of the weakly acidic HCN; acid catalysis is negligible. Lapworth concluded that the reaction hinged on a rate-limiting nucleophilic attack by cyanide ion on the carbonyl group:

$$\text{HCN} \underset{-H^+}{\rightleftharpoons} \text{CN}^-$$

$$\text{>C=O} \overset{\text{Slow}}{\rightleftharpoons} \text{>C} \overset{\text{CN}}{\underset{\text{O}^-}{<}} \overset{H^+}{\rightleftharpoons} \text{>C} \overset{\text{CN}}{\underset{\text{OH}}{<}}$$

*245*

The absence of acid catalysis indicates that protonation of the carbonyl group occurs after the rate-limiting step. This is confirmed by the kinetics of the reverse reaction, which is also subject to specific base catalysis: there is no detectable contribution from a pH-independent term (i.e.: rate $= k[>C(OH)CN]$), showing that the hydroxylic proton must be completely removed before the cyanide is expelled [78JA6119].

Another widely studied reaction is hydration [66APO1]. This is subject both to general acid and general base catalysis. Acids accelerate the reaction by protonating the carbonyl oxygen, bases by abstracting a proton from water. The fact that the catalysis is general rather than specific shows that hydrogen bonding provides sufficient activation and proton transfer occurs during the rate-limiting step.

General acid catalysis          General base catalysis

Both of these reactions, cyanohydrin formation and hydration, are readily reversible. The equilibria are markedly dependent on the nature of the groups attached to the carbonyl carbon atom of the aldehyde or ketone. Formaldehyde is almost completely hydrated in dilute aqueous solution, acetaldehyde about 60 per cent, and acetone scarcely at all (though the catalysed reaction is still fairly fast, as shown by $^{18}$O exchange). The effect is partly electronic, partly steric. Electron donors stabilise the partial positive charge on the carbonyl carbon atom; bulky groups destabilise the tetrahedral diol: both effects discourage addition. There is quite a good correlation with Taft's polar and steric parameters (§3.3):

$$\log(K/K_0) = 2.6\,\Sigma\sigma^* + 1.3\,\Sigma E_s$$

where the summations incorporate both R groups in RR'CO, and where the reference equilibrium is for acetone: $K_0 = [Me_2C(OH)_2]/[Me_2CO]$.

The formation of cyanohydrins of *meta* and *para* substituted benzaldehydes, where the steric effect is constant, is correlated by a modified Yukawa–Tsuno equation (§3.1.3), demonstrating the importance of through-conjugation between the carbonyl group and electron-donating substituents such as *p*-methoxy [79JA4678].

Hemiacetal formation (addition of ROH) is closely analogous to hydration. A widely studied variant of this reaction is the mutarotation of glucose or of its 2,3,4,6-tetra-*O*-methyl derivative, which involves equilibration of the epimeric hemiacetals via the open chain hydroxyaldehyde.

The reaction is subject to both general acid and general base catalysis and indeed was one of the earliest systems in which general catalysis was studied. In benzene solution containing mixed pyridine/phenol catalysts the reaction is third order, depending on the concentrations of substrate and of each catalyst. This was originally interpreted in terms of a termolecular transition state, but more likely it is due to conventional bimolecular catalysis by one component of the ion-pair $pyH^+ PhO^-$, since quaternary ammonium phenoxides, $R_4N^+ PhO^-$, are also effective catalysts even though they contain no acidic function. 2-Pyridone (possibly reacting via its minor tautomer, 2-hydroxypyridine) is a much more effective catalyst than the pyridine/phenol mixtures, and the kinetics are second order. This is probably a genuine example of simultaneous acid and base catalysis (figure 11.1) [83JA4767]. Carboxylic acids are more

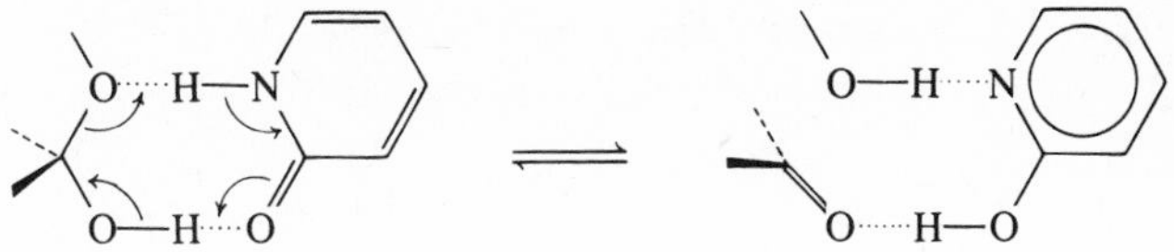

Figure 11.1. Bifunctional catalysis by 2-pyridone.

effective than phenols of the same acid strength, probably for a similar reason. Catalysts that combine acid and base properties in this way are called *bifunctional*; bifunctional catalysis may be an important feature of enzyme action [see 69JA6090, which also contains references to early work in the field, and 78Accl for studies of bifunctional catalysis in hydrogen exchange reactions].

## 11.2 Addition followed by substitution: acetals

The majority of acyclic hemiacetals, like hydrates, are stable only in solution. Further reaction with a second molecule of alcohol, however, leads to the formation of the more stable acetal (**11.1**) in which the hydroxy group has been displaced by alkoxy.

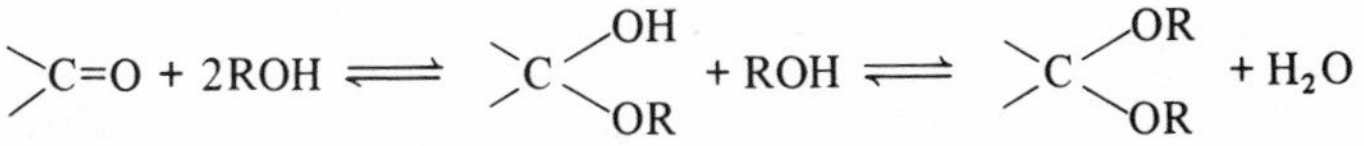

(**11.1**)

The entropy term (three molecules on the left of the equilibrium, two on the right) biases the equilibrium to the left for all but small unbranched aldehydes. Most mechanistic studies have therefore been carried out on the reverse reaction, acetal hydrolysis [74CR581].

### 11.2.1 *Acetal to hemiacetal: specific acid catalysis*

The hydrolysis of alkyl acetals (**11.1**; R = alkyl) is subject to specific acid catalysis; it is not catalysed at all by bases, and general acid catalysis is usually small or absent. This indicates that the hydrolysis is initiated by rapid protonation of the acetal. The next step must be substitution by nucleophilic water to form the hemiacetal. One could envisage four possible modes of reaction: unimolecular or bimolecular, and with attack taking place either on the alkoxy carbon atom or on that of the incipient carbonyl group. We may conveniently modify the terminology used in ester hydrolyses (chapter 12) to describe these mechanisms (figure 11.2): '*A*' denotes acid hydrolysis: '1' or '2' indicates molecularity; the subscript 'AL' or 'PC' describes the position of attack – alkyl or pro-carbonyl.

For all simple acetals the evidence points overwhelmingly to the one mechanism, $A_{PC}1$ [74CR581]:

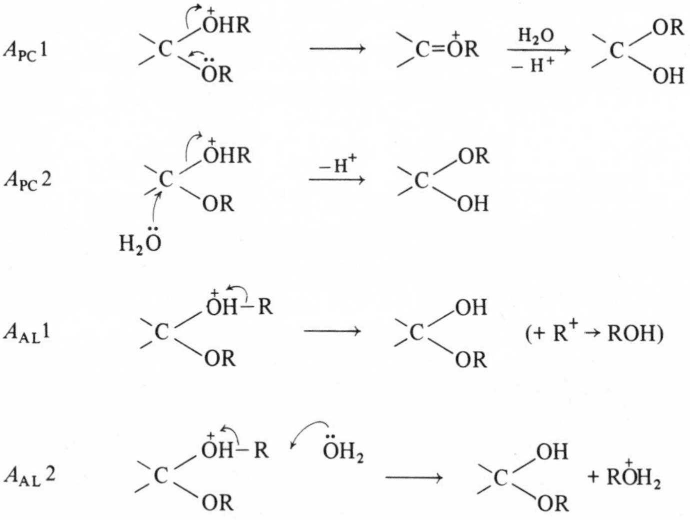

Figure 11.2. Potential modes of hydrolysis of acetals to hemiacetals.

(*a*) *Which bond is broken?* Acetals are hydrolysed in $^{18}O$-labelled water to give unlabelled ROH; conversely $^{18}O$-labelled aldehydes react with alcohols to give unlabelled acetals and $H_2{}^{18}O$. Both results show that the integrity of the R—O bond is maintained. The hydrolysis of acetals bearing chiral alkoxy groups invariably gives alcohols in which the stereochemical configuration is retained; the $A_{AL}2$-mechanism would lead to Walden inversion, and $A_{AL}1$, via $R^+$, to racemisation. The methanolysis of acetals, which presumably follows an analogous mechanism, gives the $O$-methyl acetal and ROH; the AL mechanisms would give the hemiacetal and ROMe.

(*b*) *Unimolecular or bimolecular?* The absence of base catalysis provides an initial clue in favour of the unimolecular mechanism; one would have thought that if $H_2O$ could directly displace ROH ($A_{PC}2$), then $HO^-$ could similarly displace $RO^-$. In concentrated acid solutions the hydrolysis rates relate to $H_0$ with negative Bunnett $\phi$ values (§4.3.4), characteristic of reactions in which water does not participate in the rate-limiting step.

The effects of substituent groups attached to the pro-carbonyl carbon atom point strongly to the unimolecular mechanism, the reaction being markedly accelerated by electron donation. Thus the hydrolysis rates of substituted benzaldehyde acetals correlate with $\sigma$ (or in some cases rather better with the Yukawa–Tsuno parameter: $\sigma + 0.5(\sigma^+ - \sigma)$; compare equation [3.4]). Large negative $\rho$ values, typically $-3.3$, indicate that there is substantial development of positive charge in the transition state. For aliphatic systems the order of reactivity is ketal > acetal > formal; correlation with Taft's $\sigma^*$ with large negative values of $\rho^*$ again suggests the $A_{PC}1$ mechanism. Electron donation from the non-departing alkoxy group also accelerates hydrolysis.

The entropies of activation for acetal hydrolyses are small but positive (around $+30\ J\ K^{-1}\ mol^{-1}$), typical of a dissociative transition state. The A2 mechanisms, with an extra water molecule bound into the transition state, would be expected to have large negative values of $\Delta S^{\ddagger}$.

The validity of these criteria is supported by the discovery of an exception: positive $\phi$ values, correlation with $\sigma$ ($\rho = -2.0$), and $\Delta S^{\ddagger} = -60\ J\ K^{-1}\ mol^{-1}$ are found for the hydrolyses of the cyclic acetals **(11.2)**, which are a thousand times slower than those of the unmethylated analogues. The $A1$ mechanism is inhibited because the methyl groups, which are fairly well staggered in the parent acetal, would become eclipsed in the transition state as the incipient carbon–oxygen double bond modifies the bond angles in the ring. The $A_{PC}2$ route is preferred. This can be catalysed by proton abstraction from the nucleophilic water molecule and general rather than specific acid catalysis is observed (general base-specific acid: §4.2.2) [67JA3228]:

(11.2)

## 11.2.2 Acetal to hemiacetal: general acid catalysis

If the oxocarbenium ion intermediate is more than usually stabilised it may be formed without complete prior protonation of the parent acetal, and acid catalysis becomes general rather than specific. The hydrolysis of the diethyl ketal of tropone (via the delocalised ethoxytropylium cation) is an example of this behaviour [78JA7027], and it is also the usual (but not invariable [79JA2669]) mechanism for the hydrolysis of orthoesters, the trioxy analogues of acetals (§4.2.2):

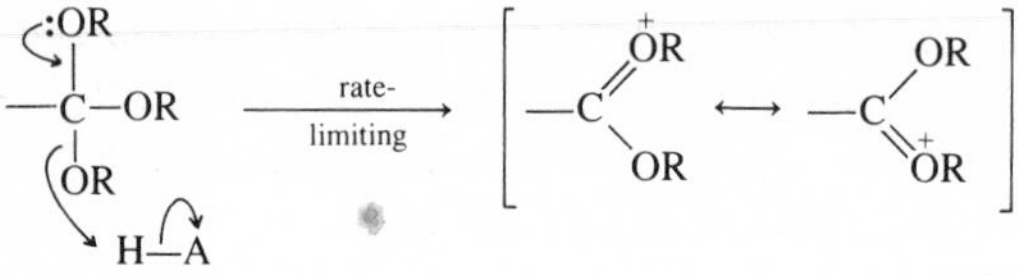

Alternatively, if the leaving group R bears strongly electron-withdrawing substituents this not only facilitates cleavage of the C—OR bond but it also discourages protonation because it reduces the basicity of the acetal. Again proton transfer is concerted with C—O bond cleavage and general acid catalysis is observed as in the following example [68JA4081]:

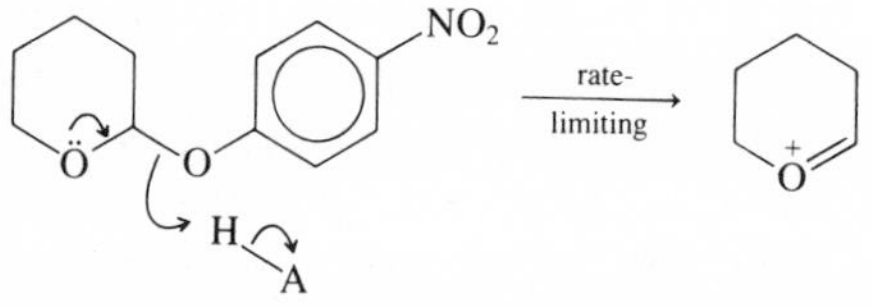

Indeed, so marked is the effect of the *p*-nitrophenoxy leaving group that in solutions with pH greater than 4 the acid catalysed route is overtaken by the unimolecular dissociation of the unprotonated acetal [70JA1681]:

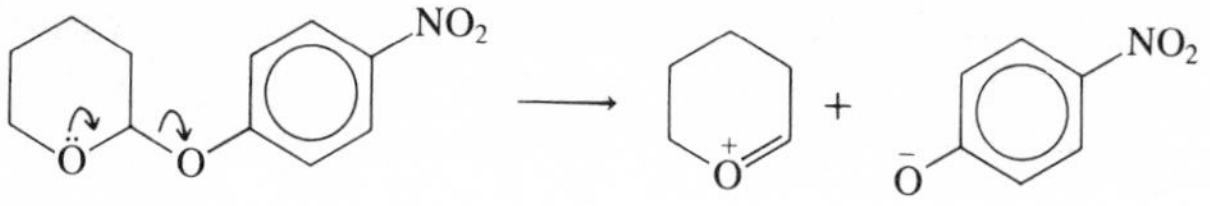

In benzaldehyde methyl phenyl acetal moderate stabilisation both of the oxocarbenium ion (PhCH=OMe$^+$) and of the nucleofuge (PhO$^-$) combine, and again the hydrolysis follows the concerted general acid catalysed route [75J(P2)1113; compare also 78JA7037; 79JA4672]:

$$\underset{\text{OAr}'}{\overset{\text{OMe}}{\text{Ar}-\text{CH}}} \underset{\text{HA}}{\rightleftharpoons} \left[\begin{array}{c} \overset{\delta+}{\underset{\underset{\underset{A^{\delta-}}{|}}{\overset{|}{H}}}{\overset{\text{OMe}}{\text{Ar}-\text{CH}}}} \\ O\!\!-\!\!^{\text{Ar}'} \end{array}\right]^{\ddagger} \rightleftharpoons \underset{+\text{ Ar}'\text{OH} + \text{A}^-}{\text{ArCH}=\overset{+}{\text{O}}\text{Me}} \underset{\text{H}_2\text{O}}{\rightleftharpoons} \underset{\text{OH}}{\overset{\text{OMe}}{\text{Ar}-\text{CH}}}$$

In this system the reaction is accelerated by electron donation in the pro-carbonyl substituent Ar (correlation intermediate between $\sigma$ and $\sigma^+$), and by increasing acid strength in the catalyst HA. Substituent changes in the aryloxy group, Ar'O, generate a curved Hammett plot; both electron donors and electron acceptors accelerate the reaction. The course of the reaction can be described in terms of the degree of C–O bond breaking and the extent of proton transfer from the catalyst to the acetal. At one extreme one could envisage a reaction in which the acetal is fully protonated before C–O cleavage begins; this would correspond to specific acid catalysis:

$$\underset{\text{OAr}'}{\overset{\text{OMe}}{\text{Ar}-\text{CH}}} \underset{\text{H}^+}{\rightleftharpoons} \underset{\overset{+}{\text{O}}\text{HAr}'}{\overset{\text{OMe}}{\text{Ar}-\text{CH}}} \rightleftharpoons \text{ArCH}=\overset{+}{\text{O}}\text{Me} + \text{Ar}'\text{OH}$$

At the other extreme the aryloxide could be expelled in an uncatalysed step and then protonated to the phenol:

$$\underset{\text{OAr}'}{\overset{\text{OMe}}{\text{Ar}-\text{CH}}} \rightleftharpoons \text{ArCH}=\overset{+}{\text{O}}\text{Me} + \text{Ar}'\text{O}^- \underset{\text{H}^+}{\rightleftharpoons} \text{Ar}'\text{OH}$$

Figure 11.3 shows schematically the potential energy surface for an intermediate reaction in which C–O breaking and O–H making keep in step. We can use the map of figure 11.3(b) to see how modifying the reaction will change its course. The top half of figure 11.3(b) relates to reactions in which C–O cleavage is ahead of proton transfer in the transition state and the leaving aryloxy group bears a partial negative charge; in the bottom half it is proton transfer that is ahead and the aryloxy group bears a partial positive charge. Note that lowering the energy in a direction perpendicular to the reaction coordinate pulls the transition state that way, but lowering it in a direction parallel to the

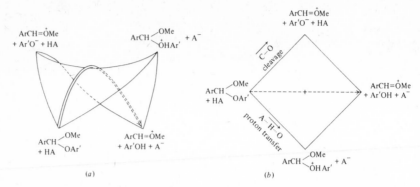

Figure 11.3. Representation of potential energy surface for general acid
catalysed acetal hydrolysis: (*a*) in perspective; (*b*) in plan. The diagrams
illustrate a process in which C—O cleavage and proton transfer keep in step
with each other; the central transition state has the C—O bond half broken
and the proton midway between the catalyst and the aryloxy oxygen atom
[*cf.* 72CR705].

reaction coordinate pushes the transition state away (compare the discussion of the *E*2 transition state, §8.1.3).*

Thus electron donation in the pro-carbonyl substituent, Ar, by stabilising the intermediate $ArCH=O^+Me$, moves a central transition state
towards the top corner of figure 11.3(*b*) and away from the right hand
corner. The result is to reduce the extent of proton transfer at the
transition state, as indicated in figure 11.4(*a*). This is reflected in the
Brønsted α values, which fall from 1.0 with *m*-nitrobenzaldehyde methyl
phenyl acetal to 0.7 with the *p*-methoxy analogue.

The effect of increasing the strength of the acid catalyst (lowering the
energy of $A^-$) is to move the transition state down and to the left (figure
11.4(*b*)). There is less C—O cleavage at the transition state and therefore
less positive charge on the developing carbon–oxygen double bond.
Again the prediction is borne out: the Hammett ρ value (calculated from
*meta* substituted benzaldehyde methyl phenyl acetals) falls from $-3.0$
with pivalic acid ($pK_a = 5.0$) to $-1.8$ with chloroacetic (2.9).

Substituent effects in the aryloxy group, Ar′, are a little more complicated. Electron withdrawal stabilises the anion Ar′O⁻, improving its
leaving group ability and placing less demand on the catalyst (figure
11.4(*c*)). Brønsted α values fall from 0.7 with benzaldehyde methyl phenyl
acetal to 0.5 for the *m*-nitrophenol derivative. The partial negative charge
on the aryloxy oxygen atom is dispersed by the substituent, the energy of

* Diagrams of this type have been used widely by More O'Ferrall to discuss
elimination reactions and by Jencks to discuss reactions of carbonyl
compounds, and are commonly called More O'Ferrall–Jencks diagrams.

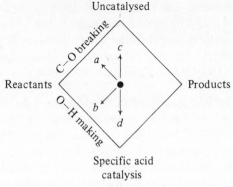

Figure 11.4. Variable transition state for general acid catalysed acetal hydrolysis (*cf.* figure 11.3). The arrows show the effect on a central transition state of the following changes: (*a*) electron donation in the pro-carbonyl substituent; (*b*) increase in the strength of the acid catalyst; (*c*) electron withdrawal, and (*d*) electron donation in the leaving group.

the transition state is thereby lowered, and the reaction is accelerated. Electron donation in Ar′ facilitates protonation of the acetal (figure 11.4(*d*)) and the value of α rises to 1.0 for the *p*-methoxyphenyl acetal. But here too the transition state energy falls because the aryloxy group now bears a partial positive charge which is stabilised by the substituent, and again the reaction is accelerated. It is as if the whole potential energy surface of figure 11.3(*a*) were balanced on the two lower corners: reducing the energy of Ar′O$^-$ moves everything down on that side; reducing the energy of ArCH(OMe)O$^+$HAr′ tips the see-saw the other way.

Extrapolating this trend from the aryloxy leaving groups to the much less stable alkoxy can be regarded as pushing the reaction coordinate right down to the lower perimeter of figure 11.4, that is, to the two-step specific acid catalysed reaction; the bottom corner of the map now represents an intermediate rather than a transition state.

When dilute hydrochloric acid (H$_3$O$^+$) is used as the catalyst instead of a carboxylic acid the pattern of aryloxy substituent effects changes; electron donation still accelerates the hydrolysis, but electron withdrawal retards it. This suggests that the strong acid has pulled the transition state so far down into the lower half of the diagram that the see-saw is fixed in one direction with the aryloxy group always bearing a positive charge.

### 11.2.3   *Hydrolysis of the hemiacetal*

The final stage in acetal hydrolysis is the hydrolysis of the intermediate hemiacetal. This is presumably the reverse of acid catalysed addition. It is generally much faster than the initial cleavage of the acetal. If it were not, there would be a build-up in the concentration of the hemiacetal during the reaction. The hydrolysis of dimethyl acetals with $D_2O$ in the presence of $CD_3OD$ reveals no sign of this. The disappearance of the acetal methoxy signal in the n.m.r. spectrum is regular, giving no sign of two different rate processes, and it is precisely matched by the appearance of the $CH_3OD$ peak, with no indication of a separate hemiacetal signal [65JA3173].

In fact there have been few studies of the hydrolysis of hemiacetals, presumably because they are hard to come by. However, the initial hydrolysis of the acylal, PhCH(OMe)OAc, is very fast, the acyloxy group being a much better leaving group than alkoxy or aryloxy, and in the overall reaction it is the hemiacetal hydrolysis which is rate-limiting [76CC871]. The n.m.r. spectra taken as the hydrolysis proceeds show the rise and subsequent decay of a signal for the hemiacetal CH proton ($\delta = 5.44$; compare $\delta = 6.48$ for the corresponding proton in the acylal and $\delta = 5.35$ for that in PhCH(OMe)$_2$). Moreover, the hydrolysis of the chloroacetyl analogue occurs at the same rate, showing that the acyl group has been expelled prior to the rate-limiting step.

This reaction is subject to general acid catalysis and, unlike most acetal hydrolyses, to general base catalysis as well. This suggests that the reason for the rapid hydrolysis of the hemiacetal intermediates during the normal acid catalysed acetal hydrolyses is the intervention by solvent or by the anion of the acid catalyst.

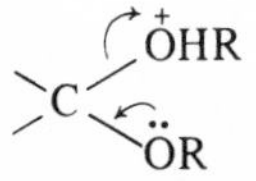

Acetal cleavage 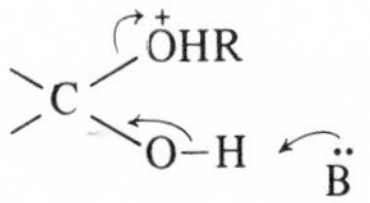          Hemiacetal cleavage

In less acidic solutions hydrolysis of the unprotonated hemiacetal can occur, also with general base catalysis. For formaldehyde hemiacetals, $HOCH_2OR$, and the corresponding deuterioformaldehyde derivatives there is a large secondary kinetic isotope effect ($k_H/k_D = 1.23$ for R = Et) pointing to substantial C–O double bond formation, and consequent $CH_2$ rehybridisation, in the transition state, in accord with general acid-specific base catalysis [80JA6472]:

## 11.3 Addition followed by elimination: imines

Primary amino compounds react with aldehydes and ketones to produce imine derivatives. The formation of *N*-aryl and *N*-alkyl imines, oximes, semicarbazones and phenylhydrazones have all been widely studied. The mechanism involves addition followed by elimination [68MI*c*].

$$RNH_2 + R_2'CO \underset{k_{-1}}{\overset{k_1}{\rightleftharpoons}} R_2'C\!\!\begin{array}{c}\diagup OH\\[-2pt]\diagdown NHR\end{array} \overset{k_2}{\rightleftharpoons} R_2'C=NR + H_2O$$

### 11.3.1   pH/rate profile: the rate-limiting step

The reaction rates vary with pH in a characteristic manner, giving a bell-shaped rate profile (figure 11.5). This is a consequence of a change in the rate-limiting step.

The dehydration step (2) is normally acid catalysed (see below) and at

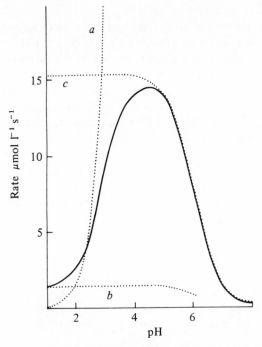

Figure 11.5. pH/rate profile for reaction of 0.0005M acetone with 0.0167M (total concentration) hydroxylamine [59JA475; 64PPO63]. The solid curve is fitted to the observed reaction rates. The dotted lines are calculated profiles for the separate steps. (*a*) The uncatalysed addition: rate = $k_1[Me_2CO][NH_2OH]$; the rate falls off with increasing acidity as the hydroxylamine is protonated. (*b*) The acid catalysed contribution to the addition: rate = $k_1'[Me_2CO][NH_2OH]$ $[H_3O^+]$. (*c*) The dehydration step, also acid catalysed: rate = $K_1 k_2[Me_2CO]$ $[NH_2OH][H_3O^+]$. In curves (*b*) and (*c*) the horizontal portions arise when the hydroxylamine is largely protonated and $[NH_2OH] \times [H_3O^+]$ is effectively constant.

low pH it is considerably faster than step 1. The rate equation is then simply:

$$\text{Rate} = k_1[\text{RNH}_2][\text{R}_2'\text{CO}]$$

The concentration of free base falls with increasing acidity as the amine becomes progressively more protonated, so that the reaction is retarded.

At lower acid concentrations the catalysis of dehydration is less effective and step 2 becomes rate-limiting. There is a region of complex kinetics with an initial build-up of the intermediate aminohydrin and a somewhat slower decay (compare figure 2.3), but at higher pH, when $k_2 \ll k_{-1}$, the first step effectively reaches equilibrium $(k_1/k_{-1} = K_1)$ and the rate equation simplifies to:

$$\text{Rate} = K_1 k_2[\text{RNH}_2][\text{R}_2'\text{CO}][\text{H}_3\text{O}^+]$$

The term $[\text{H}_3\text{O}^+]$ arises from acid catalysis, but this is actually an over-simplification as discussed in the following section.

Most early studies in the field used ultraviolet spectroscopy to follow the reaction. The aminohydrin has a transparent chromophore, but its presence could be deduced from the initial rapid decline in the carbonyl absorbance and the slower appearance of that of the imine. Recently the use of flow techniques with n.m.r. spectroscopy has allowed direct observation to be made of the aminohydrin formed during the reaction of acetone with hydroxylamine [76JA1573, 7371]. The reagents are mixed in a flowing stream within the n.m.r. tube so that at the point of observation the reaction maintains a steady state. Different stages of the reaction can be studied by varying the flow rate or by stopping the flow altogether and monitoring the subsequent change in intensity of a particular signal. In this way it has been possible to measure directly the rates of formation and dehydration of the aminohydrin.

### 11.3.2 Catalysis

(a) *The addition step* [74JA7986, 7998]. The initial attack of an amino compound on the carbonyl group produces a high-energy zwitterionic intermediate that is converted to the aminohydrin by two proton transfers (which may occur separately or, as shown below, in a concerted 'proton switch' via one or two hydrogen-bonded water molecules):

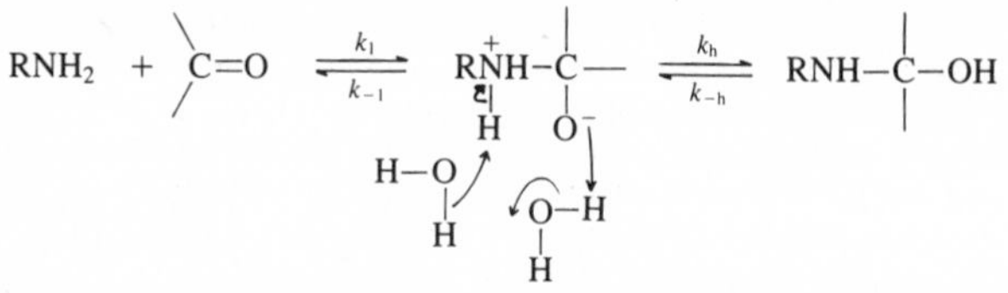

The instability of the zwitterions can be appreciated by considering the reverse of the proton transfer steps: the weakly basic nitrogen atom of the

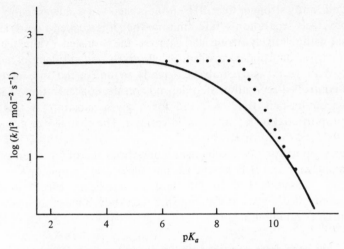

Figure 11.6. Brønsted plot for the reaction of $MeONH_2$ and $p$-$MeOC_6H_4CHO$; $k$ is the catalytic coefficient for general acid catalysts of strength $pK_a$. The dotted lines have slopes of 0 and 1, and their intersection occurs around $pK_a$ 8–9, close to the estimated $pK_a$ value for the hydroxyl group of $p$-$MeOC_6H_4CH(OH)NH_2^+OMe$ [74JA7986; see also 78JA5954].

aminohydrin has to become protonated and the weakly acidic hydroxyl ionised. A consequence is that their unimolecular dissociation back to reactants can be extremely fast: $k_{-1}$ for the expulsion of methoxyamine from $MeONH_2^+–CHAr–O^-$ is of the order of $10^8\,s^{-1}$. This is actually faster than the proton transfers ($k_h \simeq 10^7\,s^{-1}$) leading to the unusual observation of rate limitation by proton transfer between two electronegative elements [81PAC189; 73Acc41]. The proton transfers can be accelerated by the addition of acids so that the reaction is subject to general acid catalysis:

$$RNH_2^+\!-\!\overset{|}{\underset{|}{C}}\!-\!O^- \;\overset{HA}{\rightleftharpoons}\; RNH_2^+\!-\!\overset{|}{\underset{|}{C}}\!-\!OH \;\overset{-H^+}{\rightleftharpoons}\; RNH\!-\!\overset{|}{\underset{|}{C}}\!-\!OH$$

Brønsted plots are non-linear (figure 11.6). This is in accord with the so-called Eigen mechanism of proton transfer [64AG1] which visualises a three-stage process: a diffusion-controlled encounter between acid and base, the actual proton transfer occurring within the encounter-pair, and then a diffusion-controlled separation:

$$B + HA \underset{-e}{\overset{e}{\rightleftharpoons}} B\cdot HA \underset{-t}{\overset{t}{\rightleftharpoons}} BH^+\cdot A^- \underset{-s}{\overset{s}{\rightleftharpoons}} BH^+ + A^-$$

Only when $BH^+$ and $HA$ are of comparable strength ($\Delta pK_a \simeq 0$) can the middle step be the slowest of the three: encounter is rate-limiting when

HA is significantly stronger than $BH^+$ (that is, step $-e$ is slower than $t$ or $s$) and conversely separation is rate-limiting when it is weaker. Figure 11.6 shows that with relatively strong acid catalysts the Brønsted $\alpha$ value tends to zero: step $e$, diffusion-controlled, is effectively independent of acid strength. With weak acid catalysts $\alpha$ tends to unity: the rate of the diffusion-controlled separation step depends on the concentration of the second encounter-pair $BH^+ \cdot A^-$ and for a given substrate B this is linearly proportional to $K_a$ for the acid catalyst. The changeover occurs, as predicted, at around $\Delta pK_a = 0$.

Further support for this mechanism comes from the observation of a sharp maximum in the H/D kinetic isotope effect, also around $\Delta pK_a = 0$ [78JA5954]. Changing HA to DA has no significant effect on the diffusion-controlled steps, but it diminishes the catalysis when the hydron transfer itself is rate-limiting.

The catalytic behaviour described above depends markedly on the base strength of the amino compound. For bases stronger than methoxyamine (for which $pK_{BH^+} = 4.7$) the zwitterionic intermediate is less unstable, its dissociation to reactants is less rapid, and its formation more readily becomes rate-limiting. Thus with hydroxylamine ($pK_{BH^+} = 6.0$), and especially the aliphatic amines ($pK_{BH^+} \simeq 10$) the addition is effectively uncatalysed: more strictly, catalysis by water overwhelms any other. The main reaction path involves nucleophilic attack on the unactivated carbonyl group – or at least, activated only by hydrogen bonding from the solvent – followed by rapid proton exchange.

Uncatalysed route

$$RNH_2 + \;{>}C{=}O \; \underset{}{\overset{k_1}{\rightleftharpoons}} \; R\overset{+}{N}H_2 - \overset{|}{\underset{|}{C}} - O^-$$

$$H^+ \Big\updownarrow \qquad\qquad\qquad \Big\updownarrow H^+$$

$$RNH_2 + \;{>}C{=}OH^+ \; \underset{k_1'}{\overset{}{\rightleftharpoons}} \; R\overset{+}{N}H_2 - \overset{|}{\underset{|}{C}} - OH \; \overset{-H^+}{\rightleftharpoons} \; RNH - \overset{|}{\underset{|}{C}} - OH$$

Catalysed route

For acetoxime formation at 25 °C, $k_1 = 2.3 \times 10^3 \, l \, mol^{-1} \, s^{-1}$ and the catalytic coefficient $k_1' = 1.7 \times 10^5 \, l^2 \, mol^{-2} \, s^{-1}$, so that the catalysed route offers little competition until $[H^+]$ exceeds 0.01, that is until the pH is less than 2 (figure 11.5(a), (b)).

(b)*The dehydration step.* In acidic solutions and up to around pH = 9 the dehydration step is acid catalysed (figure 11.5(c)). Aminohydrins derived from the stronger bases exhibit general acid catalysis, though high values of the Brønsted α coefficient (0.7 for aliphatic amines) show there is considerable proton transfer in the transition state:

$$RNH-\overset{|}{\underset{|}{C}}-OH + HA \; \rightleftharpoons \; \left[ RNH - \overset{/}{\underset{\backslash}{C}} \cdots OH \cdots H \cdots A \right]^{\ddagger} \; \rightleftharpoons \; R\overset{+}{N}H{=}C{\overset{\diagup}{\diagdown}} + H_2O + A^-$$

Weaker bases require complete protonation of the hydroxyl group and specific acid catalysis sets in, with α values rising steadily to 1.0 for aniline and semicarbazide derivatives:

$$RNH-\overset{|}{\underset{|}{C}}-OH \; \rightleftharpoons \; R\overset{..}{N}H-\overset{|}{\underset{|}{C}}-\overset{+}{O}H_2 \; \overset{r.l.s.}{\rightleftharpoons} \; R\overset{+}{N}H{=}C{\overset{\diagup}{\diagdown}} + H_2O$$

The dehydration of $NH_2CSNHNH-CH_2-OH$ to give formaldehyde thiosemicarbazone at around pH = 7 is near the border between specific and general acid catalysis. It is clear that free protonated aminohydrin ions, $RNH-CH_2-OH_2^+$, do exist under the conditions of the reaction. However, their spontaneous dehydration (specific acid catalysis) appears to involve a transition state of higher energy than that for concerted general acid catalysis; this is revealed by a Brønsted correlation (α = 0.83) and by a substantial solvent kinetic isotope effect ($k_H/k_D$ = 2.6) [80JA6466].

In more alkaline solutions, as acid catalysis becomes less effective, different patterns of catalysis emerge. Strong base aminohydrins can dissociate without catalytic assistance:

$$R\overset{..}{N}H-\overset{|}{\underset{|}{C}}-OH \; \rightleftharpoons \; R\overset{+}{N}H{=}C{\overset{\diagup}{\diagdown}} + OH^-$$

The reaction has been studied in the reverse direction for the hydrolyses of various benzaldimines. For N-t-butylimines the rates of hydrolysis are constant over the range pH = 9 to pH = 14, and substituent effects show clearly that the absence of catalysis is due to the attack of hydroxide ion on the protonated imine and not to the attack of water on the free base [63JA2843].

With less strongly basic aminohydrins base catalysis sets in when pH ⩾ 9, and the pH rate profile bends up again. It is not known if this is general or specific catalysis, though a concerted general base reaction would avoid the intermediacy of a nitrogen anion:

$$HO^- \overset{\curvearrowright}{\phantom{}} H \qquad H_2O +$$

$$RN\overset{|}{\underset{|}{\overset{\curvearrowright}{C}}}\overset{|}{-}\!\overset{\curvearrowleft}{OH} \;\rightleftharpoons\; RN{=}C{\Big\langle} \; + HO^-$$

Again the reverse reaction is corroborative. *N*-Arylbenzaldimines hydrolyse at constant rate from pH = 9 to pH = 12, but in very strongly alkaline solution the rate rises again with onset of attack by hydroxide ion on the unprotonated (but presumably solvated) imine.

Aniline hydrochloride exerts a much greater catalytic effect on oxime and semicarbazone formation than does acetic acid, which is of comparable strength. With anilinium catalysis both reactions occur at the same rate, independent of hydroxylamine or semicarbazide concentrations. This is an example of nucleophilic (as opposed to base) catalysis. The aniline reacts rapidly to form the protonated anil, which is attacked almost instantaneously by hydroxylamine or semicarbazide, as reactions with the preformed anil confirm [62JA826]:

$$\underset{/}{\overset{\backslash}{}}C{=}O + PhNH_2 \;\overset{H^+}{\rightleftharpoons}\; \underset{/}{\overset{\backslash}{}}C{=}\overset{+}{N}HPh \;\xrightarrow[Fast]{NH_2OH}\; \underset{/}{\overset{\backslash}{}}C{=}NOH + PhNH_3^+$$

## 11.4 The aldol reaction

The aldol reaction and its many relatives are mechanistically related to aminohydrin or imine formation, depending on whether the reaction stops at the aldol stage or continues with dehydration. However, there is an additional complication because a carbonyl compound is not only the electrophile but also the source of the nucleophile, and this introduces another potentially rate-limiting step. Aldol reactions are also accompanied by a multitude of side-reactions that make kinetic analysis the more difficult. The archetypal base catalysed aldol reaction of acetaldehyde itself is actually considerably more complex than is set out in the following paragraph. Allowance has to be made for hydration of the acetaldehyde and for partial ionisation of the resulting diol, which mean that the concentration of neither the aldehyde nor the hydroxide ion is stoicheiometric. Moreover, the product is not 'acetaldol' at all, but the trimeric 2-hydroxy-4,6-dimethyl-1,3-dioxan produced by a subsequent fast reaction [64AJ953].

Making allowances for these complications one can isolate the essential aldol reaction. Under base catalysis the nucleophile is the enolate anion:

$$HO^- + CH_3CHO \underset{k_{-1}}{\overset{k_1}{\rightleftharpoons}} (CH_2CHO)^- + H_2O$$

$$CH_3CHO + (CH_2CHO)^- \underset{k_2}{\overset{}{\rightleftharpoons}} CH_3CH\underset{O^-}{\overset{CH_2CHO}{\diagdown}} \overset{H^+ \text{ (fast)}}{\rightleftharpoons} CH_3CH(OH)CH_2CHO$$

$$\text{Rate} = \frac{k_1 k_2 [HO^-][CH_3CHO]^2}{k_{-1} + k_2 [CH_3CHO]}$$

At high concentrations of aldehyde the ionisation is rate-limiting ($k_2[CH_3CHO] \gg k_{-1}$) and the kinetics simplify to $k_1[HO^-][CH_3CHO]$. General base catalysis is observed, in accord with rate-limiting proton transfer. When the reaction is carried out in $D_2O$ and stopped part way through, the unreacted acetaldehyde contains no deuterium, confirming that reprotonation of the enolate is insignificant.

As the aldehyde concentration is reduced the kinetics become complex and the reversal of step 1 can be demonstrated by deuterium exchange. At very low concentrations, when $k_{-1} \gg k_2 [CH_3CHO]$, the second step is clearly rate-limiting, the kinetics become second order in acetaldehyde, and catalysis is specific [64AJ953; 60J2983; 58J1691].

This mechanistic changeover is normally observed only with simple aliphatic aldehydes; in most other base catalysed aldol reactions, such as ketone dimerisations or the reaction of an aromatic aldehyde with a ketone anion, the addition step is generally slower than the ionisation.

Base catalysed dehydration of the aldol is usually only a minor side-reaction with aliphatic aldehydes unless the reaction is carried out in strongly alkaline solution and at elevated temperatures. With aromatic aldehydes it is much faster and the unsaturated compound is the normal product. Presumably the elimination is *E*1cB:

$$PhCHO + CH_3COPh \underset{}{\overset{EtO^-}{\rightleftharpoons}} \underset{OH}{PhCH-CH_2COPh}$$

$$\Big\Vert EtO^-$$

$$\underset{\overset{|}{\zeta OH}}{PhCH-CH=CPh} \longrightarrow PhCH=CHCOPh$$

In this situation simple kinetics cannot distinguish between rate-limiting aldol formation and rate-limiting dehydration: the observed relation (rate $\propto [PhCHO][CH_3COPh][EtO^-]$) would accommodate either [40J1295]. Studies of the base catalysed dissociation of aromatic aldols indicate that the retro-aldol reversion to reactants is normally as fast as dehydration or faster, so it is the latter that is usually rate-limiting [e.g. 59JA628].

Under acid catalysed conditions the initial stage is enolisation, followed by rate-limiting attack of the enol on a protonated carbonyl group. Acid catalysed dehydration is relatively fast and the reaction rarely stops at the aldol stage:

$$PhCOCH_3 \;\rightleftharpoons^{H^+}\; Ph\overset{\overset{\displaystyle OH^+}{\|}}{C}-CH_3 \;\rightleftharpoons^{-H^+}\; Ph\overset{\overset{\displaystyle OH}{|}}{C}=CH_2 \qquad \text{Enolisation; fast}$$

$$Ph\overset{\overset{\displaystyle :OH}{\curvearrowleft|}}{C}=CH_2$$

$$\overset{\curvearrowright}{H}\overset{\overset{\displaystyle \|}{C}Ph}{\underset{\curvearrowleft OH^+}{}} \;\rightleftharpoons^{-H^+}\; PhCOCH_2CH(OH)Ph \qquad \text{Addition}$$

$$(H^+)\Big|-H_2O$$

$$PhCOCH=CHPh \qquad \text{Dehydration}$$

The reaction rate is proportional to $[PhCHO][CH_3COPh][H_3O^+]$; in strongly acidic solution the hydronium concentration is replaced by the acidity function $h_0$ [55JA1397].

There appear to be two alternative routes for the dehydration, either a direct elimination (probably with considerable $E1$ character) or a reaction via the enol:

$$PhCOC\overset{\overset{\displaystyle H}{|\curvearrowright}}{H}-CHPh \longrightarrow PhCOCH=CHPh \qquad \text{Direct}$$

$$\underset{\curvearrowleft OH_2^+}{}$$

$$PhCOCH_2CHPh \;\rightleftharpoons^{\pm H^+}\; \;\rightleftharpoons^{+H^+}\; PhC\overset{\overset{\displaystyle :OH}{\curvearrowleft|}}{=}CH-CHPh \qquad \text{Via enol}$$

$$\underset{OH}{|} \qquad\qquad\qquad \underset{\curvearrowleft OH_2^+}{|}$$

The logarithm of the rate coefficient for dehydration of $ArCH(OH)CH_2\text{-}COCH_3$ correlates with $H_0$ for $Ar = p\text{-}MeOC_6H_4$. For $Ar = Ph$ or $p\text{-}NO_2C_6H_4$ the plot is non-linear, but a good correlation is obtained with the rate of acetophenone halogenation, a reaction known to occur via rate-limiting enolisation. The $p$-methoxy group can stabilise an $E1$ carbenium ion intermediate, and this clearly tips the mechanistic balance towards direct dehydration [58JA5539].

The rate-limiting step for the overall condensation can again be either the addition or the dehydration. In most cases dehydration is relatively more rapid with acid catalysis than with base. Benzaldehyde and butanone can give three distinct aldols, (**11.3**) and the diastereoisomeric (**11.4**). In dilute alkali all three give a single dehydration product (**11.5**), showing that rapid equilibration via a retro-aldol reaction takes place

$$PhCH(OH)CH_2COCH_2CH_3 \longrightarrow PhCH{=}CHCOCH_2CH_3$$

(11.3)  (11.5)

$$PhCHO + CH_3COCH_2CH_3$$

$$PhCH(OH)CHCOCH_3 \longrightarrow PhCH{=}CCOCH_3$$
$$\qquad\qquad\quad | \qquad\qquad\qquad\qquad |$$
$$\qquad\qquad\quad CH_3 \qquad\qquad\qquad\quad CH_3$$

(11.4)

much faster than dehydration. In acid there is no rearrangement, so dehydration is faster than dissociation [59JA628]. In either case the dehydration step is not readily reversed, so most acid catalysed reactions and those base catalysed ones for which dehydration is reasonably fast are pulled through to the unsaturated product regardless of the position of the intermediate equilibria. For example, under conditions of mild base catalysis the dimerisation of acetone produces only a few per cent of diacetone alcohol at equilibrium. With acid catalysts high yields of mesityl oxide can be obtained.

## 11.5 Further reading

General:

W. P. Jencks, *Progr. Phys. Org. Chem.*, 1964, **2**, 63.
S. Patai (ed.), *The Chemistry of the Carbonyl Group*, Interscience, London, 1966.
J. Zabicky (ed.), *The Chemistry of the Carbonyl Group*, vol. 2, Interscience, London, 1970.
A. F. Cockerill & R. G. Harrison, in *The Chemistry of Double-Bonded Functional Groups*, ed. S. Patai, Wiley, London, 1977.

Individual reaction types:

R. P. Bell, *Adv. Phys. Org. Chem.*, 1966, **4**, 1; *The Proton in Chemistry*, 2nd edn, Chapman & Hall, London, 1973 (hydration).
E. H. Cordes, *Progr. Phys. Org. Chem.*, 1967, **4**, 1; E. H. Cordes & H. G. Bull, *Chem. Rev.*, 1974, **74**, 581; B. Capon, *Pure Appl. Chem.*, 1977, **49**, 1001 (acetal hydrolysis).
P. Y. Sollenberger & R. B. Martin, in *The Chemistry of the Amino Group*, ed. S. Patai, Interscience, London, 1968 (imine formation).
A. T. Nielsen & W. J. Houlihan, *Org. React.*, 1968, **16**, 1 (aldol).

For other carbonyl addition reactions not discussed in this chapter see the general references above and:

A. Maercker, *Org. React.*, 1965, **14**, 270; S. Trippett, *Quart. Rev.*, 1963, **17**, 406; P. A. Lowe, *Chem. & Ind.*, 1970, 1070 (Wittig reaction).

E. C. Ashby, J. Laemmle & H. M. Neumann, *Acc. Chem. Res.*, 1974, **7**, 272 (Grignard reaction).

W. P. Jencks & H. F. Gilbert, *Pure Appl. Chem.*, 1977, **49**, 1021 (addition of thiols).

# 12 The hydrolysis of carboxylic esters

Carboxylic acid derivatives have a potential leaving group attached to the carbonyl carbon atom and their characteristic mode of reaction is substitution rather than addition. If the incoming nucleophile is a hydride ion or a simple carbanion the product is an aldehyde or ketone, which may then react further by addition. On the other hand, if the nucleophile is itself a potential leaving group the substitution involves the conversion of one carboxylic derivative into another, frequently a reversible process, typified by the interconversion of acid and ester:

$$R-C\underset{O-R'}{\overset{O}{\diagup}} + H_2O \rightleftharpoons R-C\underset{OH}{\overset{O}{\diagup}} + R'OH$$

## 12.1 The Ingold classification

Ingold [69MI] has classified the mechanisms of ester hydrolysis and esterification, as shown in figure 12.1, according to the following criteria: acid or base catalysis, denoted '$A$' or '$B$'; which carbon–oxygen bond is cleaved, acyl (AC) or alkyl (AL) (figure 12.2); molecularity, 1 or 2. The acid catalysed reactions are reversible, and mechanistically ester formation is equivalent to hydrolysis. Base catalysis, however, forces the equilibrium in the direction of hydrolysis (saponification) by the effectively irreversible ionisation of the carboxylic acid. In figure 12.1 most proton transfer steps have been omitted. Note in particular that the $A_{AC}1$ mechanism shows the carboxyl group protonated on the alkoxy/hydroxy oxygen atom. The carbonyl oxygen is thermodynamically the more stable site for protonation because the positive charge can then be delocalised between both oxygen atoms. Nevertheless there will be present smaller quantities of the tautomeric cation and this is the necessary intermediate in $A_{AC}1$. The two $B1$ mechanisms (of which $B_{AC}1$ is unknown) are actually not catalysed at all, being initiated by the unimolecular dissociation of the neutral ester.

$A_{AC}2$

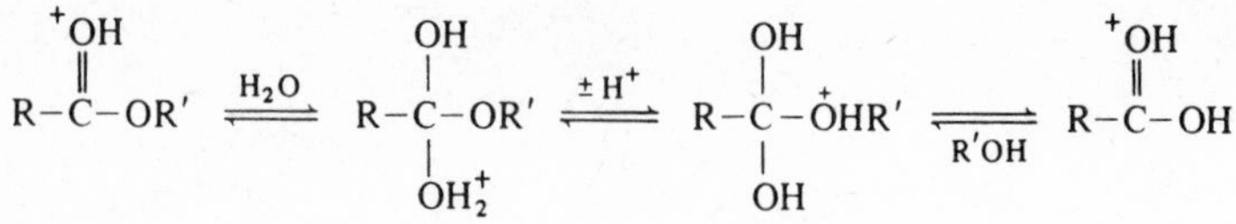

$$\underset{R-\overset{\overset{+OH}{\|}}{C}-OR'}{} \xrightarrow{H_2O} \underset{R-\overset{\overset{OH}{|}}{\underset{\overset{|}{OH_2^+}}{C}}-OR'}{} \xrightarrow{\pm H^+} \underset{R-\overset{\overset{OH}{|}}{\underset{\overset{|}{OH}}{C}}-\overset{+}{O}HR'}{} \underset{R'OH}{\rightleftharpoons} \underset{R-\overset{\overset{+OH}{\|}}{C}-OH}{}$$

$B_{AC}2$

$$\underset{R-\overset{\overset{O}{\|}}{C}-OR'}{} \xrightarrow{HO^-} \underset{R-\overset{\overset{O^-}{|}}{\underset{\overset{|}{OH}}{C}}-OR'}{} \xrightarrow{-R'O^-} RCOOH \longrightarrow RCO_2^-$$

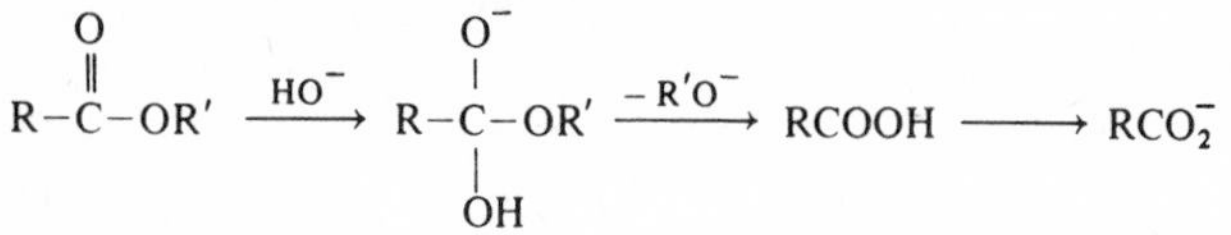

The two commonest mechanisms: nearly all saponifications are $B_{AC}2$; most esterifications and the acid catalysed hydrolyses of primary and secondary alkyl esters are $A_{AC}2$.

$A_{AC}1$ $\qquad RCO-\overset{+}{O}HR' \underset{R'OH}{\rightleftharpoons} RCO^+ \xrightarrow{H_2O} RCO-OH_2^+$

$B_{AC}1$ $\qquad RCO-OR' \xrightarrow{-R'O^-} RCO^+ \xrightarrow{HO^-} RCOOH \longrightarrow RCO_2^-$

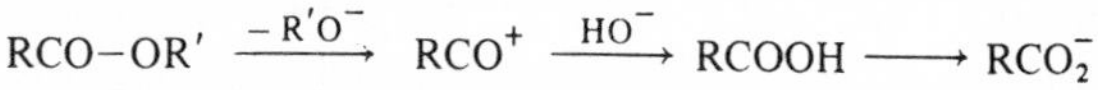

The acid catalysed mechanism is normally observed only in strongly acidic solution, especially with the esters of hindered aromatic acids. The $B_{AC}1$ mechanism is unknown.

$A_{AL}2$

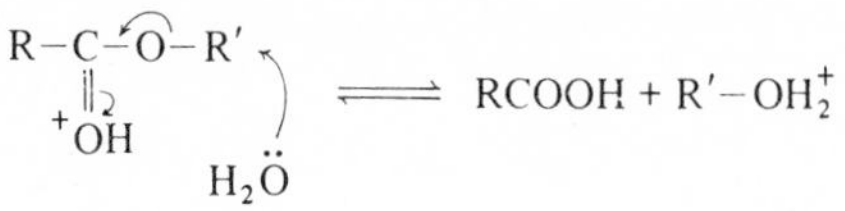

$$\left.\begin{array}{c} R-C\overset{\frown}{-O}-R' \\ \underset{+OH}{\overset{\|}{\phantom{x}}} \\ H_2\ddot{O} \end{array}\right\} \rightleftharpoons RCOOH + R'-OH_2^+$$

$B_{AL}2$

$$\left.\begin{array}{c} RCO\overset{\frown}{O}-R' \\ \\ HO^- \end{array}\right\} \longrightarrow RCO_2^- + R'OH$$

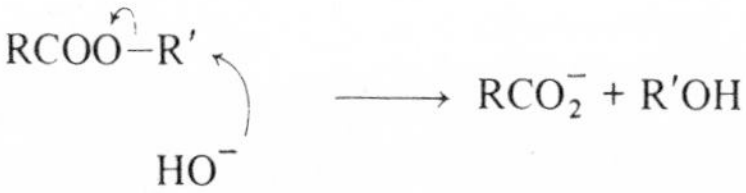

The base catalysed mechanism occurs only in the case of exceptionally severely hindered acyl groups and in the neutral hydrolysis of $\beta$-lactones. $A_{AL}2$ is unknown.

$A_{AL}1$

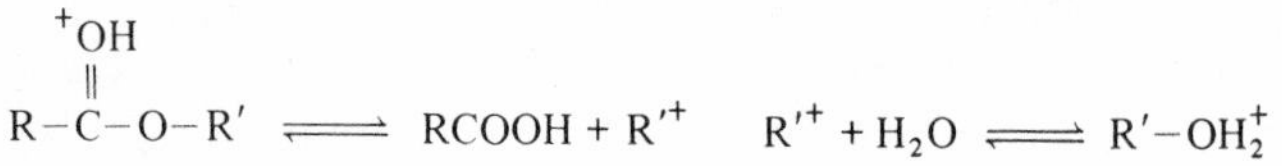

$$R-\overset{\overset{+OH}{\|}}{C}-O-R' \rightleftharpoons RCOOH + R'^+ \qquad R'^+ + H_2O \rightleftharpoons R'-OH_2^+$$

$B_{AL}1$ $\qquad RCOO-R' \longrightarrow RCO_2^- + R'^+ \xrightarrow{H_2O} RCOOH + R'OH$

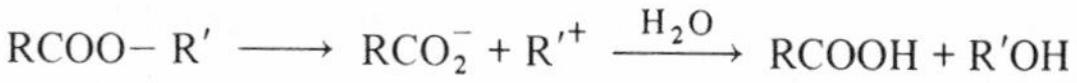

Tertiary alkyl esters in neutral or acidic solution normally react via $A_{AL}1$. $B_{AL}1$ is similar but easily overwhelmed by $B_{AC}2$ in alkaline solution.

Figure 12.1. The Ingold classification of ester hydrolyses

### 12.1.1 The $_{AC}2$ mechanisms

The normal modes of reaction for the vast majority of esters are, depending on the acidity of the solution, the $A_{AC}2$ and $B_{AC}2$ mechanisms. The essential features of the reactions are well established, although there remain some doubts about the precise role played by the catalysts.

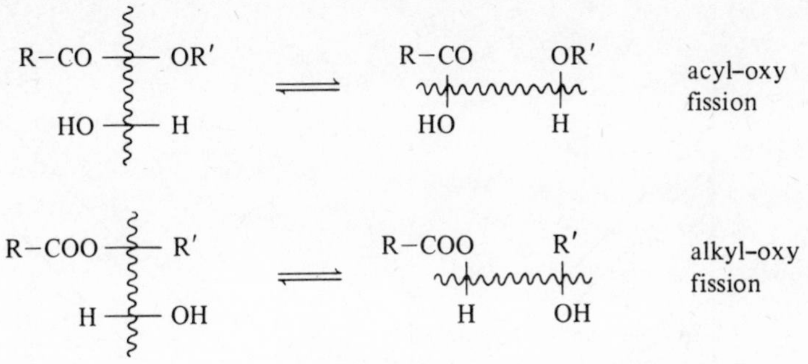

Figure 12.2. Alternative modes of carbon–oxygen bond cleavage.

Evidence for acyl-oxy cleavage is overwhelming. Hydrolysis of an ester using $H_2{}^{18}O$ leads to the labelled carboxylic acid and unlabelled alcohol. The earliest of these studies is a remarkable example of meticulous experimentation [34TF508]; the reactions were carried out on a very small scale using water enriched with only 0.35 per cent of $^{18}O$ and analysing the extent of $^{18}O$ incorporation by density measurements. The hydrolysis of chiral alkyl esters generates alcohols in which the stereochemical configuration is retained. Likewise allylic or neopentyl esters do not rearrange. Thus it is clear that the alkyl–oxygen bond remains intact during the reaction.

Esterification is first order both in carboxylic acid and in alcohol, and hydrolyses are similarly first order in ester and in hydronium or hydroxide ion (figure 12.3(a)). The evidence for a two-stage mechanism with a discrete tetrahedral intermediate comes primarily from $^{18}O$ exchange studies. The first, classic experiment was by Bender [51JA1626] who showed that during the hydrolysis of alkyl benzoates in $H_2{}^{18}O$ some of the label is incorporated into the unreacted ester. For base catalysis, for example, exchange occurs as follows:

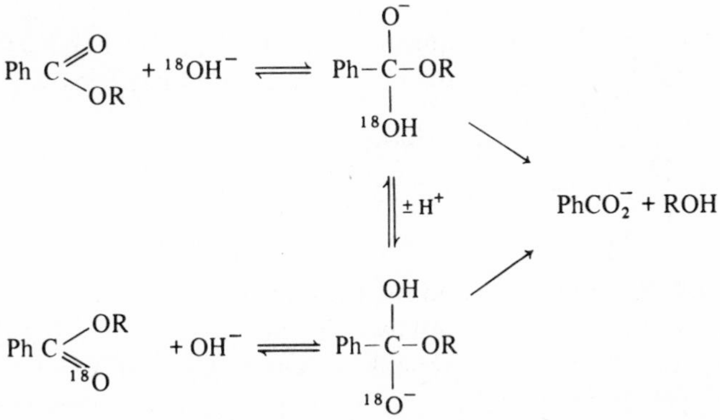

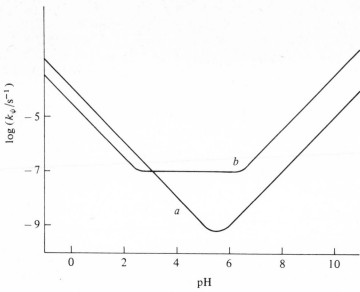

Figure 12.3. pH/rate profiles for ester hydrolyses at 25 °C: (a) ethyl acetate (*cf.* figure 4.2(b)); (b) phenyl acetate. The pseudo-first order rate coefficient $k_\psi$ is measured at any particular value of pH from the equation: rate $= k_\psi$ [ester]. Thus it incorporates the separate terms for acid and base catalysis and for the water catalysed background reaction:

$$k_\psi = k_a[H_3O^+] + k_b[HO^-] + k_w.$$

The key step is the proton transfer between the isotopically distinguished intermediates, a fast process but nevertheless one with a finite rate which could not occur within the infinitesimal confines of a transition state. Not all hydrolyses exhibit this concurrent exchange [see, for example, 68JA5848], and in any case it might be argued that the exchange process is an irrelevant side-equilibrium, running in parallel to a direct $S_N2$ type of substitution in which the carbonyl group does not participate, but an accumulation of circumstantial evidence does point towards the two-step mechanism [65APO123; 67APO237; 83Acc394]. For example, the $^{12}C/^{13}C$ and $^{16}O/^{18}O$ kinetic isotope effects shown below

$$\underset{\text{saponification}}{\overset{\displaystyle \overset{1.005}{O} \atop \displaystyle \overset{\|}{\underset{}{}} {}^{1.006}}{\underset{1.043 \qquad 1.000}{Ph-C-O-CH_3}}}$$

$$\underset{\text{hydrazinolysis}}{\overset{\displaystyle \overset{1.018}{O} \atop \displaystyle \overset{\|}{\underset{}{}} {}^{1.041}}{\underset{1.041 \qquad 1.002}{Ph-C-O-CH_3}}}$$

for the alkaline hydrolysis and the hydrazinolysis of methyl benzoate are most simply interpreted as follows. In the saponification a substantial bonding change at the carbonyl carbon and much smaller changes elsewhere suggest that the rate-limiting step is formation of the tetrahedral intermediate with a fairly early transition state. In the hydrazinolysis, on the other hand, the much larger effects at the two oxygen atoms indicate that decomposition of the intermediate is rate-limiting and that the transition state is fairly late [79JA3300]. Orthoacid derivatives with a similar general structure to that of the tetrahedral intermediates have been directly observed by generating them for highly reactive precursors – that is, under conditions where their formation is faster than their destruction – as in the following examples [81Acc306; 81JA6912]:

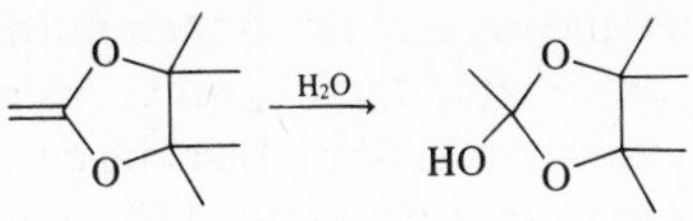

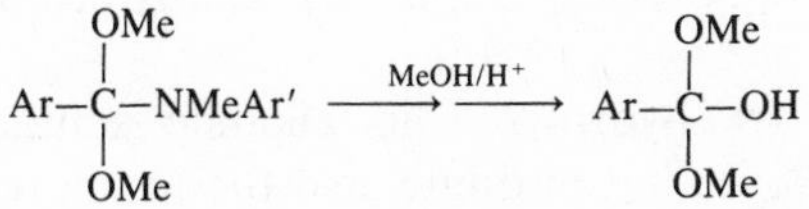

The $B_{AC}2$ hydrolysis is markedly accelerated by electron-withdrawing substituents in the carboxyl group, stabilising the negative charge in the transition state; the saponifications of substituted benzoate esters correlate with Hammett's $\sigma$ with $\rho \simeq 2.5$ (compare figure 3.1). On the other hand the $A_{AC}2$ mechanism is insensitive to substituent effects: electron donation increases the extent of protonation but retards nucleophilic attack. The difference between the two mechanisms was the basis of Taft's work on dissecting polar and steric effects (§3.3). Similarly substituent changes in the aryl group of phenyl esters are much more significant in alkaline hydrolysis (typically, $\rho = 1.2$) than in acid ($\rho \simeq -0.2$).

Acid catalysis is in general considerably less effective than base catalysis. The minimum in the pH rate profile for ethyl acetate hydrolysis is on the acid side of neutral, at about pH $= 5.5$ (figure 12.3(a)). At this point the two catalysed reactions are running at the same rate, so that $k_a[H_3O^+] = k_b[HO^-]$, where $k_a$ and $k_b$ are the catalytic coefficients. Thus $k_b/k_a = 10^{-5.5}/10^{-8.5} = 1000$. Esters with better leaving groups than ethoxide react considerably faster under base catalysis but, because of the insensitivity of the $A_{AC}2$ mechanism to polar effects, there is little difference in the rate under acid catalysis. This behaviour reveals the presence of a background water catalysed mechanism (figure 12.3(b)) which is exceptionally slow for ethyl acetate (the estimated half-life at $25\,^{\circ}$C is 75 years) but which can be quite significant with more reactive esters. These include in particular phenolic esters, because aryloxy is a better leaving group than alkoxy, and the esters of halogenoacetic and other relatively strong acids, in which the electrophilicity of the carbonyl group is enhanced. The mechanism of this background reaction is discussed later in §12.2.2.

### 12.1.2 The AL mechanisms

Hydrolyses via the AL mechanisms are simply nucleophilic substitutions at the alkyl carbon atom with a carboxylate anion ($B_{AL}$) or a carboxylic acid ($A_{AL}$) as the leaving group.

The unimolecular reactions are most commonly encountered with tertiary alkyl esters and similar substrates that can dissociate to form relatively stable carbenium ions. In aqueous acid this is their normal mode of reaction. The dissociation of a protonated tertiary alkyl ester, though rate-limiting, is extremely fast: the hydrolysis of t-butyl acetate becomes too fast to measure at acidities greater than that represented by about 20 per cent sulphuric acid, even though the ester is still only slightly protonated at that stage.

Chiral tertiary acetates are hydrolysed in alkaline solution with retention of configuration ($B_{AC}2$), but in dilute acid the resulting alcohol is largely racemised [50NA(166)680].

(ROH; $a_D = -0.63°$)

The Bunnett $\phi$ value for the hydrolysis of t-butyl acetate in dilute hydrochloric acid is $-0.2$, characteristic of a reaction in which water does not participate in the transition state (§4.3.4). When the hydrolysis is carried out with 0.01M HCl in $^{18}O$ enriched water the bulk of the label is incorporated into the t-butanol, as required for an AL mechanism, but a small proportion, decreasing with increasing temperature, appears in the acetic acid. The two mechanisms $A_{AC}2$ and $A_{AL}1$ are clearly running in parallel, and from the kinetic data and the division of the $^{18}O$ label it is possible to obtain activation parameters for the two processes [62AJ467]:

$$A_{AL}1: \Delta H^{\ddagger} = 112 \text{ kJ mol}^{-1}; \Delta S^{\ddagger} = 55 \text{ J K}^{-1} \text{ mol}^{-1}$$
$$A_{AC}2: \Delta H^{\ddagger} = 70 \text{ kJ mol}^{-1}; \Delta S^{\ddagger} = -100 \text{ J K}^{-1} \text{ mol}^{-1}$$

The positive entropy of activation for the AL mechanism is very characteristic of a dissociative mechanism. As the solvent is changed from water to aqueous dioxan with increasing dioxan concentrations, so the proportion of the AL product declines, in agreement with an ionisation mechanism.

Without labelling, of course, it is not possible to disentangle the two mechanisms. The observed activation parameters for the acid hydrolyses of virtually all tertiary alkyl esters, though varying somewhat with temperature because of the mechanistic duality, clearly point to the dominance of the $A_{AL}1$ route: $\Delta H^{\ddagger}$ is usually in the range 100 to 130 kJ mol$^{-1}$ and $\Delta S^{\ddagger}$ is 40 to 60 J K$^{-1}$ mol$^{-1}$. However, for t-butyl formate the values are 60 kJ mol$^{-1}$ and $-90$ J K$^{-1}$ mol$^{-1}$ respectively, closely similar to the values for ethyl and isopropyl formates, showing that in this one case, because of the exceptional reactivity of the formyl group, the $A_{AC}2$ mechanism takes over [60AS577].

It is less easy to study the $A_{AL}1$ hydrolyses of benzylic esters because solubility problems usually require the use of mixed solvents, which tend to suppress ionisation. Trityl (triphenylmethyl) esters ionise even in neutral solution. For benzhydryl (diphenylmethyl) esters the mechanistic balance is fairly fine, as with t-butyl acetate: the $A_{AL}1$ mechanism dominates at high temperatures, in more acidic solutions, or if electron-donating substituents are present in the benzhydryl group; the balance swings towards $A_{AC}2$ at lower temperatures, in weakly acidic solution, or at high concentrations of the non-aqueous component of the solvent mixture. Primary benzyl esters are cleaved by the $A_{AC}2$ mechanism even in fairly strong acid.

The $B_{AL}1$ mechanism (an uncatalysed reaction despite its designation) is less common. An example appears to be the hydrolysis of t-butyl trifluoroacetate in aqueous acetone for which a high value of $\Delta H^{\ddagger}$ (102 kJ mol$^{-1}$, comparable with other values for unimolecular dissociation) is observed, but for which $AS^{\ddagger} = -18$ J K$^{-1}$ mol$^{-1}$; the latter value is compatible with neither the $A_{AL}1$ nor $A_{AC}2$ mechanisms. (Part of the $\Delta S^{\ddagger}$ value for $A_{AL}1$ comes from the pre-equilibrium protonation, for which $\Delta S^{\ominus}$ is probably slightly positive.) In alkaline solution the $B_{AL}1$ mechanism is swamped by $B_{AC}2$.

Bimolecular AL mechanisms are rare; the $A_{AL}2$ mechanism is unknown and $B_{AL}2$ is found only in special cases. The carbonyl carbon atom is considerably harder than the alkoxy carbon, so that in general it is preferentially attacked by the hard base, hydroxide. Exceptionally severe steric hindrance can swing the balance: methyl 2,4,6-tri-t-butylbenzoate is hydrolysed (very slowly) by alkali in aqueous methanol. The $B_{AL}2$ mechanism is revealed by second order kinetics and by $^{18}O$ labelling [62CJ1981]. The corresponding trimethylbenzoate, however, is hydrolysed by the usual $B_{AC}2$ route.

$\beta$-Lactones are also hydrolysed in neutral or slightly acidic solution by the $B_{AL}2$ route because this offers relief of the severe angle strain. The mechanism is confirmed by $^{18}O$ labelling and by stereochemical inversion at the alkyl carbon atom [38JA2687; 41JA2459]:

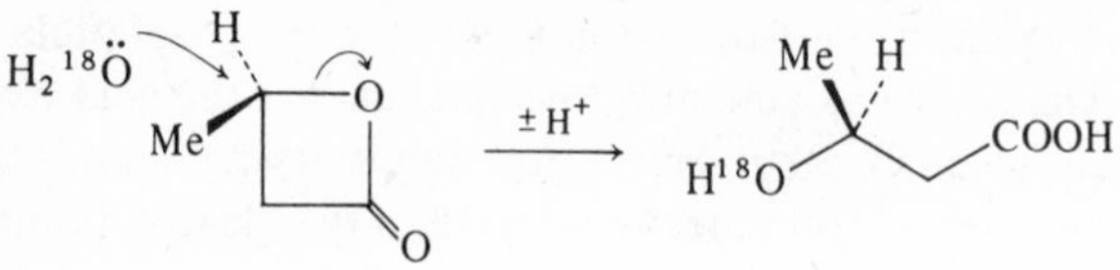

In strongly acidic solution the alkoxy oxygen atom can be protonated in a rapid pre-equilibrium and the ring can then be opened with unimolecular acyl-oxy cleavage, a special case of the $A_{AC}1$ mechanism discussed more generally in the following section:

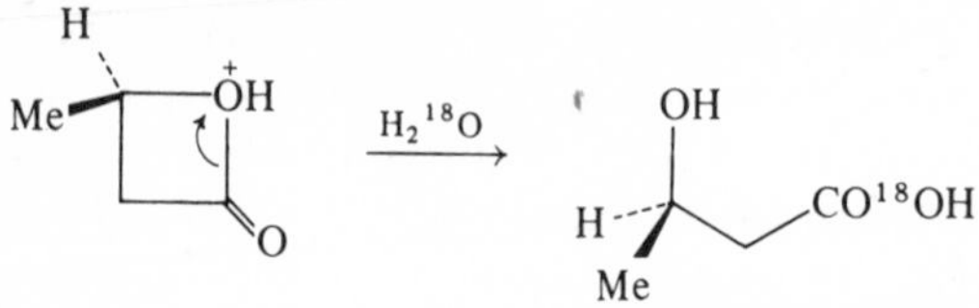

### 12.1.3 The $A_{AC}1$ mechanism

The uncatalysed dissociation of an ester into acylium cation and alkoxide anion ($B_{AC}1$) is unknown, but in acidic solution the minor alkoxy-protonated tautomer of the conjugate acid of the ester can dissociate if the resulting acylium ion is sufficiently stable. Cryoscopic measurements show that mesitoic acid (2,4,6-trimethylbenzoic acid) dissolves in 100 per cent sulphuric acid to generate four entities [38JA1708] and that methyl mesitoate produces five [45JA704]:

$$ArCOOH + 2\ H_2SO_4 \rightarrow ArCO^+ + H_3O^+ + 2\ HSO_4^-$$

$$ArCOOMe + 3\ H_2SO_4 \rightarrow ArCO^+ + H_3O^+ + MeHSO_4 + 2\ HSO_4^-$$

For synthetic purposes methyl mesitoate can be hydrolysed by dissolving it in concentrated sulphuric acid and pouring the solution into water; conversely mesitoic acid can be methylated by pouring a sulphuric acid solution into methanol.

$^{18}O$ labelling confirms that the hydrolysis of methyl mesitoate in aqueous sulphuric acid proceeds with acyl-oxy fission. Bunnett's $\phi$ value is negative ($-0.25$), and the activation parameters ($\Delta H^{\ddagger} = 120$ kJ mol$^{-1}$; $\Delta S^{\ddagger} = 70$ J K$^{-1}$ mol$^{-1}$ for 9.8M $H_2SO_4$) also accord with the unimolecular mechanism.

The hydrolyses of 4-substituted 2,6-dimethylbenzoate esters correlate with $\sigma^+$ with a large negative slope ($\rho = -3.2$), in accord with rate-limiting acylium formation [63JA37]. An interesting mechanistic transition is revealed by the hydrolyses of alkyl p-substituted benzoates in 99.9 per cent sulphuric acid. Hydrolysis of the methyl esters is retarded by electron-withdrawing substituents ($\rho \simeq -3.2$), showing the mechanism to be $A_{AC}1$. For the isopropyl esters ($A_{AL}1$) the trend is reversed; $\rho$ (estimated from only two points) is about $+3$. The ethyl esters give a curved Hammett plot, passing through a minimum around $\sigma = 0.6$: electron donors stabilise the acylium ion, encouraging $A_{AC}1$; electron acceptors destabilise it to the extent that ionisation of the ethyl–oxygen bond is preferred [56J1572; 60P84]. Under these strongly acidic conditions, where the activity of water is very low, a bimolecular reaction is unfavourable.

There is further discussion of mechanistic changes involving $A_{AC}1$ in §12.2.1.

### 12.1.4  A non-Ingold mechanism: E1cB

Esters bearing a strongly electron-withdrawing group on the $\alpha$ carbon atom of the acyl group, such as malonates or cyanoacetates, can undergo alkaline hydrolysis via an E1cB elimination–addition mechanism, with the formation of a keten intermediate:

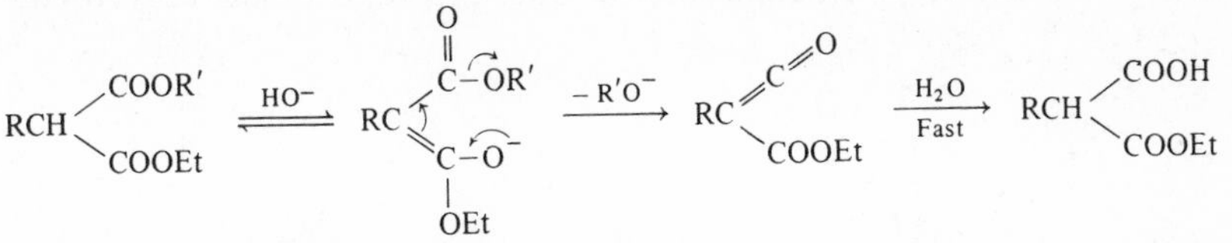

Such a mechanism was first demonstrated conclusively for the hydrolysis of ethyl nitrophenyl malonates (R = H, Me; R' = o- or p-$NO_2C_6H_4$) and related esters [69JA2993, 3003]. The hydrolysis of ethyl o-nitrophenyl dimethylmalonate, which has no $\alpha$ hydrogen atom, follows an unexceptional $B_{AC}2$ mechanism: in dilute acid there is a slow background reaction catalysed by water, and in alkaline solution the reaction rate increases linearly with hydroxide concentration. For substrates with an $\alpha$ hydrogen atom the reactions are much faster (by a factor of up to $10^4$) and the pH rate profiles are complex, in agreement with an E1cB mechanism in which the rate-limiting step changes from proton transfer (step 1) at low concentrations of hydroxide ion to phenoxide expulsion (step 2) above pH $\simeq 8$.

The E1cB reaction is very sensitive to the nucleofugality of the leaving alkoxide. The logarithms of the rate coefficients of hydrolyses of aryl acetoacetates, $ArOCOCH_2COCH_3$, correlate inversely with the p$K_a$ values of the phenols ArOH with a proportionality constant of $-1.3$: the p-nitrophenyl ester is over 3000 times more reactive than unsubstituted phenyl acetoacetate. In marked contrast the rates of hydrolysis of alkyl acetoacetates are almost independent of the p$K_a$ values of the nucleofugal alcohols (slope $= -0.05$), showing that with poor leaving groups the E1cB mechanism is overtaken by $B_{AC}2$ [70JA5956].

## 12.2    Catalysis

### *12.2.1    Reactions in strongly acidic solution*

The hydrolyses of many simple esters in dilute acid solution are very slow because so few of the weakly basic ester molecules ($pK_{BH^+}$ typically $-7$ to $-8$) are protonated. In concentrated acid, reactions are much faster and it also becomes possible to study the kinetic effects of changing the concentration of water, which is, after all, a vital reactant.

Yates and McClelland [67JA2686; 71Acc136] have investigated the hydrolyses of a series of acetates in aqueous sulphuric acid. The reactions are all first order in ester, and the pseudo-first order rate coefficients vary with acid concentration and with the structure of the acetate (figure 12.4). Clearly the different esters are reacting by different mechanisms, and for the primary and secondary alkyl esters it is apparent from the irregular rate profile that the mechanism changes with the acid concentration.

The behaviour of the tertiary alkyl esters is straightforward. They are cleaved by the $A_{AL}1$ mechanism; the rate is proportional only to the

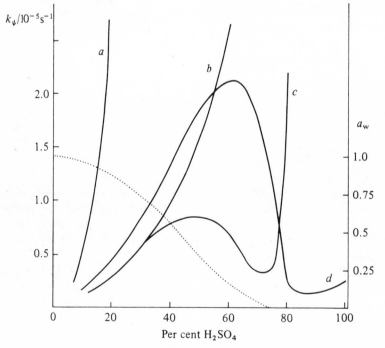

Figure 12.4. Rate profiles for acetate hydrolyses in sulphuric acid at 25 °C: (*a*) t-alkyl acetates; (*b*) aryl acetates; (*c*) secondary alkyl acetates; (*d*) primary alkyl acetates. The pseudo-first order rate coefficient, $k_\psi$, is defined by the equation: rate = $k_\psi$[ester]. The dotted line shows the change in the activity of water with concentration of sulphuric acid.

concentration of protonated ester, which increases steadily with increasing acid concentration until the reaction becomes too fast to follow.

The aryl esters also seem at first sight to follow a regular mechanistic pattern, but in fact the smooth curve is misleading. In dilute acid, aryl esters are certainly hydrolysed by the normal $A_{AC}2$ mechanism. With increasing acid concentration, this mechanism is subject to two opposing tendencies: the concentration of protonated ester increases, but that of the other reagent, water, falls off – and more important, its activity drops very sharply in the range 40 to 60 per cent sulphuric acid as competition for solvating molecules becomes intense (figure 12.4). Thus one would not expect the $A_{AC}2$ mechanism to continue its progressive acceleration through this concentration range.

Yates and McClelland sought to separate the two contributions. We may represent the hydrolysis of an ester E as:

$$E + H^+ \underset{K_{EH^+}}{\rightleftharpoons} EH^+$$

$$EH^+ + r\,H_2O \xrightarrow{k_2} \text{products}$$

where $r$ is the *change* in the number of water molecules in the rate-limiting step on going from ground to transition state. Applying the steady state approximation to $EH^+$ gives:

$$\text{Rate} = \frac{k_2}{K_{EH^+}} \cdot \frac{a_E a_{H^+}(a_w)^r}{y_\ddagger}$$

where $a_w$ is the activity of water and $y_\ddagger$ is the activity coefficient of the transition state, which has to be included when activities are used in place of concentrations in transition state theory (§2.3.3). Expanding and rearranging this equation gives:

$$\text{Rate} = (k_2/K_{EH^+}) \cdot (a_{H^+} y_E/y_\ddagger) \cdot [E] \cdot (a_w)^r$$

The term $a_{H^+} y_E/y_\ddagger$ is closely similar to $a_{H^+} y_E/y_{EH^+}$ which would define an acidity function $h_E$ appropriate for the protonation of esters (compare equation [4.4]). By making the assumption that $y_\ddagger$ is proportional to $y_{EH^+}$ – not unreasonable, in that the transition state and $EH^+$ are structurally related and both bear a positive charge – we can rewrite the rate equation as:

$$\text{Rate} = (k'/K_{EH^+}) \cdot h_E \cdot [E] \cdot (a_w)^r \tag{12.1}$$

where the change from $k_2$ to $k'$ allows for the change from $y_\ddagger$ to $y_{EH^+}$.

In practice the kinetic experiments are carried out by comparing the observed rates of reaction with the stoicheiometric concentrations of ester:

$$\begin{aligned}\text{Rate} = k_\psi[E]_{\text{stoi}} &= k_\psi([E] + [EH^+]) \\ &= k_\psi[E](1 + 1/I_E)\end{aligned} \tag{12.2}$$

where $I_E$ is the indicator ratio $[E]/[EH^+]$. Except in very concentrated acid the weakly basic ester is only slightly protonated and $I_E \gg 1$ (or $[E] \simeq [E]_{stoi}$). Thus equation [12.2] simplifies to: Rate $= k_\psi [E]$, and comparison with equation [12.1] gives:

$$k_\psi = (k' K_{EH^+}) \cdot h_E \cdot (a_w)^r$$
$$\therefore \log k_\psi + H_E = r \log a_w + \log k' + pK_{EH^+} \qquad [12.3]$$

In fact there is no established $H_E$ acidity function [see 74JA2862] but a number of acetates follow $H_0$ with a slope $n \simeq 0.62$ (compare equation [4.7], §4.3.3), leading to the approximate relation:

$$\log k_\psi + 0.62 H_0 = r \log a_w + \text{constant}$$

Plots of $(\log k_\psi + 0.62 H_0)$ against $\log a_w$ for the different types of acetate hydrolyses are shown in figure 12.5.

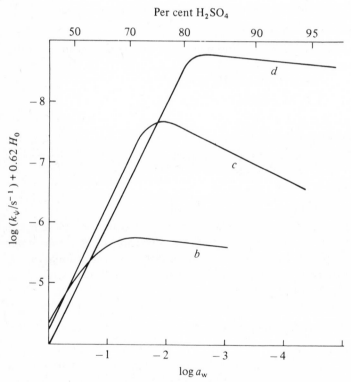

Figure 12.5. Yates–McClelland plot for the hydrolysis of acetates at 25 °C (*cf.* figure 12.4): (*b*) aryl acetates; (*c*) secondary alkyl acetates; (*d*) primary alkyl acetates. (The plot for tertiary alkyl acetates cannot conveniently be represented on the same scale, since $\log a_w$ for 20 per cent $H_2SO_4$ is only −0.06. The line is straight with slope about −9.)

These plots point very clearly to a mechanistic change in all the reactions save those of the tertiary alkyl esters; they also give an insight into the nature of the change. The initial slopes are all close to $r = +2$, suggesting that two water molecules are involved in the rate-limiting $A_{AC}2$ step. It seems likely that one molecule of water acts as a general base catalyst in faciliting nucleophilic attack by the other:

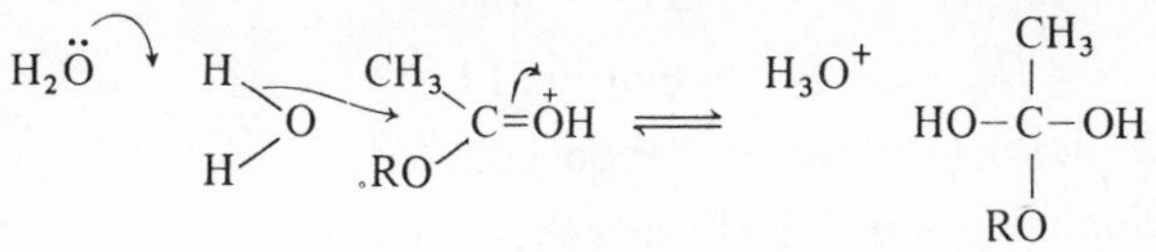

The hydrolysis would then continue with the analogous reverse process, acid catalysed expulsion of the alkoxy group (compare acetal hydrolysis, §11.2):

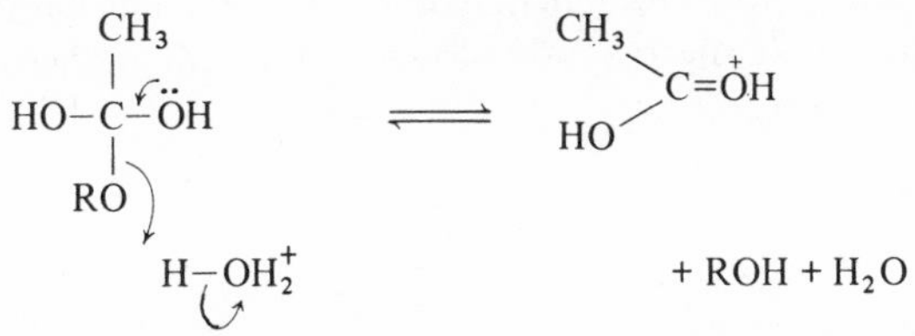

This behaviour also accords with the very high value of the Bunnett parameter for $A_{AC}2$ hydrolyses ($\phi \simeq 0.9$) and with the $H_2O/D_2O$ solvent isotope effects: the ratio $k_H/k_D \simeq 0.65$ is larger than predicted for simple nucleophilic attack on a protonated substrate ($\sim 0.33$), and suggests that the base catalysis, for which $k_H > k_D$, is exerting a moderating influence.

As the activity of water falls with increasing acid concentration so the $A_{AC}2$ mechanism becomes less viable and a unimolecular reaction takes over. For secondary alkyl acetates (and benzyl acetate behaves similarly) this is $A_{AL}1$, via the relatively stable alkyl carbenium ion pairs. For primary alkyl and aryl esters it is presumably $A_{AC}1$, with the formation of $CH_3CO^+$. The small negative $r$ values indicate differential solvation of the protonated ester compared with the dissociating transition state, but their numerical significance is obscure.

We can thus rationalise the rate profiles of figure 12.4. The tertiary alkyl acetates are hydrolysed via $A_{AL}1$ throughout. The phenyl esters follow the $A_{AC}2$ path at low acid concentrations, moving over to $A_{AC}1$ as the activity of water falls away. For primary alkyl acetates the $A_{AC}1$ mechanism does not set in till the acid concentration is considerably higher, because ROH is a poorer leaving group than ArOH; by this stage the activity of water is falling so fast that the $A_{AC}2$ mechanism is actually retarded by increasing acid concentration. With secondary alkyl and benzyl acetates the $A_{AL}1$ mechanism intervenes at a lower acid concentration than $A_{AC}1$, but still not before $A_{AC}2$ has begun to slow down.

A similar analysis has been carried out by Cox and Yates using their 'excess acidity' approach (§4.3.3) extended to deal with rates of reactions [79CJ2944]. The proportionality between $y_{\ddagger}$ and $y_{EH^+}$ discussed above is accommodated by rewriting equation [4.11] as:

$$\log I + \log[H^+] = -m^{\ddagger}m^*X - pK_{BH^+} \qquad [12.4]$$

Putting $pK_{EH^+} = H_E - \log I_E$ in equation [12.3] gives:

$$\log k_{\psi} + \log I_E = r \log a_w + \text{constant}$$

and subtracting equation [12.4] leads to:

$$\log k_{\psi} - \log[H^+] = m^{\ddagger}m^*X + r \log a_w + \text{constant}$$

For the hydrolysis of methyl benzoate esters plots of $(\log k_{\psi} + \log[H^+])$ against $X$ have the form of figure 12.6 [79CJ2960]. The curved portion closely follows the line calculated for an $A2$ mechanism (obviously $A_{AC}2$) with $r = 2$; the reaction slows down with increasing acid

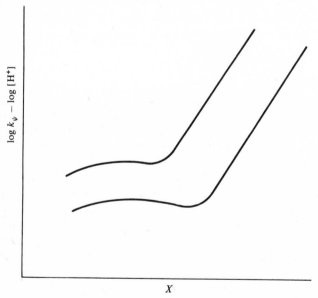

Figure 12.6. The form of plots of $(\log k_{\psi} - \log[H^+])$ for the acid catalysed hydrolyses of methyl benzoate esters against the 'excess acidity' function $X$. The upper curve illustrates how the changeover from $A_{AC}2$ (curved plot) to $A_{AC}1$ (linear plot) occurs at lower acidities when the temperature is raised or when the benzene ring bears an *ortho* methyl group [79CJ2960]. (In the original paper the ordinate is $\log k_{\psi} - \log[H^+] - \log([S]/([S] + [SH^+]))$: the last term is a small correction to allow for the difference between the stoicheiometric concentration of S and the actual concentration of unprotonated S present at any given acidity.)

concentration as the activity of water falls off. As the temperature is raised the switch to the straight line $A_{AC}1$ correlation occurs at lower acid concentrations: for the $A\,2$ mechanism, $\Delta S^{\ddagger}$ is large and negative (around $-130 \text{ J K}^{-1}\text{ mol}^{-1}$) whereas for $A\,1$ it is positive ($+40 \text{ J K}^{-1}\text{ mol}^{-1}$). An *ortho* methyl group on the benzene ring also encourages the $A_{AC}1$ mechanism, and with both *ortho* positions substituted the $A_{AC}2$ route does not offer any significant competition until the acid is very dilute (compare §12.1.3). The temperature variation of the plots reveals that this behaviour is almost entirely due to the lowering of the enthalpy of activation with *ortho* substitution. The relief of steric strain and hyper-conjugative stabilisation of the acylium ion both encourage the ionisation of the *ortho* methylated esters:

### 12.2.2 General base and nucleophilic catalysis; neutral hydrolysis

The $B_{AC}2$ mechanism, as Ingold originally conceived it, was confined essentially to alkaline hydrolyses: direct attack by hydroxide ion on the ester. Strictly speaking this is not a base catalysed reaction at all. The hydroxide ion is behaving as a nucleophile rather than a base, and it is not acting catalytically since it is consumed during the reaction. At that time relatively few hydrolyses were known to take place in neutral solution and their mechanism was uncertain, but since it seemed unlikely that they would involve the protonated ester (about one molecule in $10^{14}$ is protonated at pH $= 7$) they were included in the general $B_{AC}2$ category.

In fact, however, the hydrolysis of esters in more or less neutral solution is a fairly common reaction, catalysed by a wide variety of bases. Several different mechanisms have been suggested for these reactions [72MI*a*; 67APO237] but they all belong to two main categories: general base catalysis and nucleophilic catalysis. There remains some doubt about the precise role of the general base, but most likely it assists the attack of water on the esters:

In nucleophilic catalysis it is the catalyst that first attacks the ester, generating a more reactive intermediate which is then rapidly attacked by water:

$$\text{Nu}^- + \text{R--CO--OR}' \underset{}{\overset{\text{r.l.s.}}{\rightleftharpoons}} \text{R--CO--Nu} + \text{R'O}^-$$

$$\text{Fast} \downarrow + \text{H}_2\text{O}, - \text{H}^+$$

$$\text{R--COOH} + \text{Nu}^-$$

Neither of these catalysed routes offers serious competition to the hydroxide ion reaction in alkaline solution. A base-assisted water molecule can never be as reactive as a free hydroxide ion, so that general base catalysis is soon overwhelmed. And few other nucleophiles are very much more reactive towards an ester group than is hydroxide, so at high pH the hydroxide swamps any nucleophile that is present in only catalytic concentration. Studies of general base and nucleophilic catalysis are therefore confined in the main to the region of pH-independent hydrolysis (compare figure 12.3($b$)) where relatively small catalytic effects can be detected. This in turn means that the more reactive esters are the most easily studied, particularly phenolic esters and esters of halogeno-acetic acids. Of these, for reasons which will become apparent, only the phenolic esters normally exhibit nucleophilic catalysis.

The extent to which a particular catalyst acts as a nucleophile or as a general base depends on two main factors: the balance between the basic and nucleophilic properties of the catalyst, and the relative leaving group abilities of the catalyst and the aryloxy group of the phenolic ester. A simplified kinetic system for the two modes of catalysis operating simultaneously is shown in figure 12.7.

We assume that the base catalysed route is controlled simply by the initial catalysed attack of water; hydroxide is a much poorer leaving group than aryloxide, so the reverse of step 1$b$ is not kinetically significant. For nucleophilic catalysis we shall consider first the case when

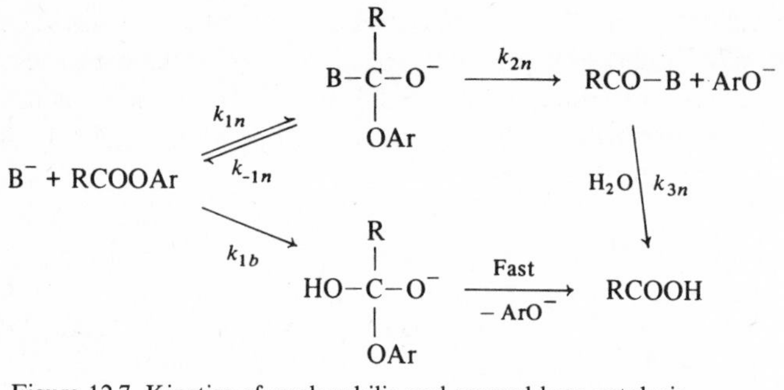

$$
\begin{array}{l}
\text{B}^- + \text{RCOOAr} \\[2em]
\end{array}
$$

Figure 12.7. Kinetics of nucleophilic and general base catalysis.

step 3n is fast. Applying the steady-state approximation to the tetrahedral intermediate gives:

$$k_n = k_{1n}k_{2n}/(k_{-1n} + k_{2n})$$

where $k_n$ is the overall rate coefficient for the nucleophilically catalysed route. Thus the balance between the two routes is:

$$\frac{k_n}{k_b} = \frac{k_{1n}}{k_{1b}} \cdot \frac{k_{2n}}{k_{-1n} + k_{2n}} \qquad [12.5]$$

The term $k_{1n}/k_{1b}$ measures the relative reactivity of the catalyst towards the carbonyl carbon atom or the proton of water, and is therefore related to softness. The second part of the expression depends on the partition of the tetrahedral intermediate: is $B^-$ or $ArO^-$ the better leaving group? If it is $ArO^-$ and $k_{2n} \gg k_{-1n}$, then the term $k_{2n}/(k_{-1n} + k_{2n})$ tends to unity and the mode of catalysis depends simply on $k_{1n}/k_{1b}$. But if $B^-$ is a much better leaving group than $ArO^-$ the nucleophilic route is suppressed. Alkoxides are among the poorest of leaving groups, so nucleophilic catalysis is rarely encountered with alkyl esters. (But note in passing the importance of nucleophilic catalysis, for example by pyridine, in reactions of acid chlorides where there is a very good leaving group attached to the acyl carbon.)

If nucleophilic attack is fast but step 3n is slow, then $B^-$ acts as an inhibitor rather than a catalyst, diverting some of the base catalysed reaction into a dead-end. In the extreme, of course, the species RCO—B is the actual reaction product, as in the preparation of amides from an ester and ammonia or a primary amine. The hydroxide ion, too, reacts this way, so that alkaline hydrolysis can be regarded as an extreme case of nucleophilic catalysis.

The two catalytic routes are clearly very different though both follow the same kinetic behaviour, first order in ester and in catalyst. General base catalysed reactions follow the Brønsted catalysis equation. The hydrolysis of ethyl dichloroacetate, for example, gives a good linear correlation ($\beta = 0.47$) with catalysts of many different types, including aniline, pyridines, carboxylate anions and the phosphate dianion [61JA1743]. Significantly, the reaction with aniline gives no dichloroacetanilide, which would have been generated by nucleophilic reaction and would not have been hydrolysed under the conditions used. The rate of the background water catalysed reaction lies on the same straight line ($pK_{H_3O^+} = -1.57$), suggesting that in this reaction one water molecule is acting as a general base in facilitating the reaction of another:

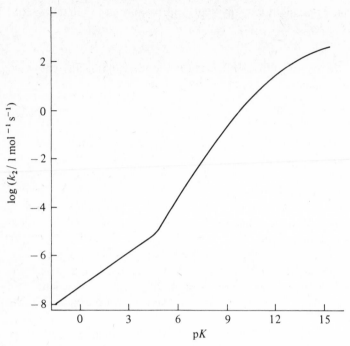

Figure 12.8. Brønsted plot for the catalysed hydrolysis of $p$-nitrophenyl acetate at 25 °C [data from 68JA2622].

In nucleophilic catalysis, on the other hand, Brønsted plots show a wide scatter of points. Phosphate ($pK_2 = 7.2$) and imidazole ($pK_{BH^+} = 7.0$) are almost equally effective in catalysing the ethyl dichloroacetate hydrolysis, but for $p$-nitrophenyl acetate the reaction with imidazole is nearly 5000 times faster than with phosphate [64JA837; 60JA1778]. The $\alpha$ effect (§7.4.3($b$)) is particularly noticeable: hydroperoxide ion is some 400 times more effective a catalyst for aryl acetate hydrolysis than is hydroxide, though its $pK_{BH^+}$ is only 11.6 [68JA2622].

If nucleophiles exhibiting the $\alpha$ effect are excluded there is a general parallel between basicity and nucleophilicity for species reacting through the same atom: oxyanions, for example. But even within this restricted field a Brønsted plot is not linear. Figure 12.8 is illustrative.

For the weakest bases nucleophilic catalysis is suppressed ($k_{-1n} \gg k_{2n}$ in equation [12.5]) and the hydrolyses are subject to general base catalysis with a linear Brønsted relation. As the catalyst approaches the ester leaving group in strength (when its $pK_{BH^+}$ is about two units less than that for $ArO^-$) nucleophilic catalysis becomes significant and the overall hydrolysis is accelerated. But with strongly basic catalysts (very poor leaving groups) control of the nucleophilic route passes to step $2n$

(figure 12.7); the rate depends mainly on the aryloxy leaving group and only secondarily on the catalyst, so the curve levels off again.

It is sometimes possible to detect the presence of the intermediate, RCO−B, formed during nucleophilic catalysis. In the imidazole catalysed hydrolysis of *p*-nitrophenyl acetate it has been possible to monitor by ultraviolet spectroscopy the formation of the *p*-nitrophenol and the build-up and decay of the acetylimidazole intermediate [57JA1652].

In the acetate catalysed hydrolysis of 2,4-dinitrophenyl benzoate the intermediate is a mixed anhydride whose presence can be revealed by $^{18}O$ labelling:

$$CH_3C^{18}O_2^- + PhCOOAr \rightleftharpoons CH_3 - \overset{\overset{^{18}O}{\|}}{C} - {}^{18}O - \overset{\overset{^{16}O}{\|}}{C} - Ph + ArO^-$$

In the hydrolytic step the acetyl group is attacked faster than is benzoyl and some of the label is thereby transferred to the benzoic acid [58JA5388].

Alternatively an anhydride intermediate may be trapped by its rapid reaction with a little added aniline [68J(B)515]:

$$AcO^- + AcOAr \rightleftharpoons Ac_2O + ArO^-$$
$$\downarrow PhNH_2$$
$$PhNHAc + AcOH$$

It is necessary to make due allowance for the relatively slow reaction of the aniline directly with the ester and for the effect of the aniline as a general base catalyst, but when this is done the formation of acetanilide can be used to monitor the presence of anhydride. As predicted by equation [12.5], esters of strongly acidic phenols such as dinitrophenols are subject to predominantly nucleophilic catalysis ($k_{2n} \gg k_{-1n}$). On the other hand no acetanilide is formed during the hydrolysis of phenyl acetate and similar esters with leaving groups that are much more basic than the acetate catalyst ($k_{2n} \ll k_{-1n}$), showing that only general base catalysis is operating.

Other criteria which may be used to distinguish the two modes of catalysis include:

(*a*) *Activation entropies.* A typical value for the termolecular general base catalysis is $\Delta S^\ddagger = -150 \text{ J K}^{-1} \text{ mol}^{-1}$; it is seldom more negative than $-80 \text{ J K}^{-1} \text{ mol}^{-1}$ for nucleophilic catalysis.

(*b*) *Solvent isotope effects.* For general base catalysis, where a proton is being transferred in the transition state, $k_H/k_D$ is usually in the range 2 to 4; it is seldom greater than 1.5 for nucleophilic catalysis.

(*c*) *Steric effects.* Steric hindrance is much more marked in nucleophilic reactions than in proton transfer. 2,6-Lutidine (dimethylpyridine) is about

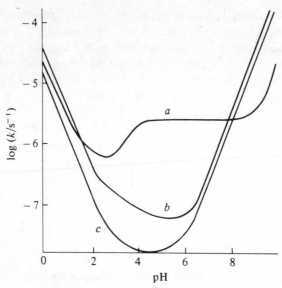

Figure 12.9. pH/rate profile for hydrolysis at 25 °C of: (*a*) aspirin; (*b*) *p*-acetoxybenzoic acid; (*c*) aspirin methyl ester.

as effective as pyridine in general base catalysis; as a nucleophilic catalyst it is almost inert.

### 12.2.3 Intramolecular catalysis

The hydrolysis of aspirin (acetylsalicylic acid) shows a marked acceleration relative to that of its *para* isomer or of its methyl ester over the the range pH = 3 to pH = 8 (figure 12.9). The sigmoid central portion of the rate profile is effectively an ionisation curve of the aspirin carboxyl group ($pK_a = 3.4$) and demonstrates that the catalysis is due to the dissociated carboxylate anion. Above pH $\simeq$ 5 the aspirin is virtually fully ionised and the rate reaches a plateau until the intramolecularly catalysed process is overtaken by direct alkaline hydrolysis at pH $\simeq$ 8.

As with intermolecular catalysis this intramolecular process could involve a nucleophilic or a general base mechanism (figure 12.10). The evidence points strongly towards general base catalysis [67JA4857]. When the reaction is carried out in $H_2^{18}O$ no label is incorporated into the salicylic acid (contrary to early reports). Solvolysis in aqueous ethanol is faster than in water whereas addition of dioxan has little effect on the rate. This suggests that a protic solvent, water or ethanol, is actually incorporated into the transition state. Moreover, in aqueous ethanol the reaction leads to the formation of some ethyl acetate but not to ethyl salicylate. Substituent effects and the solvent isotope effect ($k_H/k_D = 2$)

Nucleophilic catalysis

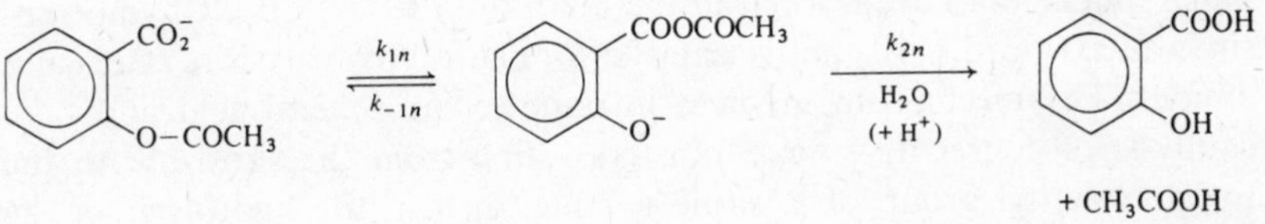

General base catalysis

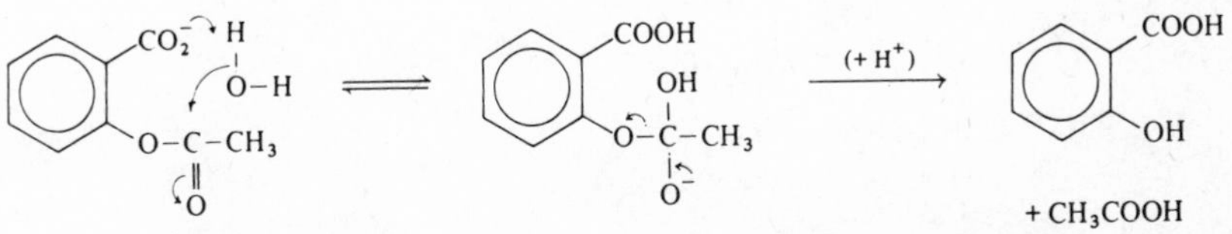

Figure 12.10. Alternative modes of intramolecular catalysis for hydrolysis of aspirin.

are comparable to those found in intermolecular general base catalysis rather than nucleophilic catalysis. The entropy of activation, $\Delta S^{\ddagger} = -95 \text{ J K}^{-1} \text{ mol}^{-1}$, is too negative for a unimolecular process.

By contrast the neutral hydrolysis of monophenyl phthalate appears to involve nucleophilic catalysis [66JA747; and compare other monophenyl esters of diacids: 60JA5858]:

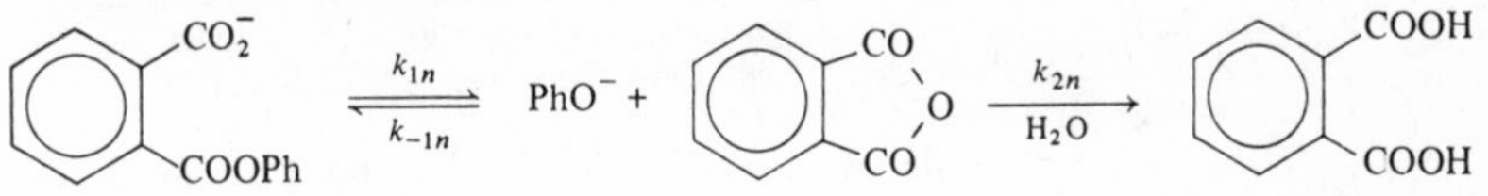

This reaction is some 10 000 times faster than aspirin hydrolysis and maybe a million times faster than the nucleophilic rate for aspirin: why the enormous difference? Application of the steady-state approximation gives:

$$k_n = k_{1n}k_{2n}/(k_{-1n} + k_{2n})$$

In aspirin hydrolysis $k_{-1n}$ is very large: it describes the intramolecular attack by a fairly strong nucleophile, phenoxide, on the highly reactive anhydride. Certainly this is very much faster than the intermolecular attack by neutral water ($k_{2n}$). Thus $k_n = K_{1n}k_{2n}$, and $K_{1n}$, the equilibrium constant between the aspirin anion and the oxyphenylanhydride, is probably about $10^{-11}$ [71MI$b$]. On the other hand the phthalate ester generates a cyclic anhydride and a separate phenoxide ion; reversion to reactants is bimolecular and $k_{-1n} \ll k_{2n}$.

Intramolecular catalysis is the key to the remarkable activity of enzymes; how it is that they can catalyse reactions under conditions much milder than are normal in the laboratory [75APO1]. Chymotrypsin is an extraordinarily active catalyst for neutral hydrolysis of esters and amides. The mechanism involves intramolecular general acid and base catalysis of consecutive acyl transfers – first from the substrate to the hydroxymethyl group of a serine residue, then in the hydrolysis of the acylserine. The tertiary structure of the enzyme is such that three particular aminoacid residues, aspartic acid (at position 102 along the chain), histidine (57) and serine (195) are ideally aligned for proton transfer:

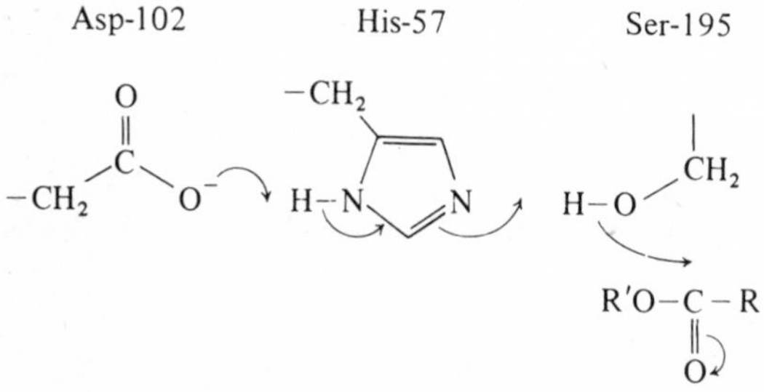

## 12.3  Further reading

A. J. Kirby, in *Comprehensive Chemical Kinetics*, eds C. H. Bamford & C. F. H. Tipper, vol. 10, Elsevier, Amsterdam, 1972.

E. K. Euranto, in *The Chemistry of Carboxylic Acids and Esters*, ed. S. Patai, Interscience, London, 1969.

A. Williams & K. T. Douglas, *Chem. Rev.*, 1975, **75**, 627 (E1cB mechanism).

K. Yates & T. A. Modro, *Acc. Chem. Res.*, 1978, **11**, 190 (acid catalysis).

S. L. Johnson, *Adv. Phys. Org. Chem.*, 1967, **5**, 237 (general base and nucleophilic catalysis).

T. C. Bruice & S. K. Benkovic, *Bioorganic Mechanisms*, vol. 1, Benjamin, New York, 1966 (see especially pp. 119–211 for discussions of intramolecular catalysis).

A. J. Kirby & A. R. Fersht, *Progr. Bioorg. Chem.*, 1971, **1**, 1 (intramolecular catalysis).

A. R. Fersht & A. J. Kirby, *Chem. in Britain*, 1980, **16**, 136 (intramolecular and enzymic catalysis).

W. J. Albery & J. R. Knowles, *Angew. Chem. Internat. Ed.*, 1977, **16**, 285; M. I. Page, *ibid.*, p. 449; F. Schneider, *ibid.*, 1978, **17**, 583; M. L. Bender & F. J. Kézdy, in *Proton-Transfer Reactions*, eds E. F. Caldin & V. Gold, Chapman & Hall, London, 1975; W. N. Lipscomb, *Acc. Chem. Res.*, 1982, **15**, 232 (enzymic catalysis).

For other substitution reactions, not discussed in this chapter, see:

J. Koskikallio, in *The Chemistry of Carboxylic Acids and Esters*, ed. S. Patai, Interscience, London, 1969 (transesterification and anhydride formation).

D. P. N. & R. S. Satchell, *ibid.* (other substitution reactions of esters and acids).

R. J. E. Talbot, in *Comprehensive Chemical Kinetics*, eds C. H. Bamford & C. F. H. Tipper, vol. 10, Elsevier, Amsterdam, 1972 (hydrolysis of amides, anhydrides and acid chlorides).

C. O'Connor, *Quart. Rev.*, 1970, **24**, 553 (amide hydrolysis).

B. C. & J. Challis, in *The Chemistry of Amides*, ed. J. Zabicky, Interscience, London, 1970 (see especially pp. 816–47 for hydrolysis and solvolysis of amides).

A. Kivinen, in *The Chemistry of Acyl Halides*, ed. S. Patai, Interscience, London, 1972 (substitution in acyl halides).

# 13 Aromatic nucleophilic substitution

## 13.1 The Meisenheimer mechanism

Nucleophilic attack on the $\pi$ system of a benzene ring can give an anion analogous to the cationic Wheland intermediate of aromatic electrophilic substitution. If a suitable leaving group is present it can then be expelled in a second step so that the overall reaction is nucleophilic substitution:

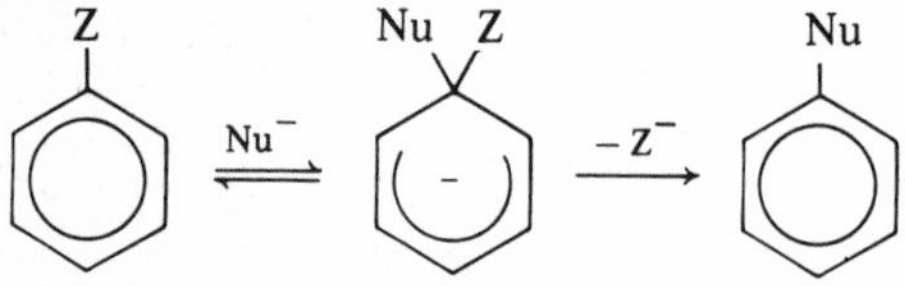

The intermediate is named after Meisenheimer, who in 1902 [02LA(323)205] first isolated salts of such a structure in systems bearing electron-withdrawing groups:

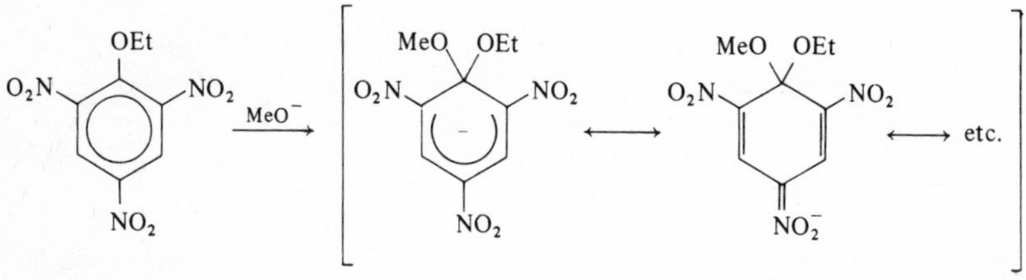

Indeed, most of these reactions *require* some activation by electron withdrawal to stabilise the anionic intermediate, because in an unactivated system the two extra electrons it bears relative to the Wheland intermediate are in an antibonding orbital. The reactions of unactivated aryl halides are facilitated by dipolar aprotic solvents such as dimethylformamide and hexamethylphosphoramide [e.g. 79JO2642; 83T193] in which the nucleophiles are only weakly solvated whereas dispersion forces allow the polarisable transition state some stabilisation (see chapter 5). Mechanistic studies have usually been carried out in hydroxylic solvents using systems such as **(13.1)** in which there is one constant activating substituent and in which the effects of changing the leaving group Z or a second substituent X can be investigated.

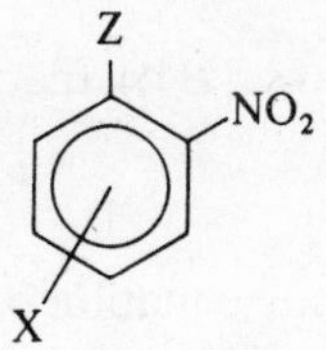

(13.1)

In accordance with the suggested mechanism the rates of these reactions correlate well with $\sigma^-$ for the substituent X, with a large $\rho$ value, typically 3.5 to 5, confirming the importance of delocalising the negative charge out of the ring and on to an electron-withdrawing substituent [e.g. 62JA1026].

### 13.1.1 Kinetics and base catalysis

Applying the steady-state approximation to the scheme:

$$ArZ + Nu^- \underset{k_{-1}}{\overset{k_1}{\rightleftharpoons}} M \overset{k_2}{\rightarrow} ArNu + Z^-$$

(where M represents the Meisenheimer intermediate) gives:

$$\text{Rate} = \frac{k_1 k_2 [ArZ][Nu^-]}{k_{-1} + k_2}$$

This represents straightforward second order kinetics with an observed rate coefficient of $k_1 k_2/(k_{-1} + k_2)$. Clearly one cannot use a simple kinetic criterion to decide whether the rate-limiting step is formation of the intermediate ($k_2 > k_{-1}$) or its dissociation ($k_{-1} > k_2$). Indeed, the kinetics would also be in accord with a concerted one-step mechanism in which the Meisenheimer complex would be the transition state rather than an intermediate. Conclusive evidence for the two-step mechanism came first from studies by Bunnett of base catalysis on reactions in which primary or secondary amines are the nucleophiles.

In these reactions the Meisenheimer intermediate **(13.2)** is a zwitterion; in the second stage of the reaction it must shed a proton as well as the

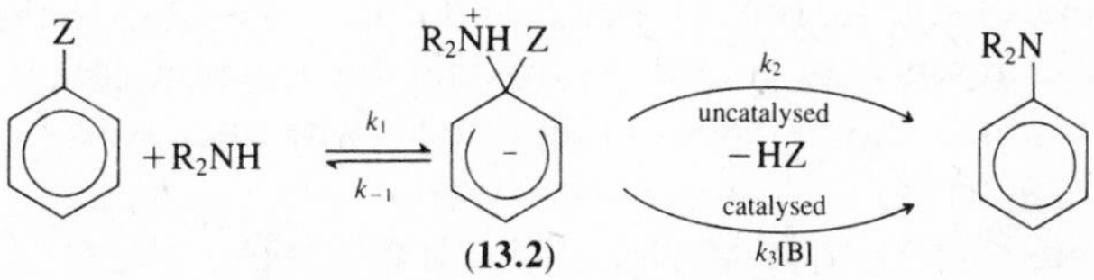

(13.2)

leaving group Z, so its decomposition can be subject to base catalysis as shown and the rate equation becomes:

$$\text{Rate} = \frac{k_1\{k_2 + k_3[\text{B}]\}}{k_{-1} + k_2 + k_3[\text{B}]} \cdot [\text{ArZ}][\text{R}_2\text{NH}]$$

When the first step is rate limiting $(k_{-1} \ll \{k_2 + k_3[\text{B}]\})$ the kinetics again reduce to second order:

$$\text{Rate} = k_1[\text{ArZ}][\text{R}_2\text{NH}] \tag{13.1}$$

At the opposite extreme $(k_{-1} \gg \{k_2 + k_3[\text{B}]\})$ the rate equation becomes:

$$\text{Rate} = \left\{\frac{k_1 k_2}{k_{-1}} + \frac{k_1 k_3[\text{B}]}{k_{-1}}\right\}[\text{ArZ}][\text{R}_2\text{NH}] \tag{13.2}$$

in which a third order term involving the base catalyst appears. This can be written more simply as:

$$k_{\text{obs}} = k + k'[\text{B}] \tag{13.3}$$

Kinetic behaviour of this type has been observed in a wide variety of substitutions when Z is a poor leaving group (so that $k_{-1} \gg k_2$). A well-studied example, due to Bunnett, is the reaction of 2,4-dinitrophenyl phenyl ether with piperidine in aqueous dioxan [65JA5209]:

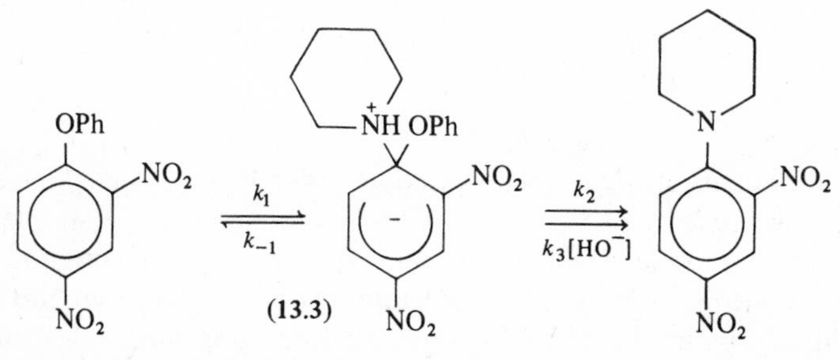

Small concentrations of hydroxide strongly catalyse the reaction but at higher concentrations the further addition of hydroxide no longer has any effect (figure 13.1). This suggests that in the absence of base the Meisenheimer intermediate (13.3) sheds a neutral piperidine molecule faster than a phenoxide ion $(k_{-1} > k_2)$. As hydroxide ion is added, the slow second step is catalysed according to equation [13.3] until eventually it becomes faster than the first step; the kinetics now approximate to equation [13.1], independent of base concentration.

Piperidine, of course, is itself a base and it too can catalyse the second step. In the absence of added hydroxide the reaction follows mixed second and third order kinetics in agreement with equations [13.2] and [13.3]:

$$\text{Rate} = k[\text{ArZ}][\text{R}_2\text{NH}] + k'[\text{ArZ}][\text{R}_2\text{NH}]^2$$

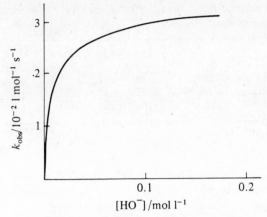

Figure 13.1. Catalytic effect of hydroxide ion on the reaction of piperidine with 2,4-dinitrophenyl phenyl ether in 10 per cent aqueous dioxan at 29.4 °C.

Subsequent studies by ultraviolet [70JA2417] and, more recently, by n.m.r. spectroscopy [77CJ1468] have conclusively revealed the formation of a Meisenheimer adduct in reactions of nitroaryl ethers with amines, and the kinetics of its build-up and subsequent decay have been analysed in terms of the mechanism described above.

Recent work has shed light on the detailed mechanisms of the base catalysis. In dipolar aprotic solvents the norm (but see §13.1.4) is specific base-general acid catalysis: rapid deprotonation of the ammonium group is followed by rate-limiting abstraction of $Z^-$ by the conjugate acid of the base catalyst, $BH^+$; that is, $k_{-z}[BH^+] \ll k_h$ in the following sequence [70JA2417; 77CJ1468]:

In water and other hydroxylic solvents, however, the uncatalysed (or more strictly, solvent-assisted) expulsion of $Z^-$ is so fast that further acceleration by acid catalysts is insignificant. With weakly nucleofugal leaving groups such as alkoxides this results in a shift from general to specific base catalysis. Extrapolation from the measured rates of alkoxide expulsion suggests that with better leaving groups the nucleofugal step is even faster than the proton transfer: $k_{-z}[H_2O] \gg k_h$ [77JA4090]. Until a few years ago it would have been almost unthinkable to suggest that the rate of a relatively slow reaction could be controlled by proton transfer between electronegative elements – a process that occurs at or near the

Table 13.1. *Second order rate coefficients* $(k_{obs}/10^{-6} \, l \, mol^{-1} \, s^{-1})$ *for some reactions of activated halogenobenzenes*

|  | F | Cl | Br | I |
|---|---|---|---|---|
| (a) $o$-$NO_2$–$C_6H_4$–Hal/MeO$^-$/MeOH/50 °C | 1810 | 2.52 | 1.77 | 0.83 |
| (b) $p$-$NO_2$–$C_6H_4$–Hal/piperidine/EtOH/90 °C | 2250 | 10.8 | 13.9 | 4.0 |
| (c) 2,4-$(NO_2)_2C_6H_3$–Hal/PhNH$_2$/EtOH/50 °C | 16800 | 269 | 405 | 131 |

(a) [55J2926] (b) [54J2109] (c) [51J3301].

diffusion limit. The reactions described here and in §11.3.2, both involving proton transfer to or from a highly unstable zwitterionic intermediate, illustrate that such a phenomenon may after all be not uncommon.

### 13.1.2 Leaving group effects

The pattern of reactivity of different leaving groups offers more evidence for the two-step mechanism. For aryl halides, sulphonates, and other substrates with fairly good leaving groups it is usually the first step that is rate-limiting. This does not entail cleaving the C–Z bond so leaving group ability is unimportant. The data of table 13.1 are typical. The high reactivity of the fluorides in no way accords with the usual pattern of nucleofugality; in aliphatic systems the fluorides are almost inert. However, it is what is expected for a Meisenheimer mechanism with the first step rate-limiting; the strongly electronegative fluorine enhances the electrophilicity of the carbon atom to which it is attached and accelerates the attack of the nucleophile.

If the nucleophile is itself also a good nucleofuge (increased $k_{-1}$) the second step can be rate-limiting. In reactions of potassium iodide with 2,4-dinitrohalogenobenzenes the relative reactivities of the fluoride, chloride and bromide are 1:17:2000 [55BB709]. The Meisenheimer intermediate has the choice of expelling the iodide ion again and reverting to reactants or of completing the substitution by expelling the other halide. Iodide is the best of the halogen leaving groups so that $k_2$ becomes rate-limiting and the well-known reactivity sequence reappears.

A particularly interesting example, again due to Bunnett [58JA6020], comes from the reactions of 2,4-dinitrohalogenobenzenes with $N$-methylaniline (table 13.2). In the absence of added salts the fluoride and chloride react at virtually identical rates, the bromide slightly faster. Sodium perchlorate accelerates all the reactions slightly: the usual electrolyte effect. Potassium acetate has a similar effect on the chloride and bromide but causes much more dramatic acceleration of the fluoride reaction. The implication is that the chloride and bromide are behaving 'normally' throughout, with the first step rate-limiting, but with fluoride,

Table 13.2. *Second order rate coefficients* $(k_{obs}/10^{-5} \, l \, mol^{-1} \, s^{-1})$ *for the reactions of N-methylaniline with 2,4-dinitrohalogenobenzenes in EtOH at 67 °C*

| Additive | Bromide | Chloride | Fluoride |
|---|---|---|---|
| None | 20.3 | 7.3 | 7.3 |
| NaClO$_4$ (0.1M) | 24.9 | 10.3 | 10.2 |
| KOAc (0.1M) | 23.2 | 9.1 | 110 |

a poorer leaving group, it is the second step that is rate-limiting in the absence of base catalysis. (It is just coincidence that the rate of fluoride expulsion is the same as the rate of amine attack on the chloride.) The acetate ion, weakly basic though it is, can catalyse the second step of the reaction. This has no effect on the overall rate of the chloride and bromide reactions, but it markedly accelerates the limiting step with the fluoride. The observed rate coefficient varies linearly with acetate concentration as predicted by equation [13.3].

It is patently impossible to draw up a standard list setting out the order of reactivity of different leaving groups; their behaviour depends on the nucleophile, the activating group, the solvent, and the possibility of catalysis (see also §13.1.4). Fairly typical leaving group behaviour is the following:

| Good | Average | Poor |
|---|---|---|
| F NO$_2$ NMe$_3^+$ OTos SMe$_2^+$ | Cl Br I OR OAr SOR SO$_2$R | NMe$_2$ H |

Note in this list the appearance of several groups not normally encountered as leaving groups in aliphatic substitutions, notably the nitro group [78T2057]; it is not uncommon to find nucleophiles reacting with, say, a polynitrochlorobenzene to displace one of the nitro groups rather than the chlorine.

### 13.1.3 Activating groups

The relative effectiveness of different activating groups follows from the correlation of reaction rates with $\sigma^-$. The diazonium group, $N_2^+$, is very strongly activating; attempts to diazotise a $p$-nitroaniline, for example, may sometimes be foiled because the nitro group is displaced by water or by the counter-ion of the diazonium salt. Almost as strongly activating is a quaternary heterocyclic nitrogen, as in 2- and 4-chloropyridinium salts. The unquaternised pyridine nitrogen is somewhat less activating, comparable to a nitro substituent. The reactions of neutral pyridines are subject to acid catalysis (protonation of the pyridine

Table 13.3. *Second order rate coefficients* $(k_{obs}/10^{-6} \; l \; mol^{-1} \; s^{-1})$ *for the reactions of 4-X-2-nitrochlorobenzenes with* $MeO^-/MeOH$ *at 50°C* [68MIa]

| X | | X | | X | |
|---|---|---|---|---|---|
| NO | 1 580 000 (est) | COMe | 5 000 (3.2 × 10^{-3a}) | Cl | 35.1 |
| SMe$_2^+$ | 796 000 | COOMe | 3 930 | CO$_2^-$ | 8.50 |
| NO$_2$ | 288 000 (8.5[a]) | CF$_3$ | 2 050 | CH$_2$OMe | 3.32 |
| NMe$_3^+$ | 53 600 | CONH$_2$ | 658 | H | 2.52 (1.2 × 10^{-10}, |
| SO$_2$Ph | 46 400 | SMe$_2$ | 54.7 | | est.[a]) |
| SO$_2$Me | 32 200 | I | 43.9 | F | 2.25 |
| COPh | 6 680 | Br | 38.7 | Me | 0.299 |

[a]Substrates with a single activating group are much less reactive. The values in parentheses are the rate coefficients for the corresponding reactions of *p*-substituted chlorobenzenes (i.e. without the *ortho* nitro group) under similar conditions.

nitrogen atom) provided the nucleophile is not so strongly basic that its demand for protons swamps that of the pyridine. With amines and other protic nucleophiles this results in autocatalysis because the reaction generates acid as it proceeds (py−Cl + XH → py−X + HCl). Table 13.3 illustrates the effectiveness of a range of activating groups.

Activating groups that function largely by mesomeric withdrawal of electrons, the nitro group for example, can only do so if they can adopt the necessary coplanarity with the benzene ring. Steric inhibition is nicely illustrated by the behaviour of 1,3-dimethyl-2,5-dinitrobenzene (**13.4**). Ammonia displaces the *more* hindered of the two nitro groups, even though the methyl groups would deactivate the *ortho* position more than the *meta*, because in the corresponding Meisenheimer intermediate (**13.5**) the unhindered nitro group is free to conjugate with the ring [23J(123)1260].

(13.4)          (13.5)

### 13.1.4   Nucleophilicity

As with aliphatic nucleophilic substitution there is no unambiguous order of nucleophilicity; variations with solvent and substrate are considerable. In the 'normal' mode of reaction, with the first step rate-

Table 13.4. *Second order rate coefficients* $(k_{obs}/10^{-6} \ l \ mol^{-1} \ s^{-1})$ *for reactions of activated fluoro- and iodobenzenes*

| | 2,4-$(NO_2)_2C_6H_3$–Hal MeO$^-$/MeOH/0 °C (a) | p-$NO_2$–$C_6H_4$–Hal PhS$^-$/MeOH/0 °C (b) | 2,4-$(NO_2)_2C_6H_3$–Hal SCN$^-$/MeOH/100 °C (c) |
|---|---|---|---|
| F | 1 740 000 | 7.79 | 3.68 |
| I | 308 | 3.40 | 3310 |

(a) [65AJ117] (b) [66J(B)310] (c) [68MI*b*].

limiting, the rate is governed by the kinetic nucleophilicity of the nucleophile: equation [13.1]. But if the nucleophile is itself a good enough leaving group to switch control to the second step then nucleophilicity, as a kinetic property, no longer influences the rate at all. Equation [13.2] shows that $k_1$ only appears as part of the ratio $k_1/k_{-1}$, the *thermodynamic* equilibrium constant for formation of the Meisenheimer intermediate. Table 13.4 illustrates the sort of variability in nucleophilic reactivity that is possible.

It is therefore perhaps sufficient to note that sulphides, hydroxide, alkoxides and phenoxides, and aliphatic amines are usually good nucleophiles, whereas anilines, ammonia and halide ions are less reactive. Surprisingly the cyanide ion, a good nucleophile in aliphatic systems, is very unreactive in the Meisenheimer mechanism.

The complicated interactions of nucleophile, nucleofuge and solvent are illustrated by the behaviour of 1,2,4-trinitrobenzene and 1-fluoro- and 1-chloro-2,4-dinitrobenzene reacting with aniline and with $CF_3CH_2NH_2$ in acetonitrile or dimethyl sulphoxide [79J(P2)1317; 81J(P2)1201]. In the following scheme B indicates that the reaction is subject to base catalysis (second step rate-limiting), U that it is uncatalysed.

| | | Nucleofuge | | |
|---|---|---|---|---|
| *Nucleophile* | *Solvent* | $NO_2^-$ | F$^-$ | Cl$^-$ |
| $C_6H_5NH_2$ | MeCN | B | B | U |
| | $Me_2SO$ | B | U | U |
| $CF_3CH_2NH_2$ | MeCN | B | U | U |
| | $Me_2SO$ | U | U | U |

The variation with nucleofuge follows the pattern discussed above (§13.1.2). The differences between the solvents arise partly because bases are stronger in acetonitrile, and partly because dimethyl sulphoxide is itself a much stronger base than acetonitrile so that catalysis by solvent overwhelms that by added bases. The difference between the two nucleophiles is at first sight surprising because they have closely similar $pK_{BH^+}$ values.

Probably it arises because the transition state for the nucleophilic attack can be stabilised by partial delocalisation of charge into the aniline ring, thereby accelerating the first step and switching control to the second:

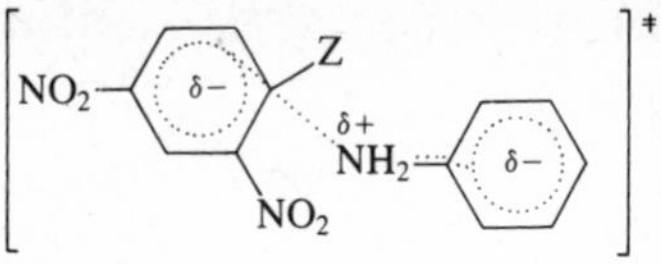

## 13.2 The $S_N1$ mechanism

There is no known example of an aryl halide reacting via an $S_N1$ mechanism, that is, dissociation to an aryl cation followed by nucleophilic attack. This is hardly surprising. The phenyl cation, $C_6H_5^+$, is an extremely high-energy species. The positive charge originates from a vacant orbital in the plane of the ring, orthogonal to the $\pi$ system and not stabilised by delocalisation other than a little hyperconjugation with the *ortho* C–H bonds. The vacant orbital ($sp^2$) has relatively high *s* character, leaving the remaining six bonding electrons with much higher average energy than, say, a t-butyl cation. Again the phenyl cation is much less strongly solvated than t-butyl, which is open on two sides to solvent molecules. Moreover, the ionisation of t-butyl halides is facilitated by steric acceleration: steric strain in the ground state is relieved as the central carbon atom eases from tetrahedral to trigonal. No such ground state destabilisation occurs with simple aryl halides; on the contrary, they are somewhat stabilised by mesomeric interaction between the ring and lone pair electrons on the halogen, and this is lost on ionisation.

### 13.2.1 *Dediazoniation via* $S_N1$

It seems surprising in view of this that any $S_N1$ reactions with aromatic substrates occur at all, but a number of displacements of nitrogen from diazonium salts do seem to proceed in this manner.

$$ArN_2^+ \underset{Fast}{\overset{Slow}{\rightleftharpoons}} N_2 + Ar^+ \overset{Nu^-}{\longrightarrow} ArNu$$

The evidence for this has been summarised and extended by Swain [75JA783, 791, 796, 799], and includes the following:

(*a*) The kinetics are first order in $[ArN_2^+]$ and independent of $[Nu^-]$.

(*b*) The addition of various nucleophiles to the reaction leads to the corresponding extra products but does not change the overall reaction rate, showing that the rate-limiting step precedes the product-forming step.

(*c*) The rate of decomposition of benzenediazonium fluoroborate at

Table 13.5. *First order rate coefficients* $(k_1/10^{-7} s^{-1})$ *for the decomposition of substituted benzenediazonium chlorides in water at 300 K* [calculated from data in 40JA1400]

| Substituent | OMe | Me | Ph | H | Cl | NO$_2$ |
|---|---|---|---|---|---|---|
| *meta* | 2500 | 2400 | 1275 | 526 | 21 | 0.45 |
| *para* | 0.08 | 64 | 27 | 526 | 1.0 | 2.2 |

constant temperature is remarkably uninfluenced by solvent. In water the rate is the same as in deuterium oxide. In 80 per cent sulphuric acid it is the same as in 105 per cent (though the activity of water decreases 1000-fold over the range 80–98 per cent). Even in such a variety of solvents as concentrated sulphuric acid, trifluoroacetic acid and methylene chloride there is only a 10 per cent change in rate. Clearly there is no solvent participation in the transition state.

(*d*) The entropy of activation for the decomposition of benzene-diazonium fluoroborate in water is $+44$ J K$^{-1}$ mol$^{-1}$. This is comparable to that for other $S_N1$ hydrolyses, whereas for $S_N2$ hydrolyses $\Delta S^{\ddagger} \simeq -50$ J K$^{-1}$ mol$^{-1}$; the difference represents the translational entropy of one mole of water.

(*e*) There is a very large $^{14}N/^{15}N$ isotope effect for Ph$-$*N$^+\equiv$N, about 1.04, compatible with virtually total cleavage of the C$-$N bond in the transition state. A small amount of Ph$-$N$^+\equiv$N* is formed during the reaction suggesting that it is initiated by the reversible loss of nitrogen: Ph$-$N$_2^+ \rightleftharpoons$ Ph$^+ +$ N$_2$. An alternative explanation, that there is an intramolecular 1,2-shift, is not consistent with later work showing that the decomposition of Ph$-^{15}$N$^+\equiv$N under an atmosphere of unlabelled N$_2$ is slower than the loss of $^{15}$N [76JA3301].

(*f*) There is a large secondary kinetic isotope effect when hydrogen atoms of the benzene ring are replaced by deuterium. A single *ortho* deuterium atom reduces the rate by a factor of 1.22. This, the largest secondary isotope effect of its type known, suggests that the *ortho* C$-$H or C$-$D bond is strongly hyperconjugated in the transition state.

(*g*) Substituents in the ring influence the reaction rate in an unusual manner (table 13.5). Substituents in the *meta* position correlate much better with $\sigma_p$ than with $\sigma_m$, and *para* substituents give a scatter plot, the reaction being retarded by both electron acceptors and electron donors.

In fact this surprising behaviour is entirely compatible with the $S_N1$ mechanism. The explanation can be found by analysing the data using a dual parameter equation to separate polar and mesomeric effects:

$$\log(k/k_0) = \rho_I\sigma_I + \rho_R\sigma_R^+$$

The values of $\rho_I$, the polar reaction constants, are $-4.5$ and $-3.7$ for *meta* and *para* substituents, respectively; the negative sign indicates deactivation by electron-withdrawing groups, which destabilise the phenyl cation. For *meta* substituents the resonance effect reaction constant $\rho_R$ is also negative, $-1.8$. It is remarkable that a *meta* interaction should involve so large a resonance contribution, emphasised by the correlation with the through-conjugation parameter $\sigma_R^+$; it arises because resonance donors generate negative charge at positions *ortho* to the cationic centre (*cf.* **3.6**, §3.2.1) which is efficiently transmitted towards it by induction and hyperconjugation. By contrast $\rho_R(para) = +2.4$; resonance donors in the *para* position can interact strongly through the $\pi$ system with the diazonium group itself, thus stabilising the parent salt and retarding its dissociation (**13.6**).

$$(13.6)$$

### 13.2.2 Related and competing mechanisms

Diazonium salts offer the only well-defined examples of $S_N1$ reactions via aryl cations. It is possible, but not established, that diaryliodonium salts, $Ar_2I^+$, can also react this way.

On the other hand it is certain that diazonium salts can also react by other mechanisms. Meisenheimer reactions may occur when there are other activating groups in the ring, and many examples of radical mechanisms are also known. Some of these are discussed in §13.4.1.

The reduction of $ArN_2^+$ in $MeO^-/MeOH$ (nominally a substitution by hydride) actually takes place by parallel radical and carbanionic mechanisms occurring in the ratio of about 2:3, as revealed by the use of $CH_3OD$ and $CD_3OH$ as solvents [83JO191]:

## 13.3 The aryne mechanism

Although unactivated aryl halides are virtually inert in Meisenheimer and $S_N1$ substitutions they are reactive towards very strong bases, notably amide ions:

$$PhCl + KNH_2 \xrightarrow[-33\,°C]{Liq.\,NH_3} PhNH_2$$

Roberts used [14]C-labelling experiments to reveal that these reactions are accompanied by *cine*-substitution; that is, the incoming nucleophile may attack the position *ortho* to that vacated by the leaving group [56JA601]:

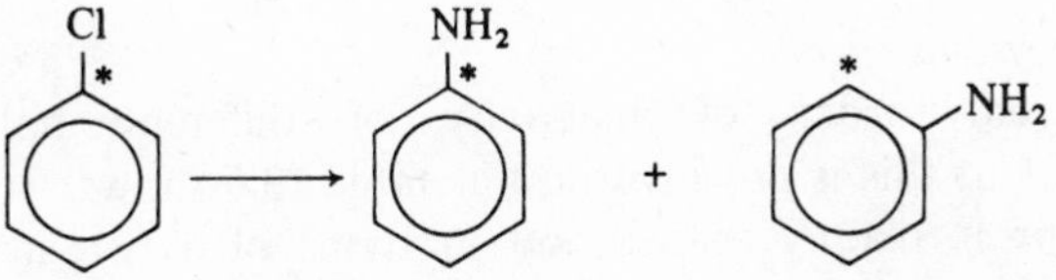

48 per cent          52 per cent

Likewise *o*-chlorotoluene reacts to give a mixture of *o*- and *m*-toluidines, *p*-chlorotoluene gives both *p*- and *m*-toluidines, and *m*-chlorotoluene gives a mixture containing all three isomers [56JA611]. Substitution only occurs at the *ipso* or *ortho* positions, never further away; moreover, if both *ortho* positions are occupied the reaction is totally suppressed.

Following an earlier suggestion of Wittig's, Roberts proposed an elimination–addition mechanism:

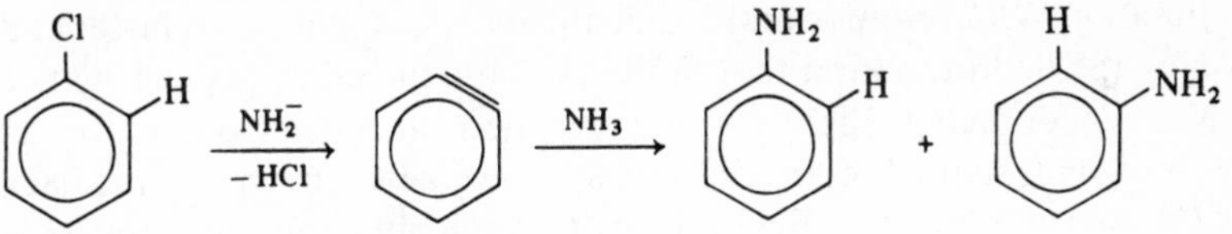

Benzyne

The intermediate benzyne (dehydrobenzene) has a weak extra bond formed by diagonal interaction of the $sp^2$ orbitals (**13.7**); the overlap is about half that of an ordinary benzene $\pi$ bond. Under some conditions the benzyne intermediate can be trapped by Diels–Alder addition with

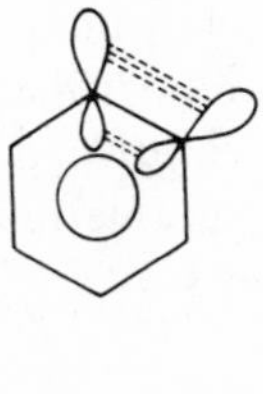

(**13.7**)

Table 13.6. *Relative rates of reaction of phenyl halides (aryne mechanism)*

|  | I | Br | Cl | F |
|---|---|---|---|---|
| $KNH_2/NH_3/-30\,°C$ [36JO170] | 8 | 20 | 1 | 0 |
| Li piperidide/$Et_2O$/20 °C [59CB192] | 0.6 | 1.6 | 1 | 4 |

dienes [75TL3371] though usually this is overwhelmed by its reaction with a nucleophilic solvent.

### 13.3.1 Leaving group effects

The usual order of reactivity of different halides is $Br \geqslant I > Cl \gg F$, but this is not invariant as table 13.6 shows.

This behaviour is readily understood in terms of a two-step $E1cB$ formation of the aryne:

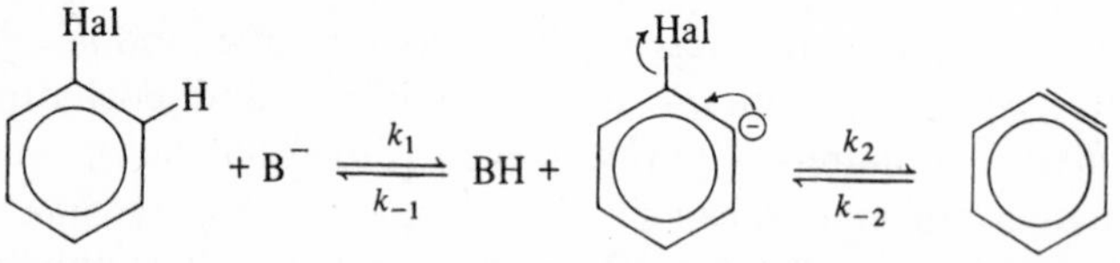

Consider first the reaction with potassamide in liquid ammonia. Fluoride is a poor leaving group and $k_2$ for this reaction is vanishingly small. Chloride is somewhat more reactive but it still leaves slowly and step 2 is rate-limiting. With bromide and iodide the second step is much faster and it is now the deprotonation that limits the rate, the more electronegative bromine encouraging slightly faster reaction than the iodine. In an aprotic solvent, step 1 becomes effectively irreversible (the rate of the back reaction depends on $k_{-1}[BH]$, and with negligible concentration of a protic compound present this is much less than $k_2$), so control passes to step 1 for all reactions. Fluorine, the most electronegative halogen, now reacts fastest (as it does in the Meisenheimer mechanism when the first step is rate-limiting). A variety of different halogen reactivity patterns can be obtained using lithium piperidide in ether/piperidine mixtures, changing the BH concentration to alter the balance between the two steps of the reaction [59CB192].

When *o*-deuteriofluorobenzene is treated with potassamide in ammonia there is a rapid exchange of deuterium even though the reaction does not continue to the substitution stage. With the chloride, deuterium exchange and benzyne formation take place at comparable rates. With the bromide no detectable exchange occurs [55JA4540].

The contrast between the normal reactivity patterns in aryne reactions (fluorides least reactive) and in Meisenheimer reactions (fluorides most reactive) is illustrated by the behaviour of the halogenonaphthalenes with sodamide in pyridine [56JA6265]. Seven of the eight compounds give an almost identical mixture of 1- and 2-naphthylamines (24:76 ± 2 per cent) via the common naphthalyne. For 1-fluoronaphthalene the ratio is 60:40. In this, the most favourable case for a Meisenheimer reaction, the two mechanisms are just about equally balanced.

### 13.3.2 Substituent effects on aryne formation

Meta substituted halogenobenzenes can generate two different arynes:

2-yne    (13.8)    (13.9)    3-yne

Which of the two predominates depends on the substituent X and on which step in the dehydrohalogenation is rate-limiting. The alternative carbanion intermediates, with negative charges ortho and para to X, experience similar resonance effects, so that the polar influence of X is dominant in differentiating them. Electron withdrawal, stabilising the carbanion, accelerates proton loss but retards the expulsion of halide. Thus m-bromoanisole reacts with potassamide to give exclusively the 2-yne; the inductive withdrawal of electrons by the methoxy group accelerates formation of the ortho anion (13.8) more than para (13.9). Alkyl groups, which inductively donate electrons, behave in the opposite manner: m-bromotoluene gives mainly the 3-yne. But with m-chlorotoluene it is the loss of chloride that is rate-limiting and now the inductive effect of the methyl group favours formation of the 2-yne [56JA611].

### 13.3.3 Substituent effects in the addition to the aryne

The rate of formation of the aryne controls the overall rate of reaction, but a substituted aryne can give two different products and the ratio in which these are formed depends on the relative rates of the alternative addition reactions.

The reaction proceeds preferentially via the more stable of the alternative anions (which presumably resemble the corresponding transition states). A substituent can influence this in two ways. The inductive or field effect dominates the interaction between substituent and anionic charge. A group that inductively withdraws electrons can stabilise an adjacent

Table 13.7. *Products of reaction of* $X-C_6H_4-Br$ *with* $KNH_2/NH_3$ [65TL963]

| X | Products from *ortho*-X (via 2-yne) | | Products from *para*-X (via 3-yne) | |
|---|---|---|---|---|
| | *ortho* (per cent) | *meta* (per cent) | *meta* (per cent) | *para* (per cent) |
| F | 0–1 | 99–100 | 20–25 | 75–80 |
| CN | 10–15 | 85–95 | 0–5 | 95–100 |
| OMe | 0–5 | 95–100 | 45–50 | 50–55 |
| Me | 55 | 45 | 60 | 40 |
| O⁻ | 85–90 | 10–15 | 100 | 0 |

negative charge and encourage a 2-yne to react via (**13.11**) to give the *meta* product. Likewise, though to a lesser degree, it can stabilise a *meta* anion relative to a *para* one, so a 3-yne reacts mainly via (**13.13**) to give the *para* product. Inductive electron donors behave in the opposite fashion. However, a secondary consideration is the possibility of *meso-meric* interaction between a $\pi$-withdrawing substituent and the incoming nucleophile if, as with an amino group, this bears a lone pair of electrons. Such an interaction tends to encourage the formation of *ortho* or *para* disubstituted products. Table 13.7 shows that this pattern is indeed borne out in practice. Fluorine, inductively a strong electron-withdrawing group, promotes almost exclusive formation of the *meta* product from the 2-yne (via anion (**13.11**)), and to a lesser extent it encourages formation of the *para* product from the 3-yne. The cyano group has a comparable inductive effect to fluorine, but it also has a strong electron-withdrawing mesomeric effect. This reinforces the stability of anion (**13.13**) from the 3-yne, but with the 2-yne it tips the balance a little way towards the less stable anion (**13.10**) and increases the proportion of the minor product. Methoxyl behaves similarly to fluorine but its influence is weaker. Methyl has a relatively small effect and considerations other than simple substituent effects become significant; for example, *o*-chlorotoluene gives about a 1:1 mixture of *ortho* and *meta* products with potassamide, but with lithamide the ratio is about 3:1 [75JO1835]. The oxide group, a strong electron donor both inductively and mesomerically, shows a pattern complementary to that of cyanide.

## 13.4 Mechanisms involving radicals

### 13.4.1 *The Sandmeyer reaction*

A number of reactions of diazonium salts, although formally nucleophilic substitutions in that a nucleophilic species replaces the nucleofugic $N_2$, actually occur by radical mechanisms. The best known of

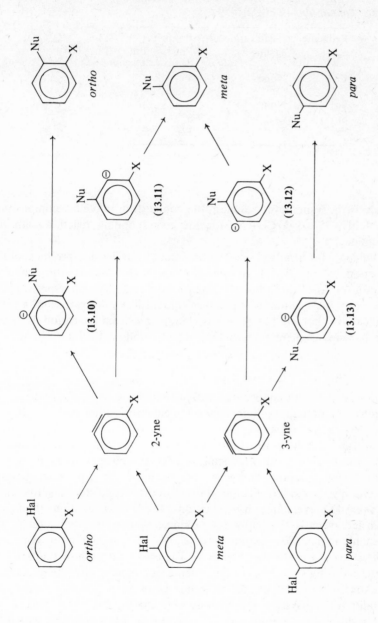

Table 13.8. *Products of the reaction of aqueous* $(PhN_2^+)_2SO_4^{2-}$ *with* $Cl^-$ *in the presence of metal ions* [81J(P2)1459]

| Reducing agent | Ligand transfer agent | Products (%) PhCl | PhH + Ph$_2$ |
|---|---|---|---|
| Cu$^+$ | None | 45 | 1 |
| Cu$^+$ | Cu$^{2+}$ | 63 | trace |
| Sn$^{2+}$ | None | >0.1 | 10 |
| Sn$^{2+}$ | Cu$^{2+}$ | 67 | >1 |
| Sn$^{2+}$ | Fe$^{3+}$ | 12 | 20 |

there is the Sandmeyer reaction, the copper(I) promoted decomposition of $ArN_2^+ Cl^-$ to $ArCl + N_2$ (or the corresponding reaction with the bromide).

Evidence for a radical mechanism comes from the side-products of the reaction, mainly Ar–H, Ar–Ar, and Ar–N=N–Ar, and from the formation of Ar–I and Ar–OH when iodine and oxygen, respectively, is added to the reaction mixture: these are all likely derivatives of an aryl radical, Ar$^{\cdot}$, but would not be expected to arise from the cation Ar$^+$. The mechanism has therefore been formulated as follows [42J266; 57JA2942].

$$ArN_2^+ + Cu^I Cl_2^- \rightarrow Ar^{\cdot} + Cu^{II}Cl_2 + N_2$$
$$Ar^{\cdot} + Cu^{II}Cl_2 \rightarrow ArCl + Cu^I Cl$$

The special role of copper was at first thought to be that it had just the right redox potential: other reducing agents were either too weak to give the aryl radical or strong enough to reduce the diazonium ion to the hydrazine. Recent work (table 13.8) has shown that its significance is rather more subtle: other reducing agents can replace copper(I) in the first step of the reaction, but copper(II) seems to be especially good at facilitating the oxidative ligand transfer in the second step. The first entry in table 13.8 exemplifies a conventional Sandmeyer reaction: the ligand transfer agent is the copper(II) produced during the reduction of the diazonium ion. The second entry shows that adding more copper(II) increases the yield of chlorobenzene. The other entries show tin(II) to be a good reducing agent, but in the absence of copper(II) the phenyl radicals are diverted into side-products. A number of other reducing agents gave similar results. When the reactions were carried out in the presence of both chloride and bromide ions the ratio of PhCl to PhBr was independent of the reducing agent used – evidence of a common intermediate, by implication the phenyl radical, in all the reactions [82J(P2)1139].

### 13.4.2 The $S_{RN}1$ mechanism

Bunnett has described how a serendipitous experiment led to his discovery of a new mechanism for aromatic substitution [78Acc413]. He found that the isomeric 2,3,5- and 2,4,5-trimethylhalogenobenzenes (**13.14** and **13.15**; Z = Cl, Br) react with potassamide in ammonia to give in each case a 3:2 mixture of the two amines (**13.14** and **13.15**; Z = NH$_2$), as expected for reaction via the common aryne (**13.16**). The iodo compounds, however, give predominantly *ipso* substitution, though some of the *cine*-substituted products are still formed. Clearly a second mechanism is running parallel to the aryne reaction.

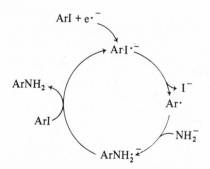

(13.14)          (13.15)          (13.16)

Bunnett [70JA7463, 7464] proposed a radical-anion mechanism, $S_{RN}1$, initiated by some electron source which is not immediately apparent and then following the cycle of steps shown in figure 13.2. Evidence for this mechanism includes the following:

(*a*) Pseudocumene (**13.14** or **13.15**; Z = H) is a side-product, presumably formed by Ar˙ abstracting a hydrogen atom from ammonia.

(*b*) If potassium metal, a good one-electron source, is used in addition to the potassamide then *cine*-substitution is largely suppressed and high yields of the *ipso*-amine together with pseudocumene are obtained.

(*c*) Conversely the addition of a radical scavenger suppresses the non-aryne reaction.

Figure 13.2. The $S_{RN}1$ mechanism.

Since these initial observations there have been many reports of other reactions that follow the $S_{RN}1$ mechanism [78Acc413]. Many are photo-initiated – further evidence for a radical mechanism, though the precise details of the activation have not been established. A range of nuc-leophiles other than $NH_2^-$ participate in $S_{RN}1$ reactions, notably thiolates, dialkyl phosphite anions, and enolates. In a competitive experiment the relative reactivities of $(EtO)_2PO^-$ and t-BuCOCH$_2^-$ towards six different substrates PhZ were in each case $1.37 \pm 0.11$ regardless of the leaving group Z [81JA7140]. This provides further evidence for the proposed mechanism, with the product-controlling step, $Ph^{\bullet} + Nu^-$, independent of Z, and is inconsistent with a possible $S_{RN}2$ alternative which would involve $PhZ^{\bullet-} + Nu^- \rightarrow PhNu^{\bullet-} + Z^-$.

Dihalogenobenzenes undergo $S_{RN}1$ reactions to give a mixture of mono- and disubstituted products but, remarkably at first sight, the monosubstituted product is not necessarily on the route to the disubsti-tuted one. For example, when $m\text{-}BrC_6H_4I$ and $(EtO)_2PO^-$ are irradiated for 7 minutes in liquid ammonia the major product (60 per cent) is the bisphosphonate, $C_6H_4\{PO(OEt)_2\}_2$, 28 per cent of the unreacted dihalide remains, and there is only 7 per cent of the monosubstituted $BrC_6H_4PO(OEt)_2$. A natural assumption might be that the monosubsti-tuted product is consumed almost as rapidly as it is formed, but in fact it reacts more slowly than the dihalide. The $S_{RN}1$ mechanism provides a simple rationalisation [78JA1873] – in the following scheme Ⓟ is an abbreviation for $(EtO)_2PO$.

$$Br-C_6H_4^{\bullet} \xrightarrow{\text{Ⓟ}^-} Br-C_6H_4-\text{Ⓟ}^{\bullet-} \xrightarrow{-Br^-} {}^{\bullet}C_6H_4-\text{Ⓟ} \xrightarrow{\text{Ⓟ}^-} \text{Ⓟ}-C_6H_4-\text{Ⓟ}^{\bullet-}$$

$$Br-C_6H_4-\text{Ⓟ} \qquad\qquad \text{Ⓟ}-C_6H_4-\text{Ⓟ}$$

monosubstituted                    disubstituted

## 13.5    Further reading

J. Miller, *Aromatic Nucleophilic Substitution*, Elsevier, Amsterdam, 1968.
J. F. Bunnett & R. E. Zahler, *Chem. Rev.*, 1951, **49**, 273.
J. F. Bunnett, *Quart. Rev.*, 1958, **12**, 1; *Acc. Chem. Res.*, 1972, **5**, 139.
S. D. Ross, *Progr. Phys. Org. Chem.*, 1963, **1**, 31.
F. Pietra, *Quart. Rev.*, 1969, **23**, 504.
C. F. Bernasconi, *Chimia*, 1980, **34**, 1.

Meisenheimer mechanism:

S. D. Ross, in *Comprehensive Chemical Kinetics*, eds C. H. Bamford & C. F. H. Tipper, vol. 13, Elsevier, Amsterdam, 1972.

C. F. Bernasconi, *MTP Rev. Sci., Org. Chem. Ser. One*, 1973, **3**, 35; *Acc. Chem. Res.*, 1978, **11**, 147.
G. Bartoli & P. E. Todesco, *Acc. Chem. Res.*, 1977, **10**, 125 (leaving group and nucleophile).

## $S_N1$ mechanism:

H. Zollinger, *Acc. Chem. Res.*, 1973, **6**, 338 (but see also [75JA783]).
H. B. Ambroz & T. J. Kemp, *Chem. Soc. Rev.*, 1979, **8**, 353.

## Aryne mechanism:

R. W. Hoffmann, *Dehydrobenzene and Cycloalkynes*, Academic, New York, 1967.
E. K. Fields, in *Organic Reactive Intermediates*, ed. S. P. McManus, Academic, New York, 1973.
J. A. Zoltewicz, *Internat. Rev. Sci., Org. Chem. Ser. Two*, 1976, **3**, 63.

## $S_{RN}1$ mechanism:

J. F. Bunnett, *Acc. Chem. Res.*, 1978, **11**, 413.
R. A. Rossi, *Acc. Chem. Res.*, 1982, **15**, 164; *J. Chem. Educ.*, 1982, **59**, 310.
R. A. Rossi & R. H. de Rossi, *Aromatic Substitution by the $S_{RN}1$ Mechanism*, Amer. Chem. Soc., Washington, 1983.
M. Chanon & M. L. Tobe, *Angew. Chem. Internat. Ed.*, 1982, **21**, 1.

## Radical mechanisms:

M. Tiecco & L. Testaferri in *Reactive Intermediates*, vol. 3, ed. R. A. Abramovitch, Plenum, New York, 1983.

# 14 Molecular rearrangements

A rearrangement involves the migration of a group from one point of attachment in a molecule to another. The commonest types of rearrangement are 1,2-shifts, where the migration is between adjacent atoms, and aromatic side-chain to ring rearrangements, and it is with these that this chapter is largely concerned:

$$\begin{array}{ccc} \underset{|}{Z} & \longrightarrow & \underset{|}{Z} \\ A-B & & A-B \end{array}$$

Pericyclic rearrangements form a third important category; examples are touched on here but detailed discussion is deferred to chapter 16.

Even within these limits the variety of rearrangements is immense and a complete treatment is quite beyond the scope of this text. The discussion is therefore centred around a few selected rearrangements that illustrate the mechanistic problems and the ways in which they can be approached.

## 14.1  Intermolecular and intramolecular rearrangements

The first mechanistic question that needs to be answered is: is the rearrangement intermolecular or intramolecular? Does the group Z become completely detached from the rest of the molecule during the migration, or is some contact maintained throughout?

### 14.1.1  Intermolecular: the Orton rearrangement

An intermolecular mechanism can most readily be detected by intercepting one of the separated fragments. The Orton rearrangement of N-chloroanilides, catalysed by hydrochloric acid, is a well-established example [27J986; 58J2982]:

$$C_6H_5-N\overset{Cl}{\underset{Ac}{\diagdown}} + HCl \underset{k_{-1}}{\overset{k_1}{\rightleftharpoons}} C_6H_5-NHAc + Cl_2 \xrightarrow[k_2]{S_EAr} Cl-C_6H_4-NHAc + HCl$$

Under reduced pressure the free chlorine can actually be sucked out of the reaction mixture, and unchlorinated acetanilide is formed. This is an unusually simple way of demonstrating intermolecularity, but even without this evidence the formation of free chlorine is easily revealed. Thus the ratio of *p*-chloroacetanilide (the major product) to its *ortho* isomer is the same as in the direct chlorination of acetanilide under the same conditions, and in the presence of *p*-cresol, a more reactive substrate than acetanilide, the chlorine is diverted into the production of dichlorocresol.

The initial equilibrium can be directly measured with a deactivated anilide for which ring chlorination is relatively slow: 2,4-dichloro-acetanilide has been used. In dilute aqueous acetic acid the equilibrium lies towards the *N*-chloroanilide; as the water concentration is reduced the equilibrium is shifted, and in glacial acetic acid it is well over to the side of free chlorine.

If the reaction is started in the middle, so to speak, by treating acetanilide with chlorine, the rates of *N*- and *C*-chlorination can be measured separately. They remain in a constant ratio throughout the reaction, confirming that the anilide is being consumed by two reactions of the same (second) order [28J998].

Kinetic studies show that the rearrangement is first order in substrate and second order in hydrochloric acid [52QR34]. This indicates that the initial equilibrium is a two-stage process:

$$Ph-N\overset{Cl}{\underset{Ac}{\diagdown}} \overset{H^+}{\rightleftharpoons} Ph-\overset{\overset{Cl}{\underset{\uparrow}{|}}}{\underset{Ac}{N^+}}-H \quad \xrightarrow{Cl^-} \rightleftharpoons \quad PhNHAc + Cl_2$$

The rate is proportional to the concentrations of hydrogen ion and chloride ion, and with the hydrogen chloride fully dissociated these are each equal to the stoicheiometric concentration of hydrogen chloride.

The behaviour of *N*-chloroanilides with other acids confirms the mechanism. Hydrobromic acid leads to the *ortho* and *para* bromo-anilines, formed in an $S_EAr$ reaction with BrCl, a brominating agent. Hydriodic acid behaves similarly, except that some of the ICl is reduced by more iodide ion to iodine, which does not attack the ring:

$$Ar-N\overset{Cl}{\underset{Ac}{\diagdown}} + 2 HI \longrightarrow Ar-NHAc + HCl + I_2$$

In sulphuric acid the N—Cl bond is hydrolysed (fairly slowly) but *C*-chlorination does not normally ensue. The chlorine is converted into hypochlorous acid $(ArN^+HClAc + H_2O \rightarrow ArNHAc + H_2O^+Cl)$, a relatively weak chlorinating agent which is reluctant to react with the anilide but whose presence can be detected by trapping with *p*-cresol [27J2761; 27JPC1192].

### 14.1.2 *Intramolecular: the benzidine rearrangements*

Hydrazobenzene rearranges in acid to give a mixture of benzidine and diphenyline.

$$\text{NH—NH} \xrightarrow[\text{H}_2\text{O/EtOH/O °C}]{\text{HCl}} \text{H}_2\text{N}——\text{NH}_2 + \text{NH}_2, \text{NH}_2$$

70 per cent benzidine      30 per cent diphenyline

Other types of product may be formed if the reaction is carried out in aprotic solution or if the phenyl rings are substituted. These include:

$$\text{NH}_2, \text{NH}_2 \qquad \text{H}_2\text{N}——\text{NH}—— \qquad \text{NH}_2, \text{NH}$$

*ortho*-benzidine      *para*-semidine      *ortho*-semidine

together with aniline and azobenzene, which come from the disproportionation of hydrazobenzene. However, these compounds are usually formed only in traces unless both phenyl rings bear *para* substituents, in which case the *ortho*-semidine may become the principle product. Even here the benzidine may be formed preferentially if the *para* substituent can be displaced; the 4,4'-dicarboxylic acid, for example, rearranges with decarboxylation.

The rearrangements are entirely intramolecular, as demonstrated by 'crossover' experiments. Two substrates, differentiated by substitution in each benzene ring, are allowed to rearrange together: no cross-products are ever detected, such as would be formed if the rings became completely separate. In this type of experiment it is important to use substrates that react at comparable rates; otherwise one rearrangement could be essentially complete while the other had hardly begun. Thus the distinguishing substitution should preferably involve closely similar groups or isotopic labelling. When a mixture of 2,2'-methoxy- and 2,2'-ethoxy-hydrazobenzenes is rearranged there is no detectable formation of the unsymmetrical benzidine (**14.1**) [33J984]. When 2-[$^{14}$C]-methylhydrazo-benzene and unlabelled 2,2′-dimethylhydrazobenzene rearrange together none of the label appears in the dimethylbenzidine (**14.2**) [52JA2282]. Similar results are found for the other rearrangement products. Moreover, there is never any sign of a product that could have been formed from attack of solvent on a hydrazobenzene fragment.

OMe    OMe    OEt    OEt      MeO    OEt

⟨ring⟩—NHNH—⟨ring⟩ + ⟨ring⟩—NHNH—⟨ring⟩  ⤳̸  H$_2$N—⟨ring⟩—⟨ring⟩—NH$_2$

(14.1)

Me    Me    Me*       Me    Me*

⟨ring⟩—NHNH—⟨ring⟩ + ⟨ring⟩—NHNH—⟨ring⟩  ⤳̸  H$_2$N—⟨ring⟩—⟨ring⟩—NH$_2$

(14.2)

    The reaction is first order in hydrazobenzene, but for a long time it proved difficult to establish its kinetic dependence on acid concentration. It is now clear that there are actually two different mechanisms, one first order in acid and one second order. With some substrates only one of the mechanisms is observed; with others the two mechanisms can run in parallel, the balance moving from first order to second as the concentration of the acid is increased. The catalysis is specific for both mechanisms. This is demonstrated not only by the absence of any rate change with a change in buffer concentration, but also, very neatly, by solvent isotope effects. As the aryl groups and the acid concentration are changed, and the balance between the two mechanisms is varied, so the observed order in acid appears to be intermediate between one and two. The D$_2$O/H$_2$O isotope effect changes in sympathy, increasing from around 2.3 with first order reactions to 4.8 with second order ones. The more or less constant proportionality points strongly to mechanisms in which all proton transfers occur in rapid pre-equilibria (for which a solvent isotope effect of 2 to 3 per proton is expected).

    No kinetic isotope effect is found when the ring hydrogen atoms are replaced by deuterium, so the final proton loss is not rate-limiting either. We can therefore put forward an outline of the mechanism as shown in figure 14.1; this illustrates the formation of benzidine itself, but similar mechanisms can be drawn for the formation of the other products.

    However, this is still only scratching at the surface of the mechanism. The rearrangement itself is established as the rate-limiting process, but how exactly does it happen? The answer is still not completely certain, but we can now make reasonably reliable suggestions, at least for the formation of benzidine itself.

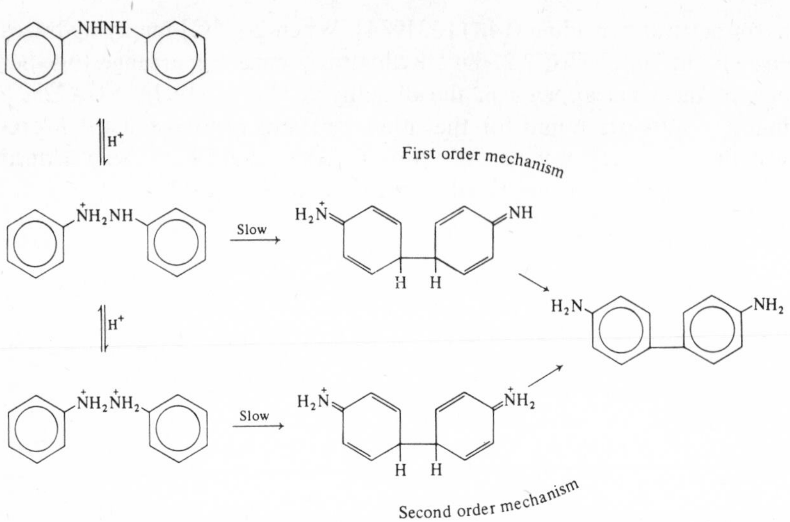

Figure 14.1. Acid catalysis of the benzidine rearrangement.

One possibility that must be considered is that there are actually two rearrangement steps: hydrozobenzene → semidine → benzidine. However, semidines do not normally rearrange to benzidines under the conditions of the benzidine rearrangement, so this can be discounted.

The mechanism that comes nearest to explaining the observed facts is the so-called 'polar transition state' mechanism [64J2864]. This envisages a partial cleavage of the N—N bond to give a transition state with one ring resembling a neutral aniline molecule (nitrogen atom positively charged, *ortho* and *para* carbon atoms negatively charged), and the other ring bearing a single or double positive charge. In the monoprotonated transition state this positive charge is extensively delocalised (**14.3**), but electrostatic repulsions confine the charges in the diprotonated transition state largely to the nitrogen and the *para* carbon atom (**14.4**).

As the N—N bond is ruptured, so the new C—C bond is being formed, preferentially between a positively charged position on one ring and a

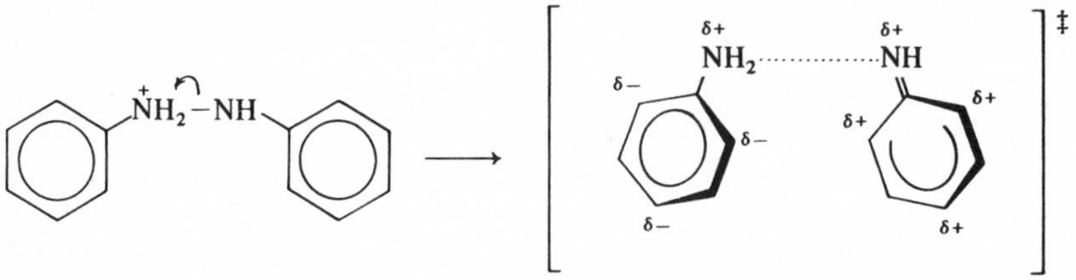

(14.3)

**(14.4)**

negatively charged one on the other. This suggests that both benzidine itself and *ortho*-benzidine could be produced in the first order reaction, but that in the second order reaction the *ortho*-benzidine should be disfavoured. The formation of diphenyline requires a mutual rotation of the two rings, and this may well be facilitated in the second order mechanism because of the mutual repulsion of the two positively charged nitrogen atoms.

To a large extent this is borne out in practice. The rearrangement of 1,1'-hydrazonaphthalene is first order in acid and the product comprises 60 per cent *para*-benzidine (**14.5**) and 20 per cent each of the *ortho*-benzidine (**14.6**) and the carbazole derived from it (**14.7**). Presumably the

mutual attraction of the two naphthalene π systems, one positively charged and the other negatively, opposes any rotation so that no dinaphthyline is formed.

In the parent system itself, on the other hand, the rearrangement is close to second order in acid and, as already mentioned, the products are benzidine and diphenyline.

This interpretation is supported and extended by the observation of heavy atom kinetic isotope effects [82JA2501]. When both nitrogen atoms are labelled the $^{14}N/^{15}N$ rate ratio is 1.0222 for the formation of benzidine and 1.0633 for the formation of diphenyline. With the *para* carbon atoms labelled the corresponding $^{12}C/^{13}C$ effects are 1.0209 and 1.0006. These values accord with a concerted mechanism for the formation of benzidine, with the N—N bond breaking as the C—C bond forms. In the formation of diphenyline, however, there is much more N—N breaking in the rate-limiting transition state and negligible C—C making, suggesting a two-step mechanism via some intermediate within which the mutual rotation of the rings takes place. The intermediate might well be a π complex (**14.8**) in which the two rings are held parallel by the mutual interaction of their π systems. Dewar [63MI] first

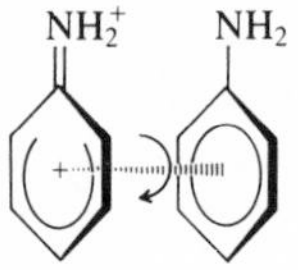

(**14.8**)

proposed such species as intermediates in the whole family of benzidine rearrangements, before it was established that some of them are concerted processes.

The semidines are produced by attack of the 'aniline-like' nitrogen on the positively charged ring. If one ring bears a strongly electron-donating *para* substituent then an *ortho*-semidine, specifically the one with the electron donor in the same ring as the primary amino group, becomes a major product. The reactions are first order in acid, and the behaviour is again in accord with the principles of the polar transition state mechanism:

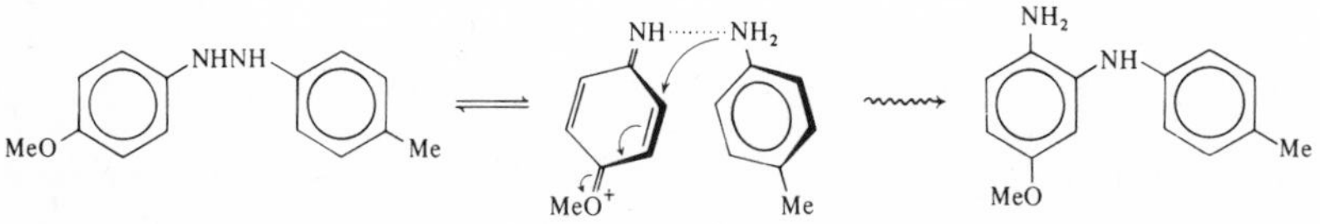

However, a relatively large $^{14}N/^{15}N$ kinetic isotope effect, comparable to that observed during the formation of diphenyline, suggests that this rearrangement too passes through some intermediate rather than being fully concerted: confirmatory evidence from carbon kinetic isotope effects is not available [82JA5181].

Until recently it was believed that *para*-semidine formation was different from the other rearrangements, either intermolecular or requiring heavy metal catalysis. This is now known not to be so; it is a normal, acid catalysed, intramolecular reaction. Moreover, the observation of significant carbon and nitrogen kinetic isotope effects suggests that, contrary to expectation, it is in fact a concerted process (actually a [1,5] sigmatropic rearrangement, §16.2.3) with the migrating nitrogen atom bridging the other nitrogen and *para* carbon atoms in the transition state, presumably with the ring buckled into a boat conformation [82JA5181]:

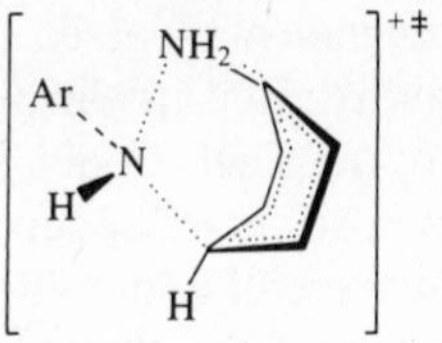

There remains the question: what determines the order of the reaction in acid? The first order mechanism requires the substrate to be weakly basic (reluctant to undergo diprotonation), yet able to disperse the positive charge in the transition state. This is the situation with the hydrazonaphthalenes; they are weaker bases than hydrazobenzene but they possess an extended π system. First order reactions are also often found with *ortho* substituted hydrazobenzenes; perhaps steric hindrance to solvation inhibits their diprotonation. The balance of effects is certainly complex and has not been treated quantitatively.

### 14.1.3 *Pericyclic rearrangements: the Claisen rearrangement*

The Claisen rearrangement of allyl aryl ethers (and allyl vinyl ethers behave similarly) is another intramolecular, side-chain to ring rearrangement:

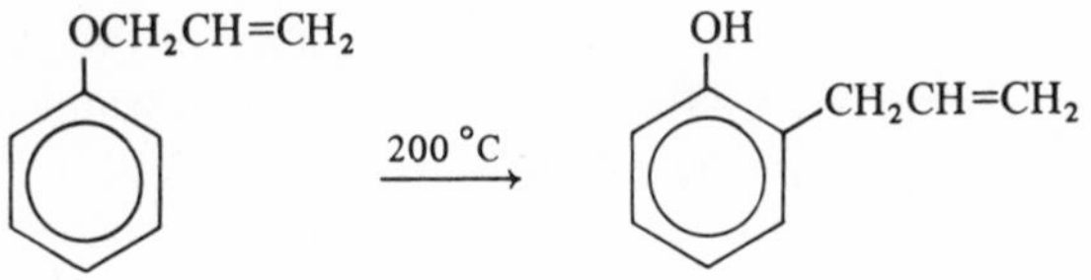

The intramolecularity is demonstrated by the absence of crossover products when mixed ethers are rearranged together [37JA107], and by the formation of a chiral product from the chiral 1,3-dimethylallyl ether [51JA4304]. Labelling, either isotopic or using substituents, demonstrates that the allyl group somersaults during the rearrangement (that is, the reaction is $S_N'$ [e.g. 52JA5866]).

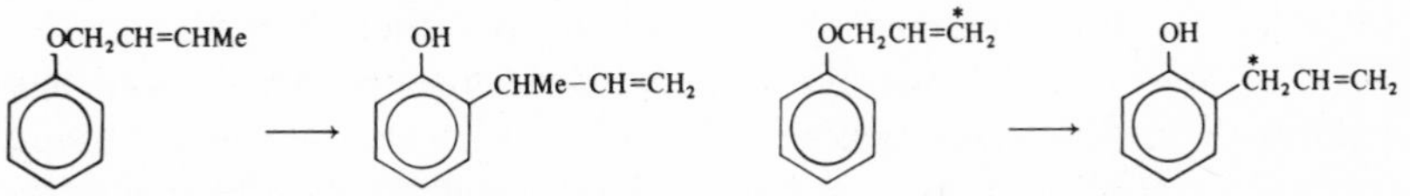

If both *ortho* positions are blocked the ether rearranges to the *para*-allylphenol. Again the reaction is intramolecular, but labelling experiments show that now the new C–C bond is formed to the same position of the allyl group as was originally bonded to oxygen; this suggests that the reaction goes through an *ortho* intermediate and the *para* product is formed via a second somersault.

Claisen himself suggested that the rearrangements passed through dienone intermediates, and this has been demonstrated in various ways. Thus the dienone (14.10) has been trapped by its Diels–Alder addition with maleic anhydride [56JA2290]. It has also been independently synthesised; when heated to 70 °C it gives a mixture of about 70 per cent of the *para* Claisen product (14.11) and 30 per cent of the allyl ether (14.9) [57JA3156]. Direct measurement by ultraviolet spectroscopy of the

steady-state concentration of the dienone shows it to be about 0.1 per cent, and this agrees with values calculated from kinetic studies. These also reveal large negative entropies of activation, typically $-50$ J K$^{-1}$ mol$^{-1}$, for both *ortho* and *para* rearrangements, in accord with the formation of a highly ordered cyclic transition state.

Thus there is little doubt about the essential nature of the rearrangement, but it has proved very difficult to obtain experimental evidence about the structure of the transition state. Indeed, it is still not irrefutably certain that the rearrangement is a one-step concerted process. Solvent and substituent effects are small. For the *ortho* rearrangement of *para* substituted ethers there is a rather surprising correlation with $\sigma^+$, but with a very small value of $\rho = -0.6$. This led to the suggestion that the transition state (or possibly an intermediate) might have some ion-pair character (14.12), though complete ionisation is ruled out by the activation entropy and by the small magnitude of $\rho$. However, the reaction is also slightly accelerated by electron donation in the allyl group, so this cannot be a true explanation [61JA3846]. There have also been suggestions that the reaction may involve homolysis rather than heterolysis; there is, however, no positive evidence for this, though Dewar has recently described experiments which imply that the analogous Cope rearrangement may involve a biradical intermediate corresponding to (14.13) [77JA4417].

However, most current opinion favours the concerted mechanism and

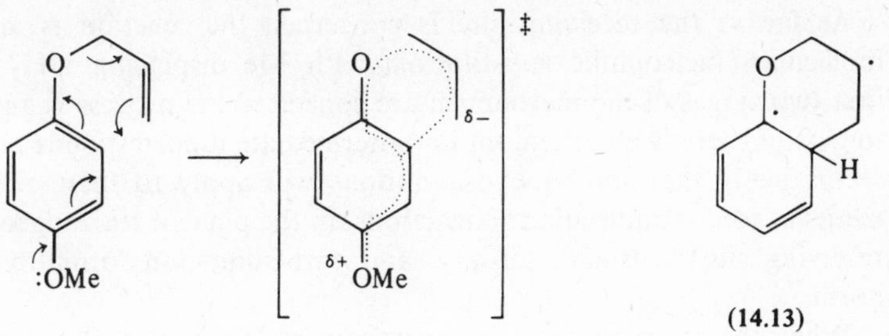

(14.12)

(14.13)

the insights into this that come from theoretical considerations are discussed in chapter 16.

## 14.2 Concerted and non-concerted rearrangements: 1,2-shifts

### 14.2.1 Pinacol and related rearrangements

An intermolecular rearrangement must be at least a two-step reaction, with the intermediate formation of the separated fragments. An intramolecular rearrangement, however, may occur in an elementary process with concerted bond making and bond breaking, though there is nothing in principle to exclude a multistep reaction.

The majority of 1,2-shifts are nucleophilic. (Electrophilic and radical mechanisms are also well known but are not discussed here; examples may be found in the books listed at the end of the chapter.) In a nucleophilic shift the migrating group moves with its bonding electrons to an adjacent electron-deficient atom. In such a rearrangement we may discern three distinct processes: the creation of the electron-deficient site, the migration,* and the restoration of electron sufficiency. There can be varying degrees of concertedness of the three stages. Thus one could envisage the pinacol rearrangement occurring in three distinct steps:

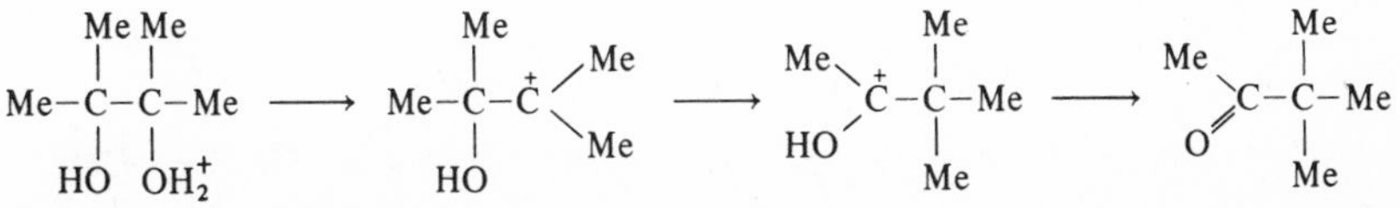

or as a single concerted process:

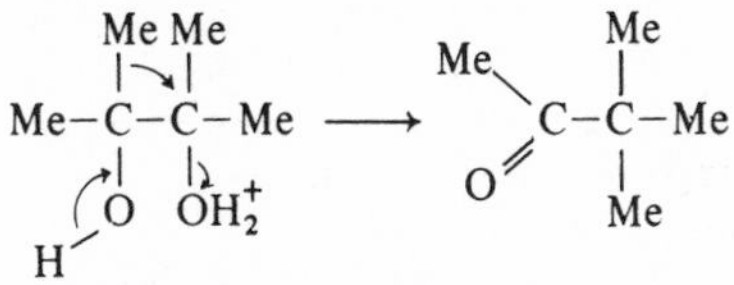

or via an intermediate mechanism.

---

* The migration itself may be concerted or it may be a two-step process; this question is discussed in §14.3.

As far as the receiving site is concerned the reaction is an intra-molecular nucleophilic substitution, with Me displacing $OH_2^+$. If the first two stages of the mechanism are concerted the process is analogous to $S_N2$; if there is a carbenium ion intermediate it corresponds to $S_N1$. It seems likely that similar considerations will apply to these rearrangements as to intermolecular $S_N$ reactions. In the pinacol rearrangement the receiving site is usually tertiary, and carbenium ion formation is the norm.

When the rearrangement is carried out in $H_2^{18}O$ some of the label is incorporated into unrearranged pinacol, pointing to the reversible formation of an intermediate prior to the migratory step [58J403].

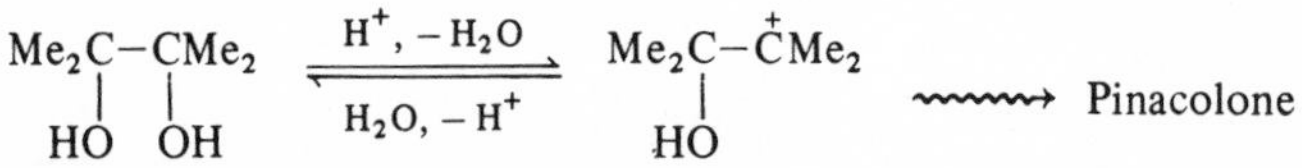

$$\underset{\underset{HO\quad OH}{|\quad\ |}}{Me_2C-CMe_2} \quad \underset{H_2O,\,-H^+}{\overset{H^+,\,-H_2O}{\rightleftharpoons}} \quad \underset{\underset{HO}{|}}{Me_2C-\overset{+}{C}Me_2} \quad \rightsquigarrow \quad Pinacolone$$

In dilute acid the exchange is about three times faster than the rearrangement; at higher acid concentrations the relative rate of exchange falls off linearly with the activity of water.

Further evidence for the intermediacy of the carbenium ion is the observation that pinacol and pinacolone are formed from a variety of other carbenium ion precursors in the same proportions as are found in the exchange experiment (figure 14.2) [59CI332].

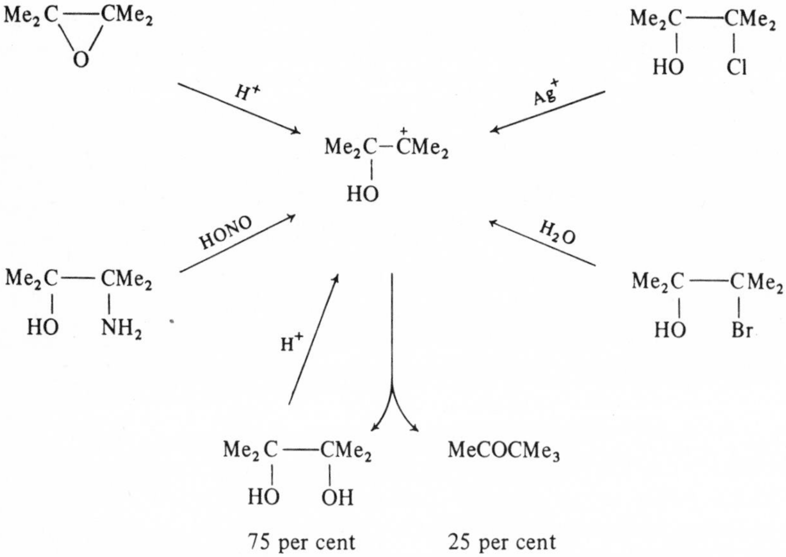

Figure 14.2. Evidence for a common carbenium ion intermediate in pinacol and related rearrangements.

Again, the rearrangements of *cis*- and *trans*-1,2-dimethylcyclohexane-1,2-diols both give predominantly 1-acetyl-1-methylcyclopentane together with a little 2,2-dimethylcyclohexanone, and the rearrangement is accompanied by $^{18}O$ exchange with solvent and by isomerisation of the two diols, all indicative of a common carbenium intermediate (**14.14**) [63J5854].

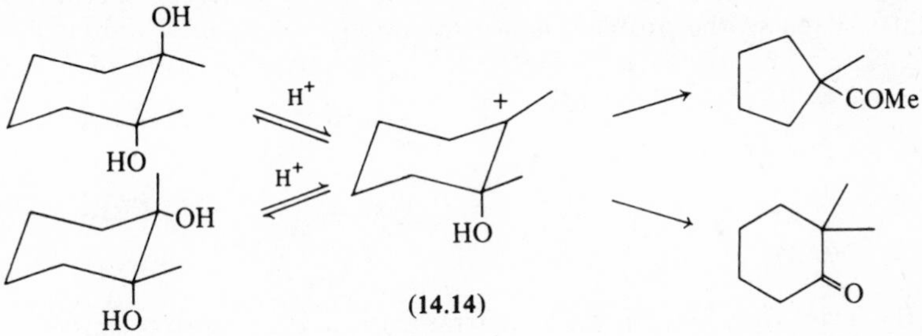

(14.14)

In pinacols that bear different groups attached to the glycol fragment the question arises as to which of them migrates. This is controlled in the first instance by which of the hydroxyl groups is expelled: migration can only occur from the other carbon atom. This problem can be studied with pinacols of the form of (**14.15**), in which this is the only consideration.

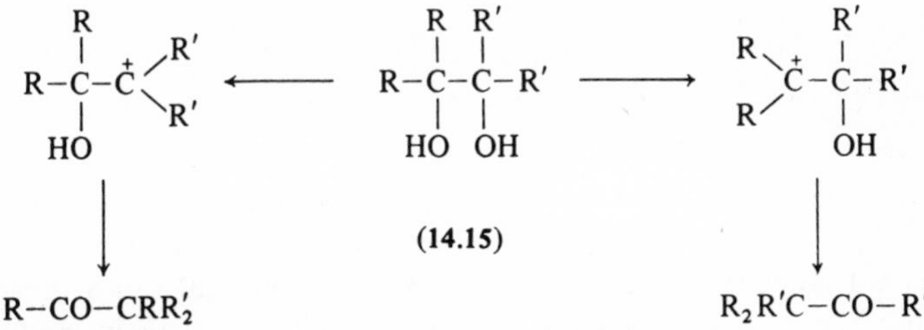

(14.15)

Normally the rearrangement occurs via the more stable of the two alternative carbenium ions. To take an extreme example, the rearrangement of $Ph_2C(OH)-CH_2OH$ gives exclusively the aldehyde $Ph_2CH-CHO$ via the highly stabilised $Ph_2C^+-CH_2OH$ cation. However, many exceptions to this behaviour are known, presumably because under some reaction conditions carbenium ion formation is reversible and the inherent migratory aptitudes of the groups R and R' become important. Thus the rearrangement of the dimethyl diphenyl glycol (**14.16**) gives different products under different conditions:

$$\underset{\substack{|\\ \text{Ph}}}{\overset{\substack{\text{Me}\\ |}}{\text{Ph}-\text{C}-\text{CO}-\text{Me}}} \xleftarrow[\text{H}_2\text{SO}_4]{\text{Cold conc.}} \underset{\substack{|\quad|\\ \text{HO}\;\;\text{OH}}}{\overset{\substack{\text{Ph Me}\\ |\quad|}}{\text{Ph}-\text{C}-\text{C}-\text{Me}}} \xrightarrow[\text{Trace H}_2\text{SO}_4]{\text{HOAc}} \underset{\substack{|\\ \text{Me}}}{\overset{\substack{\text{Ph}\\ |}}{\text{Ph}-\text{CO}-\text{C}-\text{Me}}}$$

(14.16)

Pinacols of the form (14.17) can give only one carbenium ion, so their rearrangements can give some information about inherent migratory aptitudes. Even so the problem is not so simple.

$$\underset{\substack{|\quad|\\ \text{HO}\;\;\text{OH}}}{\overset{\substack{\text{R}'\;\;\text{R}'\\ |\quad|}}{\text{R}-\text{C}-\text{C}-\text{R}}} \longrightarrow \underset{\substack{|\\ \text{HO}}}{\overset{\substack{\text{R}'\\ |}}{\text{R}-\text{C}-\overset{+}{\text{C}}}}\!\!\underset{\text{R}}{\overset{\text{R}'}{<}} \quad \begin{array}{c} \xrightarrow{\text{R migrates}} \text{R}'-\text{CO}-\text{CR}_2\text{R}' \\[2mm] \xrightarrow{\text{R}'\,\text{migrates}} \text{R}-\text{CO}-\text{CRR}'_2 \end{array}$$

(14.17)

In the transition state for the rearrangement the positive charge is shared by the original carbenium centre, the migrating group, the carbon atom from which it is migrating, and the oxygen atom (14.18). It cannot be taken for granted that electron donation is more effective in stabilising

$$\left[ \underset{\text{HO}}{\overset{\overset{\displaystyle\text{R}}{\diagup\;\diagdown}}{\text{R}-\!\!-\text{C}-\!\!\!-\text{C}}}\!\!\underset{\text{R}}{\overset{\text{R}}{<}} \right]^{+\ddagger} \qquad (14.18)$$

the charge if it is in the migrating group than if it is in the group left behind. In general it seems that it is: for aryl groups, electron-donating substituents in the *meta* and *para* positions greatly increase the migratory aptitude so that *p*-methoxyphenyl moves some 500 times more readily then phenyl. Aryl groups in general usually migrate in preference to alkyl including one example in a gas-phase rearrangement [80T1993]. Hydrogen shifts are very unpredictable.

Continuing the analogy with intermolecular $S_N$ reactions, we might expect to find that migration to a primary centre is a concerted $S_N2$-like process. There is relatively little work in this area, but there is some evidence to support the suggestion. Thus chiral deuteriated neopentyl systems (e.g. 14.19) rearrange with Walden inversion [74T1733].

Migration to a secondary centre has been extensively studied using the semipinacol rearrangement (diazotisation of a 2-aminoalcohol). Most early work pointed strongly to a concerted process. Thus the dia-

(14.19)          (+ unrearranged products)

stereoisomeric aminoalcohols (**14.20**) give different rearrangement products: regardless of its electron-donating properties the migrating aryl group is the one *anti* to the diazonium leaving group in the most stable molecular conformation, namely that with the two bulky aryl groups straddling the adjacent hydrogen [55JA355]. (Strictly one should refer to the steric interactions in the transition state rather than in the ground state conformer, but they presumably parallel one another.)

An–CO–CHMePh          (90 per cent)

(**14.20***a*)

[An = anisyl (*p*-methoxyphenyl)]

Ph–CO–CHMeAn          (90 per cent)

(**14.20***b*)

This particular study used racemic compounds, but other work indicated that similar rearrangements took place with stereochemical inversion at the receiving site [e.g. 39JA1324]. However, subsequent investigations revealed a more complex and very surprising stereochemical behaviour. The stereospecifically $^{14}$C-labelled 1,1-diphenyl-2-aminopropanol (**14.21**) rearranges with partial racemisation at the receiving site: the labelled phenyl group migrates exclusively with inversion, the unlabelled one with retention [57JA6160].

(**14.21**)          88 per cent          12 per cent

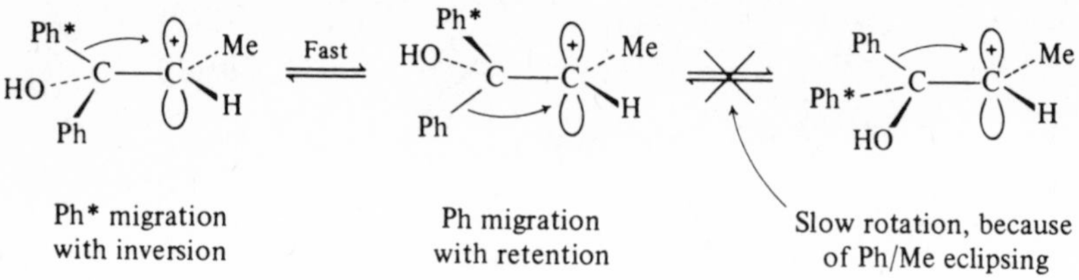

Ph* migration      Ph migration      Slow rotation, because
with inversion      with retention      of Ph/Me eclipsing

Figure 14.3. Stereospecific migrations in semipinacol rearrangement of 1,1-diphenyl-2-aminopropanol. Migrating C—Ph bond needs to be coplanar with receiving *p* orbital.

This suggests that the mechanism involves a carbenium ion (or ion-pair) in which rotation about the central C—C bond is relatively slow (figure 14.3). Even so the unequal product distribution suggests there may be a simultaneous concerted $S_N2$-like migration of Ph*; one would expect the 60° rotation of figure 14.3 to be exceedingly fast, so the $S_N1$ route should give almost equal proportions of the two products.

### 14.2.2 The Beckmann rearrangement

The rearrangement of ketoximes to amides can be catalysed by phosphorus pentachloride, concentrated sulphuric or polyphosphoric acids, or by acid chlorides. For many years the mechanism was uncertain, each new suggestion in turn being contradicted by fresh evidence. Part of the problem arose from the geometrical isomerism of the oximes: in the absence of modern spectroscopic and other physical methods it was difficult to assign their sterochemistry, and the picture was further confused because the *syn* and *anti* isomers are readily equilibrated in acid.

However, it is now established that with very few exceptions the rearrangement occurs exclusively with migration of the group *anti* to the oxime hydroxyl, and this stereochemical cohesion suggests that the migration must be concerted with the departure of the leaving group.

The nature of the leaving group has been a subject of many investigations. In general it seems the role of the catalyst is to esterify the hydroxy group, making it a better nucleofuge. Oxime sulphonates, for example, spontaneously rearrange at room temperature. The rearrangement with sulphuric acid has often been represented as taking place via simple protonation to give $RR'C=NOH_2^+$ (presumably a minor tautomer; $RR'C=N^+HOH$ is likely to be the major form), which rearranges with loss of water. This may occur with some of the more reactive oximes. That from 2,4,6-trimethylacetophenone rearranges in relatively dilute acid; migration of the mesityl group is facilitated both because it can readily accept a partial positive charge in the transition state and because there is some steric acceleration from the *ortho* methyl groups:

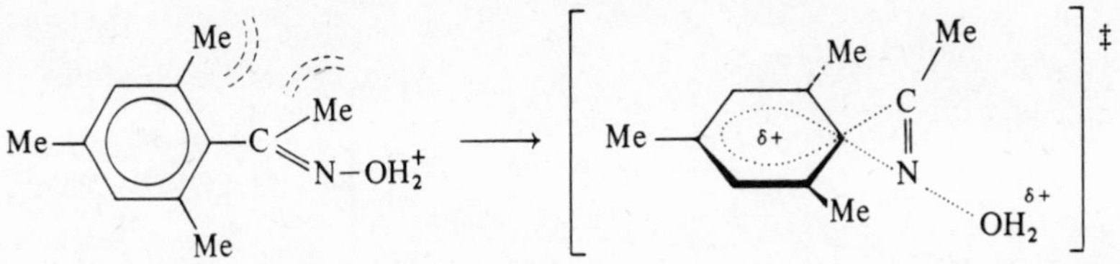

In solutions with $H_0 > -6$ (about 70 w/w per cent $H_2SO_4$) the rate of rearrangement at any particular value of $H_0$ is much the same in either sulphuric or perchloric acid. This points to a common intermediate, presumably the protonated oxime. In more concentrated acid solutions sulphuric acid is a much more effective catalyst than perchloric, suggesting that a different mechanism has set in [70J(B)338]. Under these conditions it seems likely that the acid forms the oxime hydrogen sulphate, and this is probably the reactive species in most sulphuric acid catalysed rearrangements. Many oximes give isolable salts that do not readily rearrange; the reactive species in these cases is unlikely to be the cation. Spectroscopic studies have revealed the presence both of the oxime hydrogen sulphate and of a nitrilium cation [70J(B)338; 71CC1601], suggesting that the rearrangement step is:

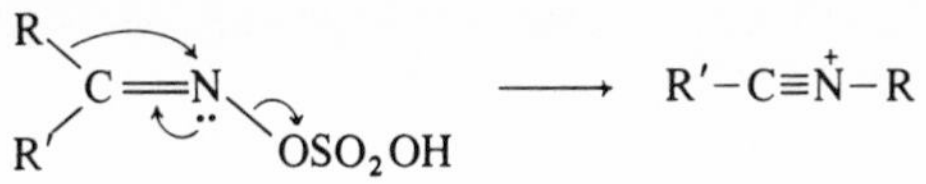

In the normal course of events the nitrilium ion is hydrated to the amide; when the rearrangement is carried out in the presence of $H_2{}^{18}O$ the amide oxygen is labelled, though neither oximes nor amides undergo direct oxygen exchange under the same conditions. In the presence of added nucleophiles the nitrilium ion can be diverted to other products: reactions in ethanol give the imidate ester, and hydrazoic acid gives a tetrazole [48J1514; 55JO726]:

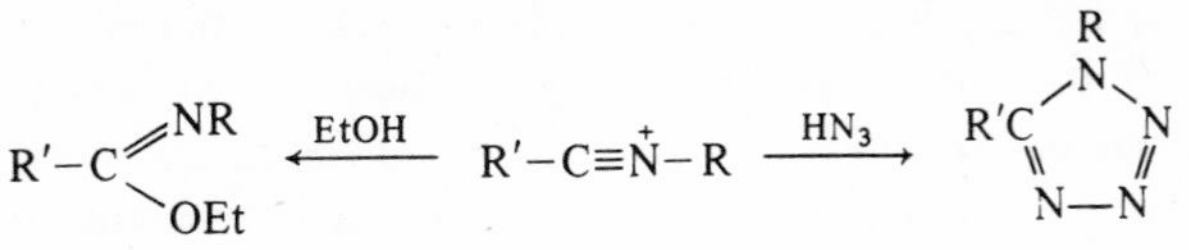

If the migrating group can form a stable carbenium ion the normal concerted rearrangement can be overtaken by an intermolecular reaction, as demonstrated by crossover experiments [63JO278]:

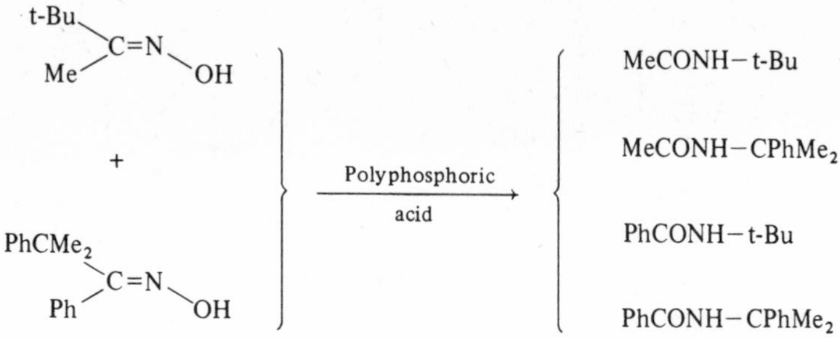

t-Bu
‚C=N
Me      OH

+

PhCMe₂
‚C=N
Ph      OH

} Polyphosphoric acid →

{ MeCONH−t-Bu

MeCONH−CPhMe₂

PhCONH−t-Bu

PhCONH−CPhMe₂

The reaction clearly involves a fragmentation (elimination) to form a nitrile, and benzonitrile has indeed been isolated during the rearrangement of phenyl t-butyl ketoxime with PCl₅ [55JA1094]:

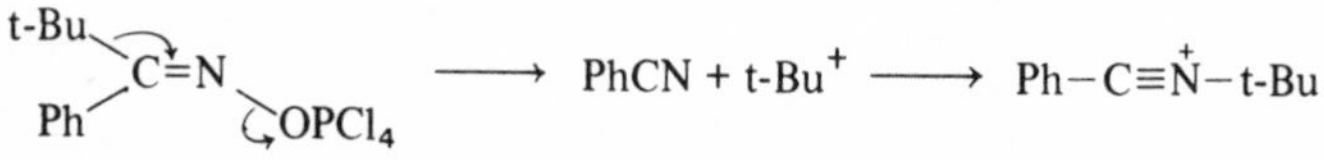

t-Bu
‚C≡N
Ph      OPCl₄     ⟶ PhCN + t-Bu⁺ ⟶ Ph−C≡N⁺−t-Bu

### 14.2.3 'Nitrene' rearrangements

When an azide is pyrolysed it can lose molecular nitrogen to generate an electron-deficient nitrene, and rearrangement takes place. The question arises: does the nitrene (**14.22**) exist as a true intermediate, or is the migration concerted with the expulsion of nitrogen?

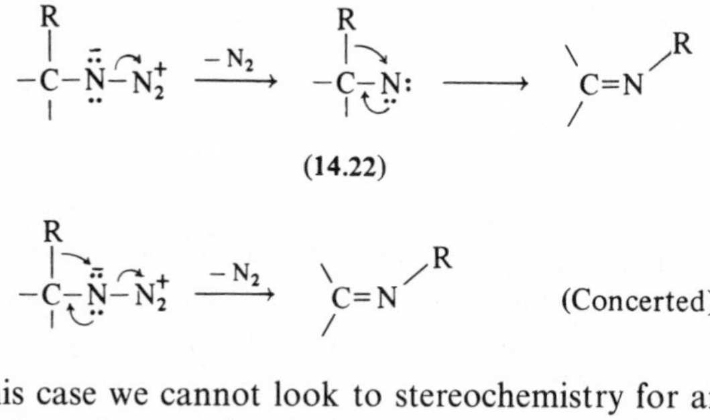

R
|
−C−N̈−N₂⁺   $\xrightarrow{-N_2}$   −C−N:   ⟶   \C=N/R
|                              |

(**14.22**)

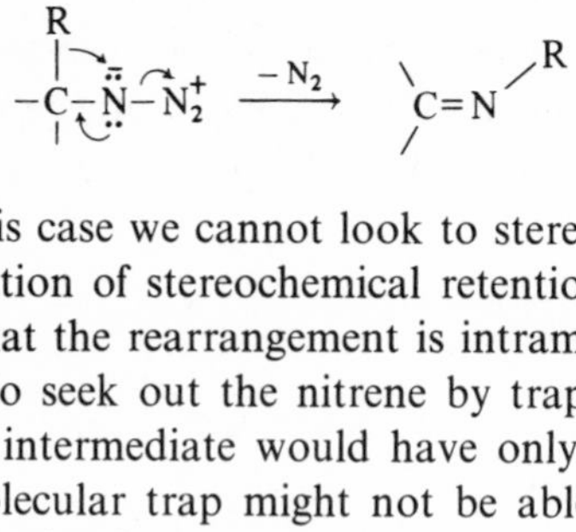

R
|
−C−N̈−N₂⁺   $\xrightarrow{-N_2}$   \C=N/R     (Concerted)
|

In this case we cannot look to stereochemistry for an answer (though observation of stereochemical retention in the migrating group R indicates that the rearrangement is intramolecular). It is therefore necessary to try to seek out the nitrene by trapping. However, it is likely that a nitrene intermediate would have only a very fleeting existence, and an intermolecular trap might not be able to compete with the rearrangement, so that intramolecular trapping is called for.

At 200 °C the tertiary azide (**14.23**) decomposes to give mainly elimination products (−HN₃; probably homolytic), but also formed, each in about 7 per cent yield, are the two rearranged imines (**14.24** and **14.25**) and the dihydrophenanthridine (**14.26**) [74JA480]. The latter is almost

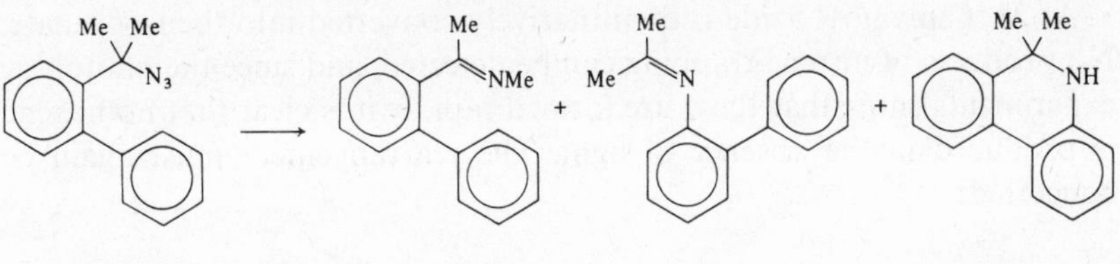

(14.23)    (14.24)    (14.25)    (14.26)

certainly formed by an intramolecular $S_E$Ar attack of the strongly electrophilic nitrene on the second benzene ring. The fact that in the rearrangements the biphenyl group migrates only about twice as readily as methyl is also evidence for a highly reactive intermediate; usually aryl migrates much faster than alkyl (compare §14.2.1).

No such insertion products can be isolated from tertiary alkyl azides in which the nitrene would be formed in less close proximity to an intramolecular nucleophilic centre; one may assume that they too rearrange via a similar mechanism, but direct evidence is lacking.

Acyl azides, $RCON_3$, likewise rearrange with the loss of molecular nitrogen to give isocyanates, $RN=C=O$ (the Curtius rearrangement). The reaction can be carried out either photolytically or thermally. Photolysis of pivaloyl azide (t-BuCON$_3$) in a range of different solvents at $-10\,^\circ$C gives in each case about 40 per cent of the rearranged t-BuNCO together with a wide variety of other products arising from reactions between the acyl nitrene and the solvent. (Acyl nitrenes are somewhat more stable than alkyl nitrenes because of delocalisation (compare **14.27**), and can therefore survive long enough to enter into intermolecular reactions.) Surprisingly this behaviour reveals that the nitrene, though it is formed under these conditions, is *not* an intermediate in the rearrangement. If it were, then the yield of isocyanate should vary, depending on the proportion of the nitrene trapped by the solvent. The simplest explanation of the facts is that there are two independent first order reactions of the azide, a concerted rearrangement, and the formation of a nitrene [67JA6308]:

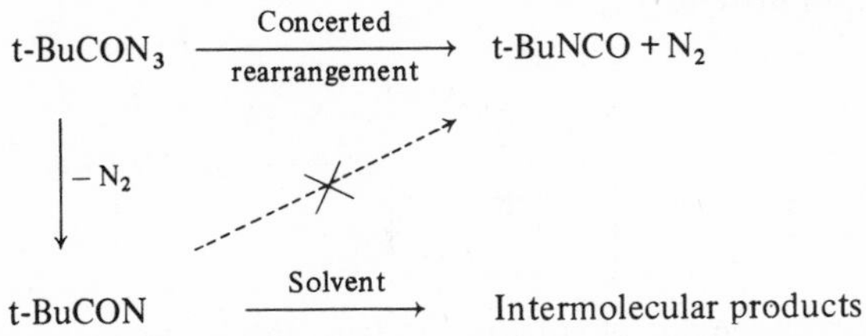

Why the acyl nitrene should not undergo rearrangement (dotted line) at a rate competitive with the other reactions is not clear.

At 28 °C pivaloyl azide is quantitatively converted into the isocyanate. No products of nitrene trapping can be detected, and since the photolytic experiments show that these are formed rapidly it is clear that no nitrene is produced in the absence of light. The rearrangement must again be concerted:

$$
\left[ \text{t-Bu}-C\overset{O}{\underset{\bar{N}-N_2^+}{\diagdown}} \quad \longleftrightarrow \quad \text{t-Bu}-C\overset{O^-}{\underset{N\overset{\frown}{-}N_2^+}{\diagdown}} \right] \xrightarrow{-N_2} \text{t-BuNCO}
$$

$$(14.27)$$

The Curtius rearrangement is one member of a series of related rearrangements of $N$-acyl compounds to isocyanates. Others include the Schmidt and Hofmann rearrangements; these are usually carried out in aqueous solution, in which the isocyanate RNCO is hydrolysed (with decarboxylation of the intermediate RNHCOOH) to the amine, $RNH_2$:

$$\text{Schmidt: } RCOOH + HN_3 \xrightarrow[H_2O]{H_2SO_4} RNH_2 + CO_2 + N_2$$

$$\text{Hofmann: } RCONH_2 \xrightarrow[H_2O]{NaOH/Br_2} RNH_2 + CO_2$$

The Schmidt rearrangement is closely similar to the Curtius. The carboxylic and hydrazoic acids react together to form a protonated acyl azide, $RCONHN_2^+$, probably in an $A_{AC}1$ reaction via $RCO^+$ (§12.1.3). This can then rearrange to the $N$-protonated isocyanate, but there is no conclusive evidence about the mechanism of the migration.

The Hofmann rearrangement has been extensively studied. It involves the sequence:

$$RCONH_2 \xrightarrow{Br_2} RCONHBr \xrightarrow{HO^-} RCO-\bar{N}-Br \rightarrow RNCO$$

The intermediate $N$-bromoamides have been isolated and shown to rearrange in alkaline solution. In the rearrangement step the migration is probably concerted with the expulsion of the bromide, without formation of an intermediate nitrene:

$$
R-C\overset{O}{\underset{\overset{\frown}{N}-Br}{\diagdown}} \quad \longrightarrow \quad \left[ \underset{N}{\overset{\delta^+}{R}}\cdots\overset{O}{\underset{\overset{|}{\underset{Br}{\delta^-}}}{C}} \right]^{\ddagger} \quad \longrightarrow \quad R-N=C=O
$$

The rates of rearrangement of substituted benzamides (R = Ar) correlate with the Yukawa–Tsuno equation ([3.4]; §3.1.3); the use of $\sigma^+$ implies that in the transition state there is some delocalisation of positive charge onto the migrating aryl group [71BJ1632]. In the rearrangement of $^{14}$C-labelled *N*-chlorobenzamide there are approximately equal $^{12}$C/$^{14}$C kinetic isotope effects for the carbonyl and 1-phenyl carbon atoms $(1.0456 \pm 0.0012$ and $1.0447 \pm 0.0006$, respectively), showing that both atoms are undergoing bonding changes in the transition state [71BJ2776].

The Hofmann rearrangements of aliphatic amides are usually assumed to be similarly concerted, but there is little evidence either way.

## 14.3 The controversy about non-classical carbonium ions

### 14.3.1 *Non-classical intermediates and σ participation*

In the preceding discussion of 1,2-shifts our attention has focussed on the possible connection between the migration and the expulsion of the leaving group: are they discrete processes or do the groups move in concert? This is another way of asking if the migrating group offers anchimeric assistance: does it participate as a neighbouring group in the departure of the leaving group (compare §7.5.3)? The answer, as we have seen, varies from one reaction to another.

We have so far given scant attention to the detailed nature of the migration itself; it could be a single elementary step or a two-step process passing through a bridged intermediate in which the migrating group is bonded simultaneously to its old site and to its new one. There is ample evidence to show that when the migrating group is aryl there can be a bridged phenonium intermediate (compare **7.2** in §7.2 and the examples in §7.5.3), although this does not mean that all aryl migrations must be two-step processes. Much more problematical are the migrations of alkyl groups or of a hydrogen atom. These reactions have engendered one of the most vigorously pursued controversies of mechanistic organic chemistry, still not fully resolved after 30 years. In essence the question is: is the midpoint of the migration (**14.28**) merely a transition state, or is it a non-classical carbonium ion intermediate?

(**14.28**; R = alkyl or H)

It has not always been made clear that there is a distinction between this question and the quite separate question as to whether the σ electrons of the migrating group can participate in the expulsion of the leaving group. There are, in fact, four essentially different mechanisms for the 1,2-shift of an alkyl group or hydrogen atom. These are illustrated in the schemes below for a Wagner–Meerwein rearrangement with substitution; analogous mechanisms can be envisaged for other 1,2-shifts.

(*a*) Ionisation and migration are two distinct elementary steps; there is no σ participation, no non-classical intermediate:

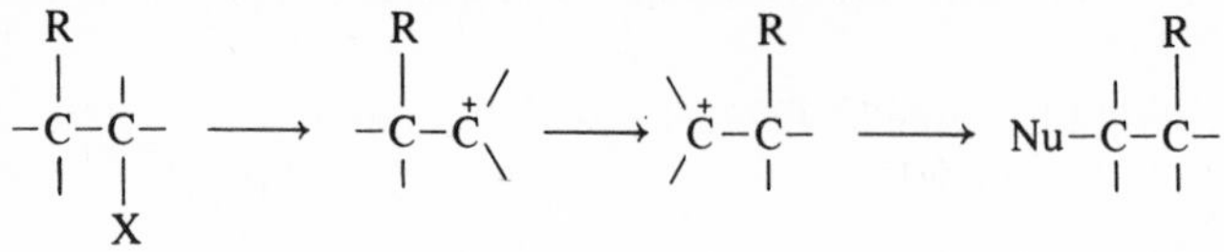

(*b*) Ionisation without σ participation is followed by a two-step migration involving a non-classical intermediate:

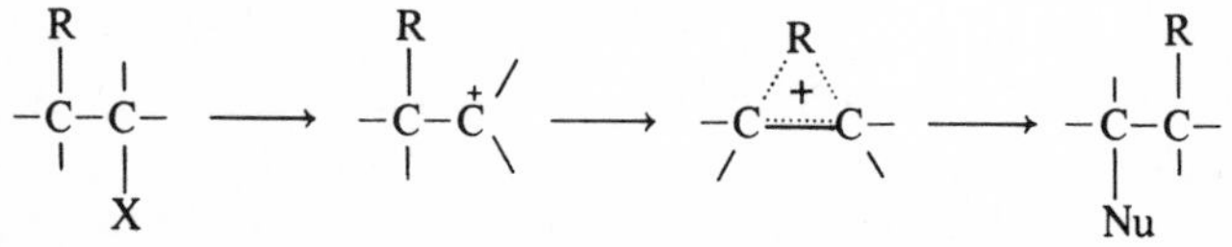

(*c*) Ionisation is assisted by the migration and concerted with it in a single step; no intermediate is formed:

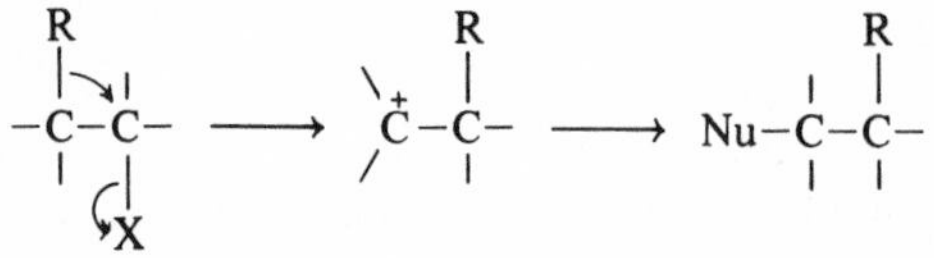

(*d*) Ionisation is assisted by and concerted with the first stage of the migration, leading to a non-classical intermediate:

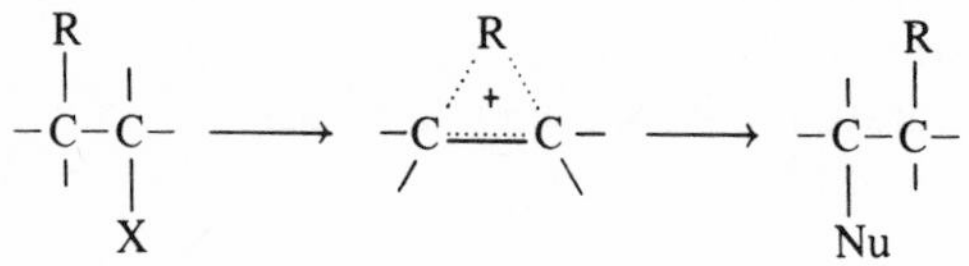

### 14.3.2 The 2-norbornyl system

The 2-norbornyl system has been at the heart of the non-classical carbonium ion controversy from its inception, and still today it remains in a key position. The acetolysis of chiral *exo*-2-norbornyl brosylate gives a racemic mixture of the *exo* acetates (**14.29**) [52JA1147, 1154].

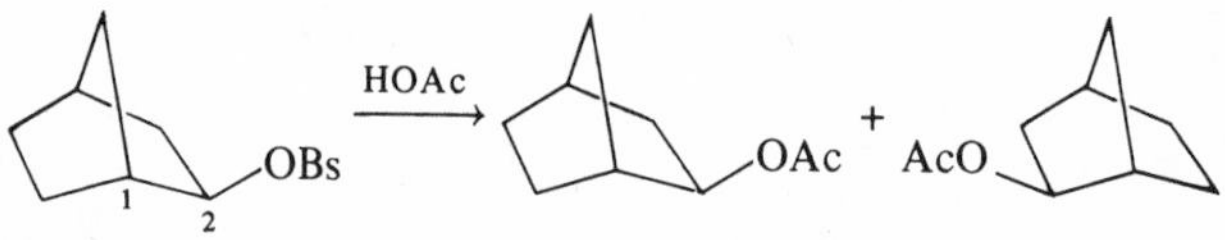

(**14.29**)

The reaction is some 350 times faster than that of the *endo* brosylate, and it involves 50 per cent rearrangement, as demonstrated by [14]C-labelling: a label at position 2 in the brosylate is equally divided between positions 1 and 2 in the acetates.

This behaviour bears all the hallmarks of neighbouring group participation with the formation of a bridged intermediate. It is closely analogous to the reaction of *trans*-2-bromocyclohexyl brosylate (§7.5.3): the enhanced reactivity is a sign of anchimeric assistance; the 50 per cent rearrangement and the characteristic stereochemistry – retention of diastereoisomeric configuration with enantiomeric racemisation – points to a symmetrical intermediate. Winstein, following an earlier suggestion by Wilson [39J1188], proposed an analogous mechanism:

(14.30)                (14.31)

The conventional way of drawing the norbornyl system is as in (14.29); the alternative (14.30) is entirely equivalent but shows more clearly the symmetry of the non-classical intermediate (14.31). It is this symmetry that leads to the 50 per cent of rearrangement with racemisation. The $\sigma$ bridging, calling for rear-side $S_N2$ attack by the solvent, leads to the retention of configuration.

The slower reaction of the *endo* brosylate also gives the *exo* product, but with rather less than 50 per cent racemising rearrangement. The *endo* geometry does not allow any neighbouring group participation, and Winstein proposed that unassisted ionisation led to a classical carbenium ion-pair which could react directly with a solvent molecule to give the unrearranged, inverted product, or else reorganise itself to the symmetrical, non-classical carbonium ion:

(Actually Winstein originally suggested that the carbenium ion would be weakly solvated (**14.32**) but recent work has shown such solvation to be insignificant. The rates of solvolysis of both *exo* and *endo* tosylates in a wide variety of solvents correlate linearly with each other and with the rates for 2-adamantyl tosylate, a model for solvolysis without nucleophilic solvation (§7.3.1). The absence of any significant rate enhancement in the presence of the strongly nucleophilic azide ion (§7.1.2) also suggests that there is negligible nucleophilic involvement in the ionisation step [78JA3139; 3143].)

H. C. Brown, on the other hand, has steadfastly maintained that there is no need to invoke σ participation or non-classical ions to account for these results [73Acc377; 76T179]. He suggests that the *exo* brosylate undergoes normal, unassisted ionisation to give an enantiomeric pair of rapidly equilibrating classical carbenium ions (or ion-pairs) (**14.33**). The symmetrical midpoint of the rearrangement is a transition state, not a non-classical intermediate.

(14.33)

The almost exclusive formation of the *exo* product, Brown asserts, does not require the presence of σ bridging; it is the natural behaviour of the norbornyl system, in which *endo* attack suffers severe steric hindrance. Thus base catalysed deuterium exchange in 2-norbornone (**14.34**) occurs 715 times faster at the *exo* than at the *endo* position, and even in camphor (**14.35**), where the *exo* side is blocked by a *gem*-dimethyl group, the *exo* exchange is 21 times faster than *endo*.

Nor does Brown believe that the greater reactivity of the *exo* substrate is evidence for anchimeric assistance. It is not the *exo* rate that is unusually fast: it is the *endo* rate that is unusually slow. Ionisation of an *endo* group is retarded as the departing anion is squeezed between the *endo*-6-hydrogen and the 2-hydrogen atom swinging towards coplanarity (**14.36**).

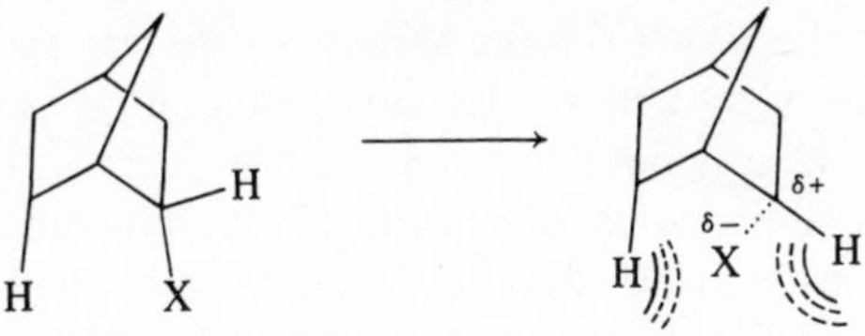

(14.36)

As further evidence against σ participation Brown points to the effects of electron-donating substituents. A 2-*p*-anisyl group promotes the un-assisted formation of a classical delocalised cation (**14.37**); compare the effect of *p*-anisyl in suppressing the vastly greater anchimeric assistance in the 7-*anti*-norbornenyl system (§7.5.3). Yet the relative reactivities of the *exo* and *endo* epimers remain more or less unchanged, so it seems that σ participation is not a necessary explanation of the rate ratios.

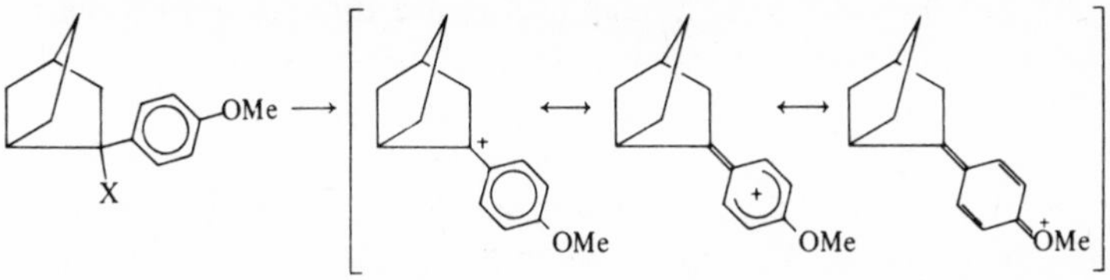

(14.37)

A 1-phenyl substituent accelerates the ethanolysis of *exo*-2-norbornyl chloride four-fold; a 2-phenyl group accelerates it about 40 million times. If the rate-limiting step led to the non-classical ion with the charge shared between carbon atoms 1 and 2 one would expect the two rates to be similar. Clearly the 2-phenyl isomer ionises directly to the classical benzylic cation (**14.38**).

1,2-Dimethyl-*exo*-2-norbornyl *p*-nitrobenzoate (**14.39**) is solvolysed to give a product mixture which is less than 50 per cent rearranged

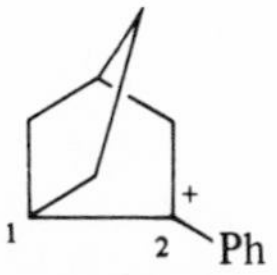

(14.38) 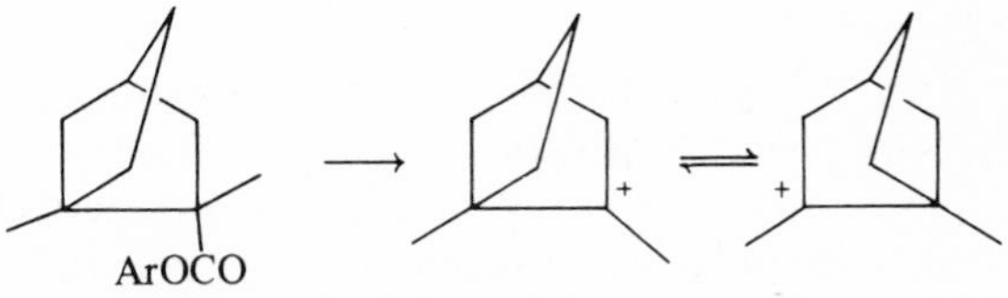　　　　(14.39)

[68JA6213; 72JA1010], suggesting that the solvent is trapping a pair of classical ions before they have fully equilibrated; the non-classical ion would, of course, be symmetrical.

Such evidence points strongly to the classical nature of *tertiary* 2-norbornyl cations [see also 78JO3667]. These must be formed without significant σ participation, yet in all cases the *exo* substrate is more reactive than the *endo*. Thus it can no longer be asserted that anchimeric assistance is necessarily the reason for the high *exo/endo* rate ratio in the parent secondary norbornyl system. On the other hand, none of Brown's studies actually prove that it is not.

Spectroscopic studies, similar to those described in the following section, confirm the largely classical nature of the tertiary cations but suggest that the 2-norbornyl cation itself is indeed non-classical [76Acc41; 80JA6867; 82JA907, 7105]. Most workers in the field now seem to accept this interpretation, though some doubts remain [74JA7638; 75APO177]. Until recently molecular orbital calculations of the relative energies of the classical and non-classical structures gave conflicting results [e.g. 77JA377, 8119], but now these also seem to point unambiguously to the greater stability of the bridged, non-classical ion [83JA5915, 6185]. However, it is not certain that the spectroscopic results, obtained on long-lived cations in super-acid solution or in the solid state, or the MO calculations, which relate to isolated molecules in the gas phase, are directly applicable to fleeting solvolytic intermediates. A linear correlation between the ionisation energies of chlorides that give stable carbocations in super-acid and the $\Delta G^{\ddagger}$ values for their solvolyses does suggest that there is substantial cation character in the solvolytic transition states [79JA522], but evidence described in the following section shows that the nature of a cation may vary with the medium.

### 14.3.3 Methyl and hydrogen migrations

Many alkyl fluorides dissociate in super-acid solutions, typically $FSO_2OH-SbF_5-SO_2$, to give carbocations which are stable and long-lived at low temperatures because the anions that are present, such as $SbF_6^-$, are negligibly nucleophilic. Spectroscopic studies of these solutions have provided new insights into the structures of carbocations [73AG173].

The $^1H$ n.m.r. spectrum of the tetramethylethyl cation (**14.40**) shows only one signal for the twelve methyl protons. Is their equivalence due to rapid equilibration (**14.40** $a \rightleftharpoons b$), or is the ion itself symmetrical (**14.40c**)? The $^1H$ spectrum cannot distinguish between these possibilities, but $^{13}C$ n.m.r. evidence points to equilibrating classical ions. The observed carbon

(14.40)

$$
\left(
\underset{(a)}{
\begin{array}{c}
\text{H} \\
\text{Me} \quad | \\
\overset{+}{\underset{\text{Me}}{\diagup}}\text{C}-\text{C}-\text{Me} \\
\quad | \\
\text{Me}
\end{array}}
\rightleftharpoons
\underset{(b)}{
\begin{array}{c}
\text{H} \\
| \quad \quad \text{Me} \\
\text{Me}-\text{C}-\overset{+}{\text{C}}\diagdown \\
| \quad \quad \text{Me} \\
\text{Me}
\end{array}}
\right)
\qquad
\underset{(c)}{
\begin{array}{c}
\text{H} \\
\quad \overset{+}{\phantom{.}} \\
\text{Me}\diagup\text{C}\text{-----}\text{C}\diagdown\text{Me} \\
\quad \text{Me} \quad \quad \text{Me}
\end{array}}
$$

chemical shift for the two central carbons is $\delta = 190$ (relative to $Me_4Si$). An average of values for a classical $C^+$ and an uncharged carbon atom, estimated from the spectrum of the t-butyl cation, is $\delta = 188$, whereas in the non-classical 7-norbornenyl cation (§7.5.3) the tetracoordinated carbons appear at $\delta = 126$. The CH coupling constant, 65 Hz, also agrees much better with the average of C–H and C–C–H couplings (64 Hz) than with the 7-norbornenyl value of 194 Hz.

Other forms of spectroscopy operate on a shorter time scale than n.m.r. and can give direct evidence about individual species even if they are undergoing rapid equilibration. The Raman spectrum of the tetramethylethyl cation shows important similarities to those of simple carbenium ions, such as t-butyl, and suggests the presence of a planar carbenium centre. X-ray photoelectron spectroscopy, reaching down into the carbon atoms to measure the binding energy of their $1s$ electrons, shows clearly that there is a single highly electron-deficient carbon atom with a very high binding energy.

Similar studies point to equilibrating, classical structures for other open chain ions, including s-butyl (**14.41**), and also pentamethylethyl (**14.42**), in which the rearrangement is a Wagner–Meerwein methyl migration rather than a hydrogen shift.

$$
\begin{array}{c}
\text{H} \\
| \\
\text{Me}-\text{CH}-\overset{+}{\text{CH}}-\text{Me}
\end{array}
\rightleftharpoons
\begin{array}{c}
\text{H} \\
| \\
\text{Me}-\overset{+}{\text{CH}}-\text{CH}-\text{Me}
\end{array}
\qquad
\begin{array}{c}
\text{Me} \\
| \\
\text{Me}-\overset{+}{\text{C}}-\text{C}-\text{Me} \\
| \quad | \\
\text{Me} \quad \text{Me}
\end{array}
\rightleftharpoons
\begin{array}{c}
\text{Me} \\
| \\
\text{Me}-\text{C}-\overset{+}{\text{C}}-\text{Me} \\
| \quad | \\
\text{Me} \quad \text{Me}
\end{array}
$$

$$\text{(14.41)} \qquad\qquad\qquad \text{(14.42)}$$

However, this does not seem to be the complete story. Labelling studies using both deuterium and carbon-13 reveal scrambling that in many cases seems to demand the intermediacy of a non-classical ion [73Acc53]. Thus n.m.r. lineshape analysis of the s-butyl cation shows that, in addition to the very rapid 2,3-hydrogen exchange, there is a slower process which scrambles all the protons, probably involving the methylbridged intermediates (**14.43**). These ions are less stable than the classical secondary ion by $10 \text{ kJ mol}^{-1}$ or more, but they are intermediates and not transition states.

Similar conclusions come from chemical evidence. The Wagner–

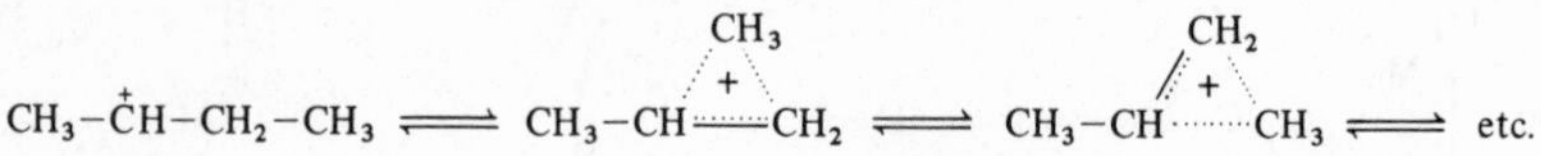

$$CH_3-\overset{+}{C}H-CH_2-CH_3 \ \rightleftharpoons \ CH_3-CH\overset{\overset{CH_3}{\diagup\,\,+\,\,\diagdown}}{\cdots\cdots\,CH_2} \ \rightleftharpoons \ CH_3-CH\overset{\overset{CH_2}{\diagup\,\,+\,\,\diagdown}}{\cdots\cdots\,CH_3} \ \rightleftharpoons \ etc.$$

(14.43)

Meerwein rearrangement of neopentyl systems frequently produces small quantities, 10 to 15 per cent, of cyclopropane products as well as the normal t-pentyl derivatives. This implies the intermediacy of a methyl-bridged ion (**14.44**), which is effectively a corner-protonated cyclopropane [compare 69CR543; 70PPO129].

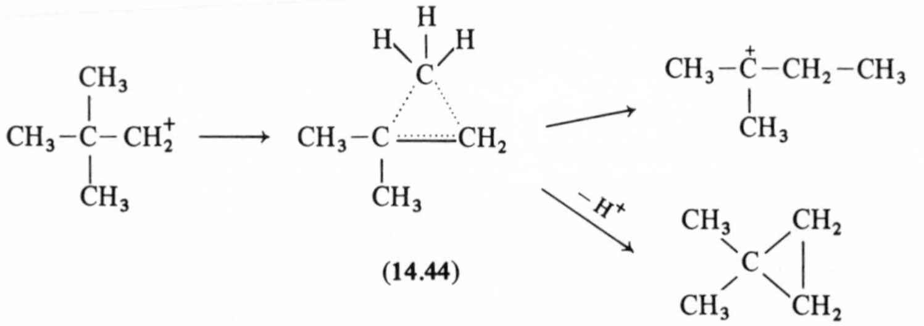

(14.44)

One further experiment indicates that the stable form of a carbocation may depend on the medium, that is, on the mode of solvation. Trifluoro-acetolysis of the labelled s-butyl tosylate (**14.45**) gives an equimolar mixture of the labelled products (**14.47**; **14.48**). This can surely arise only from the hydrogen-bridged intermediate (**14.46**). A direct $S_N1$ or $S_N2$ reaction would give no deuterium scrambling at all, and reaction via equilibrating classical ions (the stable species in super-acid) would lead to the doubly scrambled products (**14.49**; **14.50**) [72TL1241].

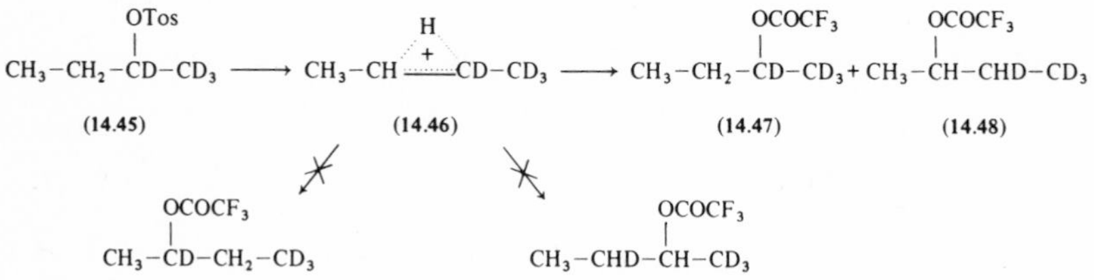

## 14.4    Further reading

P. de Mayo, *Molecular Rearrangements*, 2 vols, Interscience, New York, 1963–4.

P. de Mayo (ed.), *Rearrangements in Ground and Excited States*, 3 vols., Academic, New York, 1980.

T. S. Stevens & W. E. Watts, *Selected Molecular Rearrangements*, Van Nostrand Reinhold, London, 1973.

B. S. Thyagarajan (ed.), *Mechanisms of Molecular Migrations*, Interscience, New York (series started 1968).

H. J. Shine, *Aromatic Rearrangements*, Elsevier, New York, 1967.
D. L. H. Williams, in *Comprehensive Chemical Kinetics*, eds C. H. Bamford &
C. F. H. Tipper, vol. 13, Elsevier, Amsterdam, 1972 (aromatic
rearrangements).
H. J. Shine, *Internat. Rev. Sci., Org. Chem. Ser. Two*, 1976, **3**, 87 (aromatic
rearrangements).
D. V. Banthorpe, in *Chemistry of the Azido Group*, ed. S. Patai, Interscience,
New York, 1971 (azide rearrangements).

## Carbocations:

G. A. Olah, *Angew. Chem. Internat. Ed.*, 1973, **12**, 173; *Acc. Chem. Res.*, 1976, **9**,
41.
H. C. Brown, *Acc. Chem. Res.*, 1973, **6**, 377; *Tetrahedron*, 1976, **32**, 179.
R. E. Leone & P. v. R. Schleyer, *Angew. Chem. Internat. Ed.*, 1970, **9**, 860.
H. C. Brown & P. v. R. Schleyer, *The Non-classical Ion Problem*, Plenum, New
York, 1977.
W. J. Hehre, *Acc. Chem. Res.*, 1975, **8**, 369.
G. M. Kramer, *Adv. Phys. Org. Chem.*, 1975, **11**, 177.
C. A. Grob, *Angew. Chem. Internat. Ed.*, 1982, **21**, 87.
See also a series of papers in *Acc. Chem. Res.*, 1983, **16**: C. A. Grob, p. 426; H.
C. Brown, p. 432; G. A. Olah, G. K. S. Prakash & M. Saunders, p. 440; C.
Walling, p. 448.

# 15 Aliphatic radical substitution

## 15.1 Characteristics of radical reactions

One of the main distinctions between radical and ionic reactions is that radicals seldom occur in other than very low concentrations. Anions and cations can coexist in high concentrations, each stabilised by solvation. Radicals are less strongly solvated and therefore tend to recombine with the formation of a covalent bond: even if they bear polar substituent groups that facilitate solvation these will still be present in the combined product and the solvation will not specifically stabilise the dissociated radicals. A few radicals exist as such, for example the nitroxide $t\text{-Bu}_2\text{NO}^\bullet$ which is sterically crowded and for which the dimer is not a stable compound, but even the 'stable' $\text{Ph}_3\text{C}^\bullet$ (the first radical to be recognised: Gomberg [00JA757]) is largely dimerised at room temperature:

$$2\,\text{Ph}_3\text{C}^\bullet \underset{K}{\rightleftharpoons} \qquad K \approx 5000\ \text{l mol}^{-1}$$
(25 °C in benzene)

A consequence is that most radical reactions require *initiation* – the generation of radical species in a preliminary step. Initiation usually occurs by the homolysis of a relatively weak bond, either thermally or photolytically, for example:

$$(\text{PhCOO})_2 \xrightarrow{\Delta} 2\,\text{PhCO}_2^\bullet$$

$$\text{Cl}_2 \xrightarrow{h\nu} 2\,\text{Cl}^\bullet$$

The thermolysis of azo compounds is also a source of radicals, the decomposition being driven by the stability of the newly forming molecular nitrogen:

$$\text{CH}_3\text{--N=N--CH}_3 \xrightarrow{\Delta} 2\,\text{CH}_3^\bullet + \text{N}_2$$

Radicals may also be generated by the transfer of a single electron to or from an electrode or a suitable transition metal ion, from a dissolving metal (for example, with Na/EtOH), or to or from another radical (for example, in *homosolvolysis* using a stable nitroxide or other radical solvent [80J(P2)260]). The existence of suitable conditions for initiation, and especially the inhibition of a reaction when such conditions are removed, is often the first pointer to a radical mechanism.

The converse of initiation is *termination*, the mutual destruction of two radicals. This commonly occurs by *disproportionation* (for example, $2C_2H_5^{\bullet} \rightarrow C_2H_6 + C_2H_4$) or *colligation* ($R^{\bullet} + R'^{\bullet} \rightarrow R-R'$). The colligation of two atoms often occurs only in a three-body collision or on the walls of the reaction vessel. The third body is needed to absorb some of the energy of reaction that would otherwise appear in the new diatomic molecule as vibrational excitation and is, of course, exactly equal to the bond dissociation energy: the two atoms simply bounce off each other. In some atom–atom colligations (of halogens, for example) the excess energy can also be lost directly in emitted radiation. In the colligation of polyatomic radicals the energy can be transferred to other vibrational and to rotational modes.

Termination seldom follows directly upon initiation. The instability of most radicals means not only that they are present in low concentrations so that contact between two radicals is infrequent, but also that they are highly reactive and tend to react rapidly either with the much more abundant even-electron species that surround them or in a unimolecular decomposition. Such reactions fall into three main categories:

Substitution: $R^{\bullet} + X-Y \rightarrow R-X + Y^{\bullet}$
(abstraction)

Addition: $R^{\bullet} + X=Y \rightarrow R-X-Y^{\bullet}$

$\beta$ fission: $R-X-Y^{\bullet} \rightarrow R^{\bullet} + X=Y$

In each case a radical enters into the reaction and another is formed as a product. Such reactions therefore continue until eventually two radicals destroy each other in a termination step. Often a great many different radical–molecule reactions occur in parallel and a complicated mixture of products is formed. For example, when azomethane is pyrolysed the first-formed methyl radicals can undergo substitution and addition reactions with undissociated azomethane molecules, and the resulting new radicals can enter into a series of further reactions with each other, with methyl radicals, with other azomethane molecules, or with some of the products of earlier reactions:

$$CH_3^{\bullet} + CH_3-N=N-CH_3 \rightarrow CH_4 + {}^{\bullet}CH_2-N=N-CH_3 \rightsquigarrow \text{etc.}$$
$$CH_3^{\bullet} + CH_3-N=N-CH_3 \rightarrow (CH_3)_2N-N^{\bullet}-CH_3 \qquad \rightsquigarrow \text{etc.}$$

In other cases the reaction is cleaner, as in the radical addition of HBr to an alkene (§9.2). Two (or sometimes more) radical–molecule reactions occur repetitively, being energetically more favourable than possible alternatives, and thereby establish a *chain-reaction*. In the reaction of HBr with ethylene the chain *propagation* steps are:

$$\begin{cases} Br^{\textbf{·}} + CH_2{=}CH_2 \rightarrow BrCH_2CH_2^{\textbf{·}} \\ BrCH_2CH_2^{\textbf{·}} + HBr \rightarrow BrCH_2CH_3 + Br^{\textbf{·}} \end{cases}$$

The addition of $Br^{\textbf{·}}$ to the double bond is much faster than the alternatives – abstraction of hydrogen atoms from ethylene or bromo-ethane molecules, both of which involve cleaving the strong $C{-}H$ bond. The bromoethyl radical preferentially attacks HBr, though there is some competition from addition to ethylene as well.

The chain steps may be repeated many times before the radicals are finally consumed. In a photo-initiated reaction the occurrence of a chain mechanism may be revealed by a high *quantum yield* (molecules of reactant transformed per photon absorbed): a single photon can initiate a chain that is anything from two or three steps to a million or more steps long.

### 15.2 $S_H2$ reactions

The reaction described above as substitution is the radical analogue of a bimolecular nucleophilic or electrophilic substitution and is sometimes designated $S_H2$, H for homolytic.

The direct displacement of one radical by another at a saturated carbon, the radical analogue of the normal $S_N2$ reaction, has never been unambiguously demonstrated. The reason is not hard to find. Estimates of the energy of an $S_N2$-like transition state (15.1) for the attack of a deuterium atom on the carbon atom of methane suggest that the activation energy for such a process would be around $160\,kJ\,mol^{-1}$: the observed value for the alternative attack at hydrogen (15.2) is only about $40\,kJ\,mol^{-1}$.

$$\begin{bmatrix} H \\ | \\ D\cdots\cdots C\cdots\cdots H \\ \diagup \\ H \quad H \end{bmatrix}^{\textbf{·}\ddagger} \qquad \begin{bmatrix} H \\ \diagup \\ D\cdots\cdots H\quad C \\ \diagdown H \\ H \end{bmatrix}^{\textbf{·}\ddagger}$$

$$(15.1) \qquad\qquad (15.2)$$

The difference is largely steric, though with many attacking radicals there is also an electronic contribution in that bonds to hydrogen are often stronger than those to carbon:

$$Cl^{\textbf{·}} + CH_4 \rightarrow Cl{-}H + CH_3^{\textbf{·}} \qquad \Delta H^{\ominus} = +3\,kJ\,mol^{-1}$$
$$Cl^{\textbf{·}} + CH_4 \rightarrow Cl{-}CH_3 + H^{\textbf{·}} \qquad \Delta H^{\ominus} = +84\,kJ\,mol^{-1}$$

(Bond dissociation enthalpies: $CH_3{-}H$ 435, $CH_3{-}Cl$ 351, $H{-}Cl$ 432 kJ $mol^{-1}$)

The normal mode of reaction of a radical with a saturated hydrocarbon grouping is therefore substitution at hydrogen not at carbon, a process that is more readily pictured as a hydrogen atom transfer or as the abstraction of a hydrogen atom from the substrate by the radical reagent.

The radical produced by such an abstraction can react further with an even-electron species, often that which first initiated the reaction as in the chlorination of an alkane, where the propagation steps are:

$$\begin{cases} Cl^{\bullet} + R-H \rightarrow HCl + R^{\bullet} \\ R^{\bullet} + Cl_2 \rightarrow RCl + Cl^{\bullet} \end{cases}$$

The overall reaction is then substitution at carbon, the abstracted hydrogen being replaced by, in this example, chlorine.

## 15.3    Kinetics of aliphatic halogenation

A simple example of the kinetic analysis of a chain-reaction was presented in §2.2.3, where the bromination of an alkane was shown to follow the rate equation: Rate $= k[\text{alkane}][Br_2]^{\frac{1}{2}}$. In practice the kinetics of many radical reactions are more complicated than this. Indeed, even for alkane bromination the above relation applies only at the start of the reaction. As the concentration of HBr builds up the reverse of the hydrogen abstraction step becomes significant. Using the symbols of §2.2.3 we should write:

$$Br_2 \underset{k_4}{\overset{k_1}{\rightleftharpoons}} 2\, Br^{\bullet}$$

$$\begin{cases} Br^{\bullet} + R-H \underset{k_{-2}}{\overset{k_2}{\rightleftharpoons}} HBr + R^{\bullet} \\ R^{\bullet} + Br_2 \overset{k_3}{\rightarrow} RBr + Br^{\bullet} \end{cases}$$

Applying the steady state approximation first to $R^{\bullet}$ gives:

$$k_2[Br^{\bullet}][RH] = [R^{\bullet}](k_3[Br_2] + k_{-2}[HBr])$$
$$\begin{aligned} \text{Rate} &= k_2[Br^{\bullet}][RH] - k_{-2}[HBr][R^{\bullet}] \\ &= k_2[Br^{\bullet}][RH]\{1 - k_{-2}[HBr]/(k_3[Br_2] + k_{-2}[HBr])\} \\ &= k_2 k_3 [Br^{\bullet}][RH][Br_2]/(k_3[Br_2] + k_{-2}[HBr]) \end{aligned}$$

Setting $k_1[Br_2] = k_4[Br^{\bullet}]^2$, as in §2.2.3, gives:

$$\text{Rate} = k_2 k_3 \cdot (k_1/k_4)^{\frac{1}{2}} \cdot [RH][Br_2]/(k_3[Br_2] + k_{-2}[HBr])$$

and in agreement with this it is found that bromination is retarded by the build-up of HBr as the reaction proceeds, and also by the addition of HBr to the reaction mixture [44JCP(12)469].

If the reaction is photo-initiated then, provided the light is intense enough to penetrate the entire reaction mixture, $k_1 = I\varepsilon$ where $I$ is the intensity of the light and $\varepsilon$ the absorption coefficient of bromine. The rate of reaction is therefore proportional to $I^{\frac{1}{2}}$. If the light is less intense and is totally absorbed – that is, some of the bromine is unaffected by the light – then the rate of initiation no longer depends on $[Br_2]$ but only on $I\varepsilon$. The kinetics of the initiation/termination equilibrium become: $I\varepsilon = k_4[Br^{\bullet}]^2$, and in the overall rate equation the term $[Br_2]^{\frac{1}{2}}$ drops to $[Br_2]^1$.

For chlorination with molecular chlorine the kinetic analysis is further complicated because the concentration of chlorine atoms may be comparable to that of alkyl radicals and termination no longer occurs via a single predominant reaction. The observed kinetics are therefore a composite of several different rate expressions.

In practice the accurate measurement of the absolute rates of radical reactions is not easy and most kinetic analyses are based on competitive experiments (§2.2.2), either intermolecular (that is, using a mixture of two different substrates) or intramolecular (for example, by comparing the relative reactivities of the methyl and methylene hydrogen atoms in propane). In fact all the absolute rates reported for gas-phase chlorinations of alkanes derive from a series of competitive experiments built up from measurements of the absolute rates of the reaction of chlorine with hydrogen.

### 15.4    Hydrogen abstraction reactions

In the reaction of an alkane with a halogen atom (or other radical) the site of substitution is controlled by the first of the propagation steps, hydrogen abstraction. In almost all cases this also controls the rate of reaction because it involves cleavage of the strong carbon–hydrogen bond. Indeed, when the second propagation step is slower than the first the chain sequence is seldom established and the substitution does not proceed cleanly. For example, methyl radicals can be produced by the thermolysis of iodomethane, for which the bond dissociation enthalpy is similar to that of $Cl_2$. Subsequent reaction with an alkane, however, does not lead to the analogous substitution: $MeI + RH \rightarrow\!\!\!\times\!\!\!\rightarrow MeH + RI$. The chain propagation would have to be:

$$\begin{cases} Me^{\bullet} + RH \rightarrow MeH + R^{\bullet} & \text{exothermic} \\ R^{\bullet} + MeI \rightarrow RI + Me^{\bullet} & \text{endothermic} \end{cases}$$

In such a sequence the more stable of the chain-carrying radicals ($R^{\bullet}$ – see below) propagates the chain very slowly and is consumed mainly by other reactions such as disproportionation or colligation.

We shall therefore focus attention now on the factors influencing the hydrogen abstraction step of the chain.

### 15.4.1    The structure of the substrate

In the reactions of alkanes with radicals, tertiary hydrogen atoms are usually abstracted fastest, primary slowest (table 15.1). The main

Table 15.1. *Second order rate coefficients for*
$X^\bullet + RH \rightarrow XH + R^\bullet$ *at* *164°C*
$(k_2/l \ mol^{-1} \ s^{-1})$ [calculated from data in
71CR247; 82T313]

|           |       | R | Me  | Et    | i-Pr   | t-Bu      |
|-----------|-------|---|-----|-------|--------|-----------|
| X = Me:   | $k_2$ |   | 10  | 330   | 2800   | 23000     |
| X = Br:   | $k_2$ |   | 5   | 3000  | 20000  | 3500000   |

Table 15.2. *Second order rate coefficients* (*per*
*hydrogen atom abstracted*) *for*
$Me^\bullet + CH_3-X \rightarrow MeH + {}^\bullet CH_2-X$ *at* *164°C*
$(k_2/l \ mol^{-1} \ s^{-1})$ [calculated from data in
71CR247]

| X: | H  | OMe | COMe | CMe=CMe$_2$ | Ph   |
|----|----|-----|------|-------------|------|
| k: | 10 | 600 | 700  | 400         | 2000 |

reason is that a tertiary C−H bond is weaker than a primary one: indeed, there is a linear correlation between the activation energy for hydrogen abstraction and the bond dissociation enthalpy. The relative weakness of the tertiary C−H bond is to a small extent due to hyperconjugative stabilisation of the resulting tertiary radical, but the main reason is the release of steric compression that accompanies the abstraction of the tertiary hydrogen atom [70AG830]. In similar fashion cyclopentane is much more reactive towards radical attack than is cyclohexane, because torsional strain is relieved in the cyclopentyl radical (compare $S_N$ reactions of cycloalkyl substrates, §7.4.1).

Conjugation with neighbouring $\pi$ bonds or with lone electron pairs (e.g. $MeO-CH_2^{\bullet+} \leftrightarrow MeO-CH_2^-$) can also stabilise a radical, but the effect of this is less than is often supposed as shown by a comparison between tables 15.1 and 15.2 (see also §15.4.3).

*15.4.2 The effect of the attacking radical*

Table 15.1 shows a marked difference in selectivity between methyl radicals and bromine atoms, and such variations are common in radical substitutions. A branched alkane is brominated almost exclusively at tertiary positions, chlorinated with much less selectivity ($k_{tertiary}/k_{primary} \simeq 3$–10), and fluorinated with virtually no selectivity at all. The difference arises from the different degrees of bond making and breaking in the transition state (figure 15.1). In the very exothermic reaction of

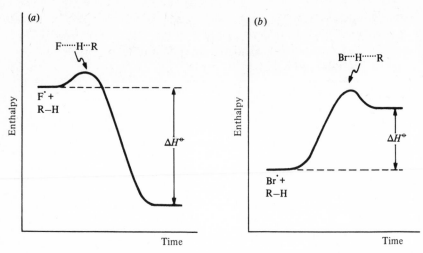

Figure 15.1. Reaction profiles for hydrogen abstraction by (a) fluorine atoms and (b) bromine atoms. $\Delta H^{\ominus}$ values for RH = $CH_4$ are: (a) $-133$ kJ mol$^{-1}$, (b) $+69$ kJ mol$^{-1}$.

fluorine with an alkane the transition state (according to the Hammond postulate, §3.4) is reactant-like: there is relatively little C—H bond breaking and structural changes in the alkane have little effect. Conversely in the endothermic reaction of bromine there is much more C—H cleavage in the product-like transition state and the reaction rate is much more dependent on the C—H bond strength.

The reactions of the methyl radical with alkanes are thermoneutral (with $CH_4$) or slightly exothermic; this accords with their observed selectivity (table 15.1) intermediate between fluorine and bromine. The apparent anomaly that the endothermic bromine reactions occur at rates similar to those of the methyl reactions is an entropic effect: $\Delta H^{\ddagger}$ for bromine attack is indeed greater than for methyl attack (78 and 60 kJ mol$^{-1}$, respectively, for reaction with methane) but the methyl group must reorganise its geometry during reaction and this lowers the value of $\Delta S^{\ddagger}$ for attack at a single hydrogen atom of methane from $-50$ J K$^{-1}$ mol$^{-1}$ for the reaction of bromine to $-90$ J K$^{-1}$ mol$^{-1}$ for methyl.

If a radical substitution is to be used synthetically it is obviously desirable to have as high a selectivity as possible. Chlorinations are often carried out with other reagents of the general type X—Cl rather than molecular chlorine. The chain propagation steps are:

$$\begin{cases} X^{\bullet} + RH \rightarrow R^{\bullet} + XH \\ R^{\bullet} + X-Cl \rightarrow RCl + X^{\bullet} \end{cases}$$

so that the requirements are that $X^•$ should be more selective (less reactive) than $Cl^•$ but that $X-Cl$ should not be so unreactive that $R^•$ fails to abstract chlorine from it. t-Butyl hypochlorite (t-BuOCl) is commonly used, typically giving a tertiary/primary selectivity around 50 [64JA3368].

### 15.4.3 Polar interactions between radical and substrate

As in the addition of radicals to carbon–carbon double bonds (§9.2) the site of attack preferred by one radical may not be that preferred by another. For example:

$$^•CH_2CH_2COOH \xleftarrow{Cl^•} \overset{\beta}{CH_3}\overset{\alpha}{CH_2}COOH \xrightarrow{Me^•} CH_3\dot{C}HCOOH$$

97 per cent                                    90 per cent

The carbonyl group slightly activates the adjacent $\alpha$ position towards attack by $Me^•$ (compare table 15.2) but markedly deactivates it towards attack by $Cl^•$ to the extent that the normally less favoured primary site is preferred by a factor of some 30:1.

The reason lies in the electrophilic character of the chlorine radical (using the term in its broadest sense [83PAC1281]). As the $Cl-H$ bond forms it begins to develop its dipolar character, leaving the hydrogen atom with a partial positive charge in the transition state. This is destabilised by the electron-withdrawing carbonyl group and $\alpha$ attack is thereby discouraged:

$$\overset{\delta-}{Cl}\cdots\overset{\delta+}{H}\cdots\overset{|}{\underset{|}{C}}-COOH$$

Conversely an organometallic radical such as $Me_3Sn^•$ behaves nucleophilically: it abstracts halogen in preference to hydrogen (via $\overset{\delta+}{Me_3Sn}\cdots\overset{\delta-}{Hal}\cdots CR_3$) and its reactions are accelerated by electron withdrawal in the substrate.

Polar effects are particularly important in exothermic, early transition state reactions in which there is little $C-H$ cleavage (figure 15.1(a)). They are still present in late transition state reactions, but may be counteracted by the effects of steric acceleration and delocalisation as exemplified by the different effects of substituents in 1-substituted butanes on gas-phase chlorination and bromination (table 15.3). In these reactions the effect of the substituent is minimal beyond the $\beta$ position: the relative reactivities of the $\gamma$ and $\delta$ hydrogen atoms are virtually independent of substituent. The $\delta$ position is therefore taken as a reference, and reactivities at the $\alpha$ and $\beta$ positions can be determined from relative product ratios. In

Table 15.3. *Reactivities of α and β hydrogen atoms of 1-substituted butanes in gas-phase halogenations relative to the reactivity of a δ hydrogen atom* [82T313].

$$\overset{\alpha}{X-CH_2}-\overset{\beta}{CH_2}-\overset{\gamma}{CH_2}-\overset{\delta}{CH_3}$$

| X | Chlorination (50 °C) $\alpha/\delta$ | $\beta/\delta$ | Bromination (160 °C) $\alpha/\delta$ | $\beta/\delta$ |
|---|---|---|---|---|
| H | 1 | 3.6 | 1 | 80 |
| $CF_3$ | 0.04 | 1.2 | <1 | 7 |
| MeOCO | 0.4 | 2.4 | 40 | 35 |
| CN | 0.2 | 1.7 | 25 | 8 |
| F | 0.9 | 1.7 | 9 | 7 |

chlorinations the polar effect is dominant: electron-withdrawing substituents deactivate both α and β positions. The compensating effect of an adjacent π bond (MeOCO and CN) and especially of an adjacent lone pair (F) can be seen by comparing their effects on the α reactivity with that of $CF_3$. In the brominations the polar effect is still present – witness the deactivation of all the β positions. At the α positions, however, delocalisation and steric decompression are dominant and all the substituents except $CF_3$ accelerate hydrogen abstraction.

The significance of polar interactions can also be seen in the reactions of radicals with the methyl hydrogen atoms of substituted toluenes:

$$X-C_6H_4-CH_3 + R^{\cdot} \rightarrow X-C_6H_4-CH_2^{\cdot} + RH$$

For halogenations a linear correlation is found between $\log(k/k_0)$ and the through-conjugation substituent parameter $\sigma^+$, the $\rho$ values of $-1.8$ for bromination and $-0.7$ for chlorination reflecting the lower selectivity of chlorine discussed above. The correlation with $\sigma^+$ indicates that in the transition state a positive charge is developing at the reaction site (**15.3**) [63JA3142; 64JA2357]. On the other hand, reaction with the nucleophilic t-butyl radical correlates with $\sigma$ with a $\rho$ value of $+1.0$, pointing to the build-up of negative charge (**15.4**) [73JA4754].

$$\overset{\delta+}{\overbrace{Ar-CH_2}}\cdots H\overset{\delta-}{\cdots Br} \qquad \overset{\delta-}{\overbrace{Ar-CH_2}}\cdots H\overset{\delta+}{\cdots CMe_3}$$

(**15.3**)      (**15.4**)

Methyl and phenyl radicals generate non-polar transition states and the rates of their reactions with substituted toluenes are virtually independent of the substituent ($\rho \simeq 0$) [63JA3754; 69JO2018].

There have been many attempts to isolate this polar effect from the inherent effect of the substituent on the stability of the tolyl radical, mostly by analysing radical reactions in terms of a dual substituent parameter equation of the type:

$$\log (k/k_0) = \rho\sigma + \rho^{\bullet}\sigma^{\bullet}$$

in which the $\rho\sigma$ term measures the polar effect and $\sigma^{\bullet}$ describes the substituent effect on the radical stability. It is assumed that *meta* substituents do not interact significantly with the radical site (that is, $\sigma^{\bullet} = 0$ for all *meta* substituents), and hence a $\rho$ value is derived for whatever is selected as the reference reaction (for which $\rho^{\bullet}$ is arbitrarily defined as unity). Then for *para* substituents any deviations from the simple Hammett plot of log $(k/k_0)$ against $\rho\sigma$ are taken as a measure of $\sigma^{\bullet}$. The difficulty with this approach is that the $\rho^{\bullet}\sigma^{\bullet}$ term is often much smaller than $\rho\sigma$ and it is not easy to distinguish true radical stabilisation effects from variations in the polar effect. Even in non-polar systems ($\rho \simeq 0$) the residual rate changes are so small that their interpretation is uncertain. It is not surprising, therefore, that the mutual consistency of the various $\sigma^{\bullet}$ scales that have been proposed is not high.

One point, however, is clear. Radicals are stabilised by both $\pi$ electron-withdrawing and $\pi$ electron-donating substituents (compare table 15.2). Almost every $\sigma^{\bullet}$ scale allots positive values to all groups. For example, one recently proposed set of $\sigma^{\bullet}$ values (based on the rate of homolysis of $(ArCH_2)_2Hg \rightarrow ArCH_2^{\bullet} + ArCH_2Hg^{\bullet}$) correlates well with $(\sigma^- - \sigma)$ for $\pi$ electron-withdrawing substituents and with $- (\sigma^+ - \sigma)/2$ for $\pi$ electron donors [81J(P2)1127]: the $(\sigma^{\pm} - \sigma)$ term is taken to measure the ability of the substituent to enter into through-conjugation with the benzylic radical site (compare equation [3.4]). Figure 15.2 suggests an explanation. Interaction between a single non-bonding electron and a $\pi$ bond produces a radical with an electronic structure similar to that of the allyl radical (compare figure 6.8); the odd electron remains in a non-bonding

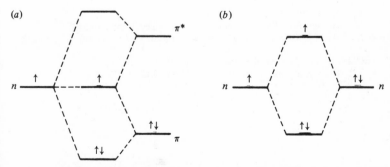

Figure 15.2. Interactions between a non-bonding unpaired electron and (*a*) a $\pi$ bond and (*b*) a lone electron pair.

orbital and the stabilisation of the system arises from the lower energy of the $\pi$ bonding orbital. To a first approximation this is the same as for the corresponding anion (whence the correlation with $(\sigma^- - \sigma)$ or cation. Interaction between a radical site and a lone pair, however, forces one electron into an antibonding orbital and results in the net lowering of the energy of only one electron: stabilisation of the corresponding cationic centre, as measured by $(\sigma^+ - \sigma)$ would be twice as effective [compare 52JA3353].

### 15.4.4   Solvent effects

In general, radical reactions are insensitive to solvent changes, a consequence of the weak solvation of most radical species (§15.1). However, reactions of electrophilic radicals, such as Cl$^\bullet$, are more influenced than most by the nature of the reaction medium. For example, chlorine is surprisingly selective in its reaction with ethylbenzene in the liquid phase: a hydrogen atom is abstracted 14.5 times faster from the $\alpha$ position than from $\beta$. Solvation of the chlorine atom by the aromatic $\pi$ system reduces its reactivity leading to a later transition state and enhanced selectivity (§15.4.2). When the ethylbenzene is diluted with nitrobenzene, a much weaker $\pi$ donor, the selectivity is reduced, tending towards a limiting $\alpha/\beta$ ratio of 2.0 in pure nitrobenzene [63JA2976]. Likewise in the chlorination of alkanes the tertiary/primary selectivity ratio rises from its normal value of 3–10 when the alkane substrate is also the solvent, to 50 in benzene and to 225 in a high concentration of $CS_2$ which complexes chlorine atoms very strongly, probably with the formation of the covalently bonded ClS–ĊS [73MIc].

### 15.5   Further reading

D. C. Nonhebel, J. M. Tedder, & J. C. Walton, *Radicals*, Cambridge University Press, Cambridge, 1979.

J. M. Hay, *Reactive Free Radicals*, Academic, London, 1974.

R. L. Huang, S. H. Goh, & S. H. Ong, *The Chemistry of Free Radicals*, Arnold, London, 1974.

J. K. Kochi (ed.), *Free Radicals*, 2 vols, Wiley, New York, 1973.

E. S. Huyser, *Free Radical Chain Reactions*, Wiley, New York, 1970.

J. M. Tedder, *Tetrahedron*, 1982, **38**, 313; *Angew. Chem. Internat. Ed.*, 1982, **21**, 401.

C. Rüchardt, *Angew. Chem. Internat. Ed.*, 1970, **9**, 830.

E. S. Huyser, in *The Chemistry of the Carbon–Halogen Bond*, ed. S. Patai, Wiley-Interscience, New York, 1973 (halogenations).

A. L. J. Beckwith, *Tetrahedron*, 1981, **37**, 3073.

C. Rüchardt & H. D. Beckhaus, *Angew. Chem. Internat. Ed.*, 1980, **19**, 429.

# 16 Pericyclic reactions

## 16.1 Cyclical reactions: concerted and non-concerted

Many important reactions of organic chemistry involve a cyclical reorganisation of bonds. The pyrolytic eliminations discussed in §8.6 are examples; others include the Diels–Alder reaction, 1,3-dipolar cycloadditions, the Claisen rearrangement (§14.1.3), and the closely related Cope rearrangement (see below). There is a great variety of possible reaction types [for a rationalised classification see 74AG47], but three classes have been particularly studied:

(a) *Cycloadditions*, typified by the Diels–Alder reaction (**16.1**), and their inverse fragmentations, cycloeliminations.

(b) *Electrocyclic reactions* (**16.2**), in which a conjugated polyene cyclises to a cycloalkene with one double bond fewer and the inverse ring opening.

(16.1)                                      (16.2)

(c) *Sigmatropic rearrangements*, in which a σ bonded group is transferred from one end of a conjugated π system to the other, as in the 1,5-hydrogen shift of a conjugated diene (**16.3**), or the Cope rearrangement (**16.4**) in which the σ bond moves across two allyl fragments.

(16.3)                                      (16.4)

All of these reactions could, in principle, proceed in one elementary step, with a concerted migration of electrons, passing through closely

*347*

similar cyclic transition states which can be represented respectively as follows:

These particular examples involve six-membered hydrocarbon rings, but other ring sizes are possible, and so are heteroatomic systems: again, the pyrolytic eliminations provide examples.

The term *pericyclic* has been coined to describe all such processes, but pericyclic reactions need to be distinguished from apparently similar non-concerted reactions. In a cycloaddition, for example, the two ring-closing bonds could be formed in separate steps, via a zwitterionic (**16.5**) or biradical (**16.6**) intermediate.

It is not always easy to make this distinction in practice. A number of criteria may be applied, but they are frequently ambiguous. A characteristic of the concerted mechanism is its stereospecificity. A loss of stereochemical integrity points to a two-step mechanism with conformational changes intervening. 1,1-Dichloro-2,2-difluoroethene reacts with the isomeric hexa-2,4-dienes to give mixtures of stereoisomeric vinylcyclobutane adducts (figure 16.1). The results accord with a biradical intermediate in which the allylic π system maintains the stereochemistry about the vinyl double bond, but rotation about single bonds leads to stereoisomerisation in the cyclobutane ring [64JA622, and see 68SC(159) 833].

$$
\begin{array}{c}
\overset{a\quad b}{CH_3CH=CH-CH=CHCH_3} \\
+ \\
CF_2=CCl_2
\end{array}
\Bigg\}
\xrightarrow[80\,°C]{}
\begin{array}{c}
\overline{CH_3CH-CH-CH-CHCH_3} \\
\downarrow \\
CF_2CCl_2
\end{array}
\longrightarrow
\begin{array}{c}
CH_3 \quad \overset{x}{\underset{y}{\quad}} CH=CHCH_3 \\
\square \\
F_2 \quad Cl_2
\end{array}
$$

| Stereochemistry: | Hexadiene | | | Vinylcyclobutane | |
|---|---|---|---|---|---|
| | a | b | | x | y |
| | trans | trans | ⟶ | trans | cis + trans |
| | cis | cis | ⟶ | cis | cis + trans |
| | trans | cis | ⟶ | cis + trans | cis + trans |

Figure 16.1. Stereochemical evidence for two-step cycloaddition.

However, although the loss of stereospecificity certainly demonstrates that a two-step mechanism is operating (provided it is shown that the reactants or products do not independently isomerise), the converse is not always true. Stereospecificity can arise even in two-step reactions if the ring closure is faster than rotation; in the example of figure 16.1 the variation in isomer proportions shows that the two processes are of comparable speed, and some demonstrably zwitterionic reactions are highly stereospecific [e.g. 66JA5254].

The zwitterionic mechanism may frequently be detected through its sensitivity to solvent polarity and to substituent effects. Tetracyanoethylene reacts almost instantly with enol ethers and with styrenes bearing electron-donating substituents to form coloured complexes, the colour fading again as cycloaddition ensues. The rate is markedly dependent on solvent polarity: a reaction which is complete in a minute in acetonitrile can take a month in cyclohexane. For a wide range of solvents plots of $\log k$ against the Dimroth solvent polarity parameter (table 5.5; §5.2) are approximately linear. The extent of isomerisation also depends on solvent polarity, since in a zwitterionic intermediate internal rotation, which increases the degree of charge separation, is suppressed in a non-polar solvent [73JA5054, 5055, 5056]. In the reactions with styrenes the effects of *para* substituents are very large with $\rho = -7$ [70QR473]. Clearly these reactions all involve zwitterionic intermediates such as (**16.7**).

$$
\left.
\begin{array}{c}
ArCH{=}CH_2 \\
+ \\
(NC)_2C{=}C(CN)_2
\end{array}
\right\}
\longrightarrow
\begin{array}{c}
Ar\overset{+}{C}H{-}CH_2 \\
\big| \\
C(CN)_2 \\
(NC)_2\bar{C}
\end{array}
\longrightarrow
\begin{array}{c}
ArCH{-}CH_2 \\
\big|\quad\big| \\
(NC)_2C{-}C(CN)_2
\end{array}
$$

(**16.7**)

In less extreme cases, however, such effects provide less clear-cut criteria. A reaction may be concerted yet not synchronous: the bond changes occur within a single elementary step, but do not keep in time with each other. Then charge separation appears in the cyclic transition state and the reaction begins to show zwitterionic characteristics. The pyrolytic eliminations exemplify this: table 8.7 (§8.6) points to early C–H cleavage in the Cope elimination, early C–O cleavage in acetate pyrolysis. On the other hand many of these reactions are very insensitive to the nature of the solvent, as exemplified by the Diels–Alder dimerisation of cyclopentadiene (§5.2), but such behaviour could accord either with a concerted pericyclic mechanism or with reaction via a biradical.

It is often possible to make fairly accurate estimates, based on bond energies, of the activation enthalpies for gas-phase biradical reactions. In some cases there is close agreement with experimental values, strongly pointing to the biradical route. In others the observed enthalpies are considerably less than the calculated values and this indicates a concerted mechanism in which bond breaking at the transition state is partly compensated by bond making [e.g. 68JPC1866]. Similar, though necessarily cruder, calculations can be used for reactions in solution.

Conversely the entropy of activation for a concerted reaction, with its highly ordered transition state, is often large and negative. This applies particularly to cycloadditions: a typical value for the Diels–Alder reaction is $\Delta S^{\ddagger} = -150 \text{ J K}^{-1} \text{ mol}^{-1}$.

The secondary kinetic isotope effect can also provide evidence for a concerted mechanism. In the cycloelimination of the methylfuran–maleic anhydride adduct (16.8) the H/D rate ratio for each of the three hydrogen

(16.8; X, Y, Z = H, D)

atoms denoted X, Y, Z is 1.08. This indicates that in the transition state all three carbon atoms to which they are bonded have undergone about the same degree of rehybridisation and suggests a more or less symmetrical transition state structure. Changing the methyl group to $CD_3$ has much less effect on the rate, confirming that there is little build-up of positive charge or radical density on the adjacent carbon atom [65JA1534]. Conversely, unequal H/D kinetic isotope effects point to a two-stage ring opening in the following example [82JA6836]:

(X,Y=H,D. $k_H/k_D(X)=1.208$. $k_H/k_D(Y)=0.944$)

The carbon atoms of the four membered ring are intermediate between $sp^2$ and $sp^3$: the observed kinetic isotope effects, one greater than unity and one less, accord with a rate-limiting step in which one atom is moving towards $sp^2$ hybridisation and the other towards $sp^3$ (§2.4.2).

Nevertheless, despite these various mechanistic criteria, and despite intensive study over two decades and more, there remain doubts about the nature of even some of the best known cyclic reactions. The Cope rearrangement has long been held to be a typical concerted process and it is discussed in this context later in the chapter, but recent work indicates that it may well be a two-stage reaction via a biradical intermediate [77JA4417; 79JA6693]:

Likewise there is intense controversy about the nature of 1,3-dipolar cycloadditions. Huisgen, the father of 1,3-dipole chemistry, interprets the abundant experimental evidence in terms of a concerted process [76JO403; 81PAC171]; Firestone argues from the same data that the reactions follow a biradical route and suggests that this applies to many other cycloadditions, including the Diels–Alder reaction, that have long been regarded as concerted [77T3009]. Theoretical calculations of transition state structures are no more helpful. Both for 1,3-dipolar cycloadditions [82T1847; 80JA1763; 83JA719] and for Diels–Alder reactions [78JA5650; 76JA2190] there are predictions of unsymmetrical transition states associated with two-step reactions and of symmetrical, concerted transition states [compare also 84JA203, 209].

This inability to get to the heart of these reactions is the more frustrating in that recent years have seen a revolution in our theoretical understanding of concerted pericyclic processes. The vagaries of these reactions had for long defied the probes of physical organic chemists, to the extent that they were called 'no-mechanism' reactions. Why, for example, should Diels–Alder cycloaddition occur so readily when the concerted dimerisation of alkenes to cyclobutanes is rare? It cannot only be a question of strain; in the unambiguously two-step reaction of dichlorodifluoroethene with hexadiene, discussed above, the cyclobutane is actually formed in preference to the Diels–Alder adduct.

Looking back through the literature one can see that a number of workers were beginning to gain insights into possible reasons for such behaviour, but the explosive development of the subject sprang from the theory developed by Woodward and Hoffmann, and especially from a comprehensive account of their work published in 1969 [69AG781]. Their conclusions are summarised in the *Woodward–Hoffmann rules*, tables 16.1 to 16.3.

Various other approaches to the same problems have been developed, notably theories based on frontier molecular orbital concepts and on ideas related to aromaticity. The predictions of these different approaches are essentially equivalent [75JA2431].

The Woodward–Hoffmann rules divide pericyclic processes into two categories: those which are 'allowed' have a low activation energy, those which are 'forbidden' have a high activation energy. A 6-electron cyclo-addition, as in the Diels–Alder reaction, is allowed; the 4-electron cyclodimerisation of alkenes is forbidden. With remarkable accuracy the rules have rationalised the behaviour of many previously puzzling cyclical reactions, and have successfully predicted the outcome of hitherto unknown ones. Their dramatic success has tended to blind us to the underlying uncertainty about the reaction mechanisms. The theoretical approaches establish that *if* the Diels–Alder reaction is a concerted, pericyclic process it can take place readily via a low energy cyclical transition state. It does not follow that because the Diels–Alder reaction does indeed take place readily it necessarily *must be* concerted. The biradical route may be still more favourable.

The rest of this chapter is concerned with true pericyclic processes, and with the theories that describe them. The discussion is couched in terms that imply that reactions such as the Diels–Alder are pericyclic, but in view of the cautionary comments of the preceding paragraphs the reader may wish to regard this assumption with an element of scepticism. There is not space enough here to describe the applications of each theoretical method to every type of pericyclic reaction: detailed expositions are available in numerous reviews and monographs, a selection of which is listed at the end of the chapter. The discussion that follows is therefore illustrative rather than exhaustive.

## 16.2 Conservation of orbital symmetry

### 16.2.1 Electrocyclic reactions

The essence of the Woodward–Hoffmann approach is that during the course of a pericyclic reaction, the symmetries of the changing orbitals relative to the overall symmetry of the system must be retained from reactants through to products.

Consider for example the electrocyclic interconversion of cyclobutene and butadiene. The reaction leaves the $\sigma$ framework of the butadiene largely unchanged, but the delocalised $\pi$ orbitals are permuted to form the localised $\pi$ and $\pi^*$ orbitals of cyclobutene together with the $\sigma$ and $\sigma^*$ orbitals of the new ring-closing single bond (figure 16.2).

It is clear that during the reaction the terminal C–C bonds must be rotated. If both rotations are in the same sense the process is called *conrotatory*; if in the opposite sense, *disrotatory* (figure 16.3). During the conrotatory process the only molecular symmetry element that is retained throughout is the $C_2$ axis. In the disrotatory process the $C_2$ axis is lost but a mirror plane ($m$) is retained.

Consider first the conrotatory process, and the relationship of the eight

Butadiene                               Cyclobutene

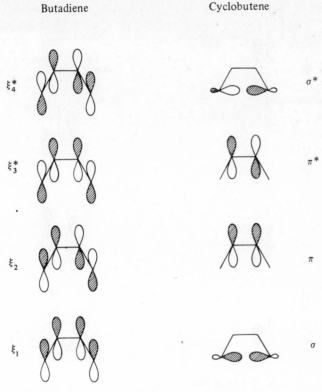

Figure 16.2. Molecular orbitals involved in electrocyclic interconversion of butadiene and cyclobutene. The molecular orbitals are represented by their constituent atomic orbitals; the relative phases are indicated by shading, but no attention is paid to the magnitudes, the symbol $\xi$ is used for delocalised $\pi$ orbitals.

*Conrotation*:

*Disrotation*:

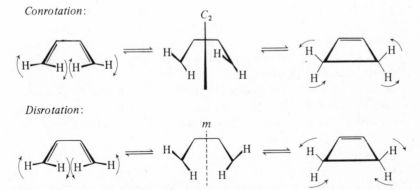

Figure 16.3. Alternative modes of electrocyclic interconversion of butadiene and cyclobutene, showing the symmetry elements retained throughout the reaction.

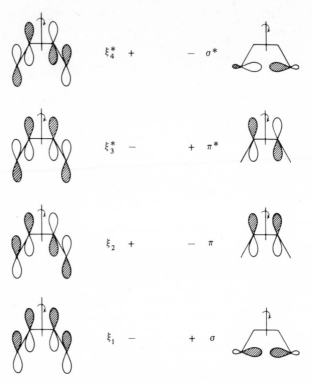

Figure 16.4. Symmetry relations relative to $C_2$ axis in conrotatory interconversion of butadiene and cyclobutene. Orbitals that are symmetrical relative to $C_2$ are indicated ' + ', those that are antisymmetrical are indicated ' − '.

molecular orbitals to the $C_2$ axis (figure 16.4). The lowest energy $\xi$ orbital ($\xi_1$), being antisymmetric with respect to $C_2$, cannot transform itself into the lowest energy cyclobutene orbital ($\sigma$) because that is symmetric; it can transform itself into the $\pi$ orbital. Similarly $\xi_2$ can become $\sigma$, $\xi_3^*$ must become $\sigma^*$ and $\xi_4^*$ becomes $\pi^*$. These relations are indicated by a *correlation diagram* (figure 16.5).

In the ground state of butadiene the four $\pi$ electrons occupy $\xi_1$ and $\xi_2$. These electrons are redistributed during a conrotatory process to occupy the $\pi$ and $\sigma$ orbitals of cyclobutene, which is also the ground state. The reaction takes place readily.

In the disrotatory process the orbital symmetries must relate to the mirror plane (figure 16.6). The ground state of butadiene now correlates with a high-energy excited state of cyclobutene, with two electrons in the antibonding $\pi^*$ orbital. Clearly the disrotatory process is energetically unfavourable.

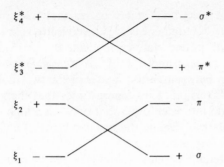

Figure 16.5. Correlation diagram for conrotatory interconversion of butadiene and cyclobutene.

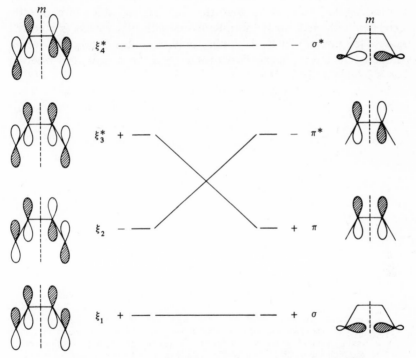

Figure 16.6. Symmetry relations relative to mirror plane *m* in disrotatory interconversion of butadiene and cyclobutene, and resulting correlation diagram.

Actually the situation is a little more complicated than the previous paragraph indicates. Instead of considering simply the correlation between individual orbitals we should look at the molecule as a whole, taking into account all the occupied orbitals, to discover symmetry relations between different electronic states of reactant and product molecules. The ground state of butadiene is $\xi_1^2\xi_2^2$ (two electrons in each of the orbitals $\xi_1$ and $\xi_2$), and ground state cyclobutene is $\sigma^2\pi^2$. Similarly the first excited states are $\xi_1^2\xi_2^1\xi_3^{*1}$ and $\sigma^2\pi^1\pi^{*1}$. A *state diagram* (figure 16.7) shows the correlation between various states of the two molecules for the disrotatory interconversion. This demonstrates that the ground states can, in fact, transform into each other, but only via a high energy transition state in the 'crossover' region, the reaction remains unfavourable relative to the conrotatory process.

The interconversion of butadiene and cyclobutene can be carried out photochemically as well as thermally. In this case we are concerned not with the ground state but with excited states of the molecules, produced by the absorption of a photon. The state diagram for the disrotatory process (figure 16.7) shows that the first excited states are directly correlated with each other; no interaction with other states is involved because the non-crossing rule applies only to states of the same overall symmetry. In the photochemical process it is the conrotatory reaction that is inhibited (figure 16.8).

There is abundant experimental evidence for these conclusions. Thus thermolysis of *trans*-3,4-dimethylcyclobutene leads exclusively to *trans*-

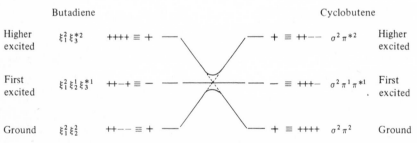

Figure 16.7. State diagram for the disrotatory interconversion of butadiene and cyclobutene. Not all higher states are shown.

The straight-line correlations (including the dotted sections) follow the correlations of the individual orbitals (figure 16.6). Thus $\xi_1$ correlates with $\sigma$ and $\xi_2$ with $\pi^*$, so the ground state of butadiene, $\xi_1^2\xi_2^2$, correlates with one of the high energy excited states of cyclobutene, $\sigma^2\pi^{*2}$.

The total symmetry of each state is the product of the separate symmetries of the individual electrons. In ground state butadiene the two $\xi_1$ electrons are symmetric relative to $m$ (figure 16.6) and the two $\xi_2$ electrons are antisymmetric. The resultant is a symmetric state, shown on the figure as:
$+ + - - \equiv +$.

Electronic interactions prevent two reaction paths of the same symmetry from crossing each other; the actual paths are those shown by solid lines. Thus the two ground states can transform one into the other because both are overall symmetric, but the reaction path reaches up into the high energy 'crossover' region.

Butadiene                                                                 Cyclobutene

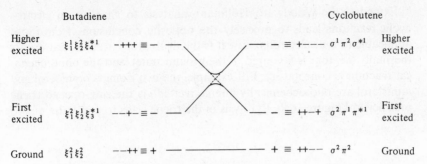

Higher
excited      $\xi_1^1 \xi_2^2 \xi_4^{*1}$   $-+++ \equiv -$  — — — — — — $-- \equiv +---$ $\sigma^1 \pi^2 \sigma^{*1}$   Higher
excited

First
excited      $\xi_1^2 \xi_2^1 \xi_3^{*1}$   $--+- \equiv -$  — — — — — — $-- \equiv ++-+$ $\sigma^2 \pi^1 \pi^{*1}$   First
excited

Ground       $\xi_1^2 \xi_2^2$   $--++ \equiv +$ — — — — — — — — — — $+ \equiv ++--$ $\sigma^2 \pi^2$   Ground

Figure 16.8. State diagram for conrotatory interconversion of butadiene and
cyclobutene. Not all higher states are shown. The direct (crossing) interactions
follow from the correlation diagram of figure 16.5.

*trans*-hexadiene, which in turn can be photolysed to *cis*-dimethylcyclo-
butene [65TL1207; 68JA4498].

Note that there is an alternative conrotatory ring opening which is
disfavoured not because of symmetry considerations, but because of the
steric interactions involved:

   This example, though illustrating the validity of the Woodward–
Hoffmann approach, also reveals one of its weaknesses. The introduction
of the methyl groups reduces the symmetry of the system, and during the
conrotatory process the $C_2$ axis is lost. Yet despite this the predictions of
the theory are upheld – indeed, they remain valid even in systems in
which the molecules lack all symmetry. One might regard this as an
unimportant criticism; a mechanism that is applicable to a totally
symmetrical situation is not likely to be drastically altered by introducing
substituents into the molecule that interfere with the symmetry but not
with the essential character of the reaction. On the other hand there are
those who maintain that these weaknesses pervade the whole foundation
of the Woodward–Hoffmann approach: a key assumption is the non-
correlation of states of opposite symmetries, and this is demonstrably
only valid for systems of perfect symmetry.

Extending the Woodward–Hoffmann analysis to six-electron electro-cyclic reactions leads to precisely the opposite conclusions from those applicable to four electrons: now it is the disrotatory process that occurs thermally (i.e. that is favoured in the ground state), and the photochemical reaction is conrotatory. For example, the two isomers ergosterol and lumisterol are photochemically interconverted via the ring-opened triene precalciferol (figure 16.9). Pyrolysis of the triene leads to a mixture of two

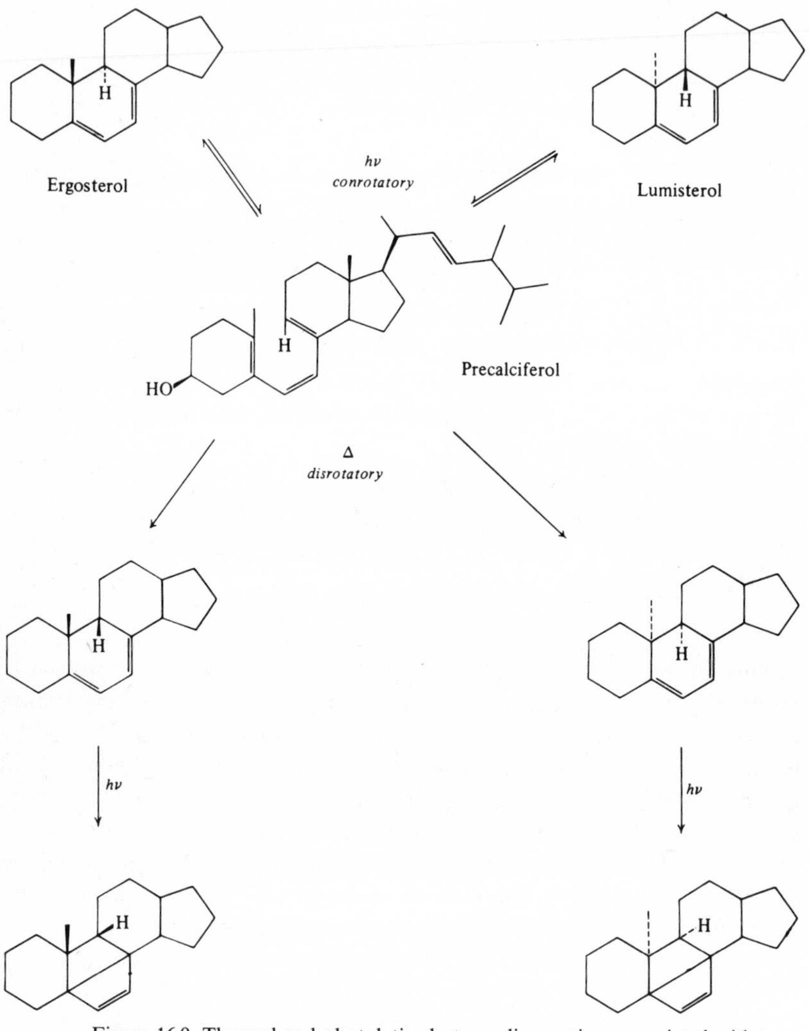

Figure 16.9. Thermal and photolytic electrocyclic reactions associated with vitamin D chemistry (see also §16.2.3). The complete substitution pattern is shown only for precalciferol [61T(16)146].

Table 16.1. *Woodward–Hoffmann rules for electrocyclic reactions*

|            | $4n$ electrons | $4n + 2$ electrons |
| ---------- | -------------- | ------------------ |
| Thermal    | Conrotatory    | Disrotatory        |
| Photolytic | Disrotatory    | Conrotatory        |

different isomers, and if these are photolysed they do not revert to the triene because the six-electron disrotatory process by which they were formed is photochemically forbidden and the conrotatory process involves impossible molecular contortions; instead they undergo a symmetry-allowed *four-electron* disrotatory ring closure to give cyclobutene derivatives.

The principles established in this way can be extended to systems with more than six electrons. The general rules are summarised in table 16.1.

### 16.2.2 Cycloadditions and cycloeliminations

(*a*) $[2 + 2]$ *Cycloadditions.* The dimerisation of ethylene to cyclobutane involves the transformation of four $\pi$ orbitals (two bonding, two antibonding) into four $\sigma$ orbitals. The orbitals of a single ethylene molecule do not relate to the symmetry of the reaction as a whole, nor does a single $\sigma$ bond of the cyclobutane. We therefore need to combine the separate orbitals in appropriately symmetrical ways. This is illustrated in figure 16.10, together with the resulting correlation diagram which indicates that the $\pi_2$ orbital pair of ethylene transforms into the high energy $\sigma_3^*$ orbital pair of cyclobutane. This suggests that the ground state thermal reaction is symmetry-forbidden, and the state diagram (figure 16.11) confirms this, revealing a high activation energy reaching up into the crossover region. The photolytic reaction involves no such barrier.

The theory again accords with observation: olefin dimerisation is almost invariably a photocatalysed process unless the double bonds bear substituents that facilitate the two-step zwitterionic or biradical routes.

So far the discussion has been couched in terms of a face-to-face approach of the two ethylene molecules in a mutual *syn* addition; the term that is generally used in the context of pericyclic reactions is *suprafacial.*

*Suprafacial*                                        *Antarafacial*

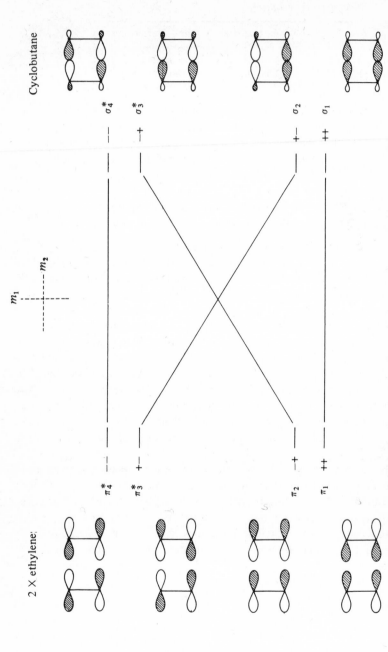

Figure 16.10. Symmetry orbitals and correlation diagram for interconversion of two ethylene molecules with cyclobutane. Orbitals $\pi_1$ and $\pi_2$ are combinations of the bonding $\pi$ orbitals of ethylene and are therefore of equal energy; $\pi_3^*$ and $\pi_4^*$ are combinations of the antibonding $\pi^*$ orbitals and are also of equal energy. Likewise $\sigma_1$ and $\sigma_2$ are bonding, $\sigma_3^*$ and $\sigma_4^*$ antibonding. The two symmetry symbols associated with each pair of orbitals describe their relation to the vertical and horizontal mirror planes $m_1$ and $m_2$, both of which are retained throughout the reaction.

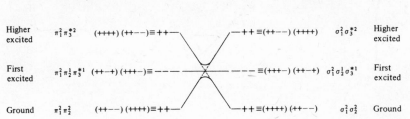

Figure 16.11. State diagram for interconversion of two ethylene molecules with cyclobutane. The symbols in parentheses show the symmetries of the individual electrons relative to the two different mirror planes: thus in ground state cyclobutane all four electrons are symmetrical with respect to $m_1$, and the two $\sigma_1$ electrons are also symmetrical with respect to $m_2$, but the two $\sigma_2$ electrons are antisymmetrical with respect to $m_2$. The symmetries of the separate electrons are then combined to give the overall symmetry of each state, but one cannot further combine the $m_1$ and $m_2$ symmetries: the symmetry about the two mirror planes must be independently conserved.

At first sight the alternative *antarafacial* attack seems impossible, but it might occur if the molecules were to approach each other at right angles. In figure 16.12 the lower ethylene molecule is regarded as stationary, lying coplanar with the paper, and it undergoes suprafacial attack from above. The upper molecule is initially in a plane perpendicular to that of the paper, and it reacts antarafacially; the 'north' end forms its new $\sigma$ bond from the left while the 'south' end forms it from the right. Notice that this requires a mutual twisting of the two ends of the upper molecule; the two hydrogen atoms shown in bold type begin the reaction *cis* to each other in the ethylene molecule, but they are *trans* in the cyclobutane.

Correlation and state diagrams can be drawn for this process too, having regard now to the $C_2$ axis bisecting the two C–C bonds which is the only symmetry element of the system. (It is a good exercise to work them out!) They show the reverse pattern from supra-supra addition, namely that the thermal reaction is allowed, the photochemical one forbidden. However, it is doubtful if any $[2s + 2a]$ reactions have been observed: some alkenes possessing severely strained double bonds do dimerise when heated, but these are probably biradical reactions [e.g. 71JA3633].

(*b*) $[4 + 2]$ *Cycloadditions.* The $[4 + 2]$ Diels–Alder reaction is the archetypal cycloaddition. For the mutually suprafacial attack of diene on dienophile the only symmetry element involved is a mirror plane *m* (figure 16.13).

The interacting orbitals, their symmetry relations with *m*, and their correlation diagram are shown in figure 16.14. All the bonding orbitals of

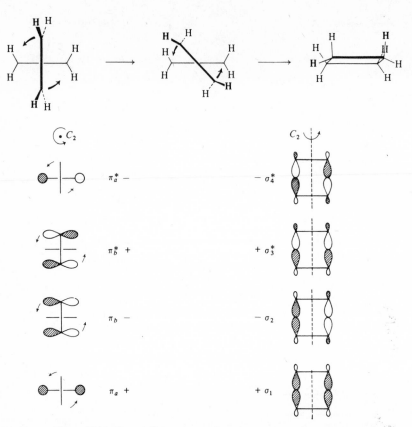

Figure 16.12. [2s + 2a] Interconversion of two molecules of ethylene with cyclobutane. In this process the separate π orbitals of the ethylene molecules are directly related to the $C_2$ axis; the suffix $a$ refers to the lower ethylene molecule whose orbitals are shown in plan so that only the top lobe is indicated, and the suffix $b$ refers to the upper ethylene molecule. The cyclobutane σ orbitals are combined into symmetry-related pairs, as in figure 16.10.

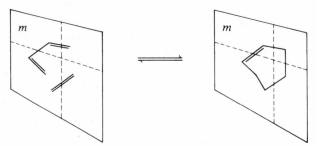

Figure 16.13. Plane of symmetry in Diels-Alder interconversion of ethylene + butadiene with cyclohexene.

Diene/dienophile                                                   Cyclohexene

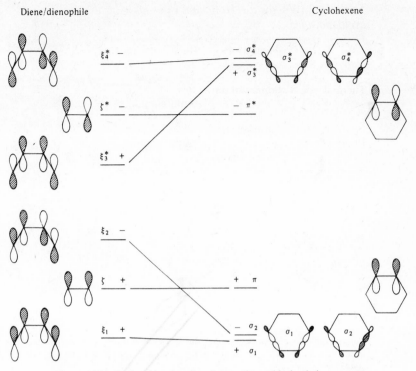

Figure 16.14. Diels–Alder reaction. Interacting orbitals, their symmetry relation to the mirror plane of the system, and their correlation. The symbols $\xi$ and $\zeta$ are used for the $\pi$ orbitals of the diene and dienophile respectively.

the reactants correlate directly with bonding orbitals of the products and vice versa, and we can determine without drawing a full state diagram that the ground state thermal reaction is symmetry allowed. The photochemical reaction of the first excited state is, on the other hand, forbidden.

Again the pattern is reversed if one of the reactants undergoes the geometrically less favoured antarafacial attack (figure 16.15($a$)); for these processes the photochemical reaction is allowed, the thermal one forbidden. One could also envisage a $[4a + 2a]$ reaction in which both components react antarafacially (figure 16.15($b$)). This should show a second reversal of behaviour and become thermally allowed, but no examples of this are known.

($c$) *Other cycloadditions.* The patterns of reactivity established for four- and six-electron reactions can be generalised according to table 16.2. For

Table 16.2. *Woodward–Hoffmann rules for cycloadditions*

|  | 4n electrons | 4n + 2 electrons |
|---|---|---|
| Thermal | Supra-antara | Supra-supra *or* Antara-antara |
| Photolytic | Supra-supra *or* Antara-antara | Supra-antara |

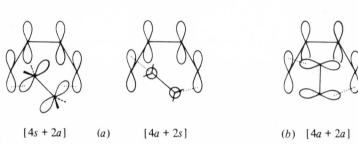

[4s + 2a]    (a)    [4a + 2s]    (b)  [4a + 2a]

Figure 16.15. Six-electron cycloadditions: (*a*) supra-antara, symmetry-allowed in the first excited state; (*b*) antara-antara, symmetry allowed in the ground state.

example, heptafulvene undergoes cycloaddition reactions which can be classed as [8s + 2s]:

Heptafulvalene, on the other hand, undergoes a cycloaddition which is shown by the stereochemistry of the adduct to be [14a + 2s]:

It is important to notice that the rules apply to the total number of *electrons* involved in the cyclical reorganisation, not the total number of orbitals, nor the size of the ring. Thus 1,3-dipolar cycloadditions also fall within their scope. In ozonolysis, for example, the three-centre $\pi$ orbitals of ozone contain a total of four electrons, so the reaction with an alkene is $[4s + 2s]$, entirely analogous to the Diels–Alder reaction:

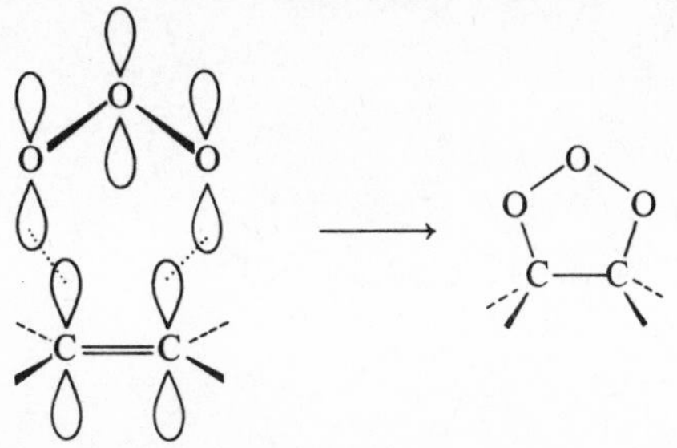

Similar is the reaction of an alkene with osmium tetroxide (addition across the O=Os=O fragment) which leads, via hydrolysis of the resulting osmate ester, to *syn* hydroxylation: a useful counterpart to the *anti* addition using peracids.

Not all 1,3-dipolar additions are as symmetrical as these; indeed, many of them lack any semblance of symmetry:

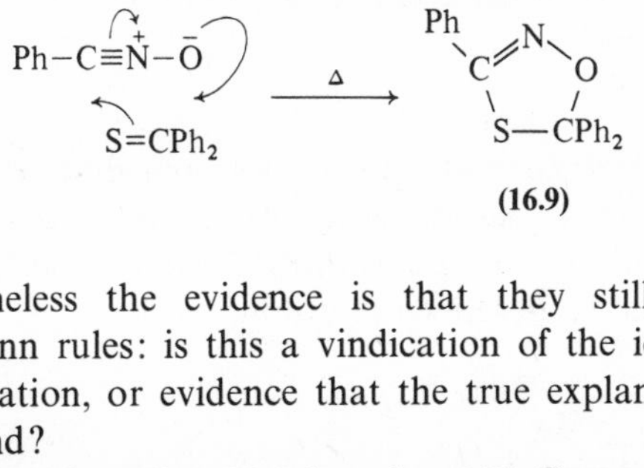

(16.9)

Nevertheless the evidence is that they still obey the Woodward–Hoffmann rules: is this a vindication of the ideas of orbital symmetry conservation, or evidence that the true explanation is something more profound?

The reactions that Woodward and Hoffmann call *chelotropic* – the ring closure of a polyene with the inclusion of a single additional atom – are also a type of cycloaddition. The corresponding cycloelimination is usually called an extrusion reaction. Some examples are shown in figure 16.16.

Most cycloadditions involve only two reacting systems, but examples are known of $[2 + 2 + 2]$ cycloadditions. The Woodward–Hoffmann rules are readily extended to accommodate them: $[2s + 2s + 2s]$ and $[2s + 2a + 2a]$ are the thermally allowed and photochemically forbidden processes; vice versa for $[2a + 2a + 2a]$ and $[2a + 2s + 2s]$.

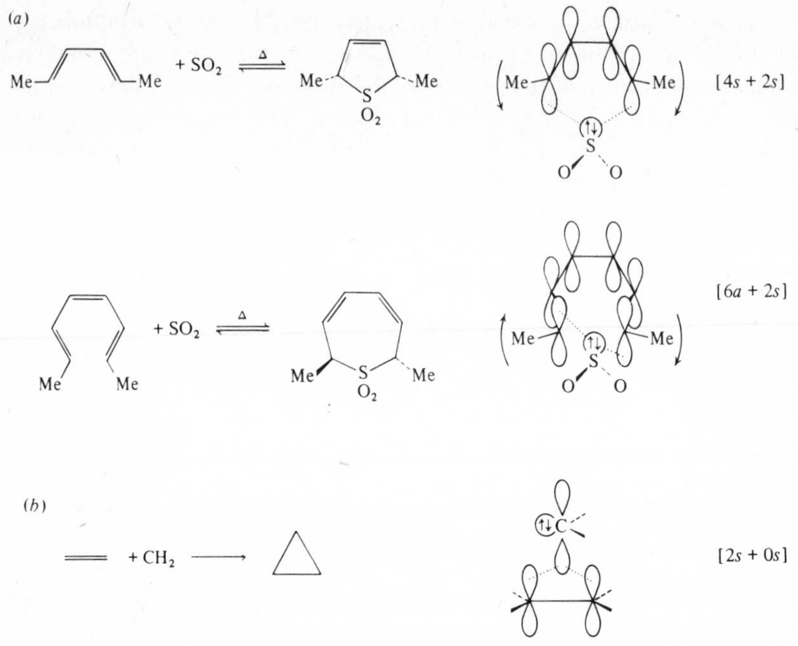

(a)

Me + SO$_2$ ⇌ (Δ) Me—S—Me [4s + 2s]
O$_2$

Me—Me [4s + 2s]

Me + SO$_2$ ⇌ (Δ) Me—S—Me [6a + 2s]
O$_2$

Me—Me [6a + 2s]

(b)

+ CH$_2$ ⟶ △ [2s + 0s]

Figure 16.16. Chelotropic reactions. (a) Stereochemical consequences of orbital symmetry control in cycloadditions of SO$_2$ to polyenes. (b) Cycloaddition of singlet carbene to alkene involves a vacant carbene orbital; triplet carbenes have no low-lying vacant orbitals and cannot react in such a concerted process.

## 16.2.3 Sigmatropic rearrangements

The Woodward–Hoffmann treatment of sigmatropic rearrangements differs somewhat from that of electrocyclic and cycloaddition reactions in that frequently only the transition state, and not the reactants and products, can be described in symmetry terms:

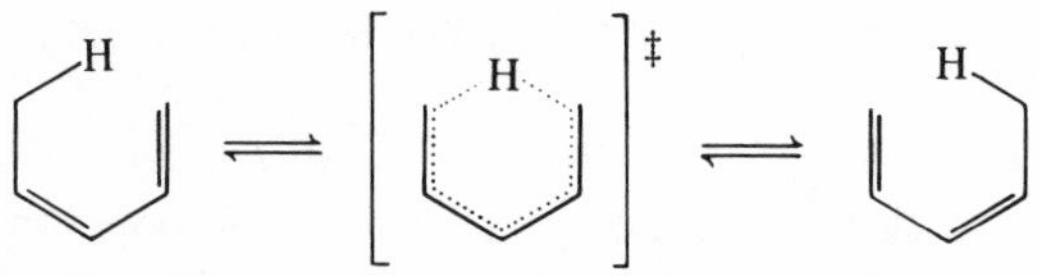

This is an inherently different situation from those in previous examples of unsymmetrical systems. Even the nitrile oxide + thione reaction (**16.9**) can be regarded as a (highly) perturbed version of the symmetrical ozone + alkene reaction; the fundamental nature of the interacting orbitals is the same in both.

For this reason only the summarising rules are given here; their rationale is discussed via an alternative theoretical approach in §16.4.

Table 16.3. Woodward–Hoffmann rules for sigmatropic hydrogen shifts

|  | 4n electrons | 4n + 2 electrons |
|---|---|---|
| Thermal | Antara | Supra |
| Photolytic | Supra | Antara |

A hydrogen shift can occur either suprafacially or antarafacially:

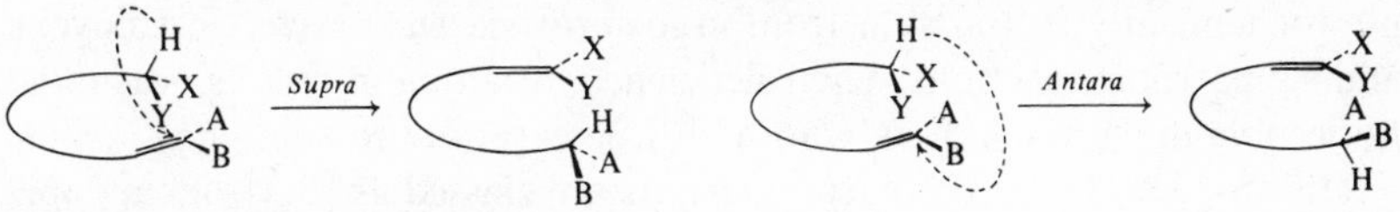

The allowed processes are listed in table 16.3.

Conjugated dienes readily undergo suprafacial 1,5-hydrogen shifts; in cyclopentadiene the reaction occurs even at room temperature [63T1939; 69CJ1555]:

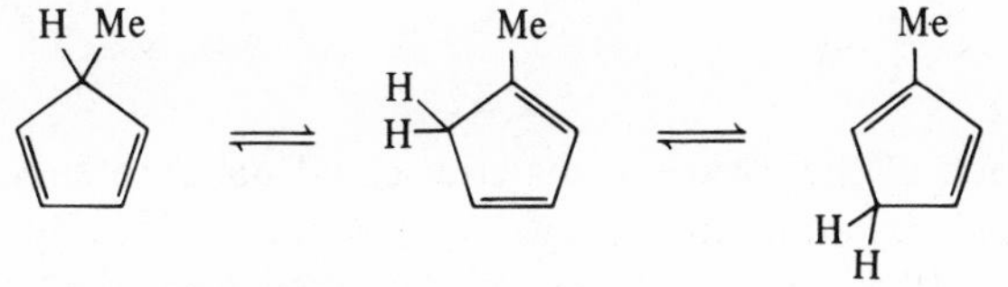

Simple alkenes do not undergo antarafacial two-electron 1,3-shifts which would involve the highly strained transition state (16.10), but when photolysed they do undergo suprafacial 1,3-shifts. Notice that antarafacial hydrogen transfer involves a conrotatory displacement of the bonds, and suprafacial transfer (16.11) involves disrotation. Thus the table of allowed hydrogen shifts is entirely analogous to the table 16.1 for electrocyclic reactions.

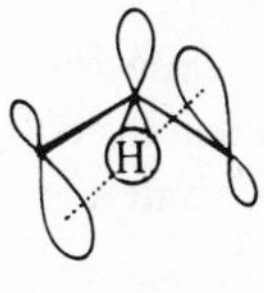

(16.10)

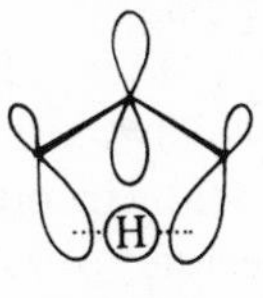

(16.11)

1,7-Shifts can occur either thermally (antarafacial) or photolytically (suprafacial) as in the equilibrium between vitamin D and precalciferol:

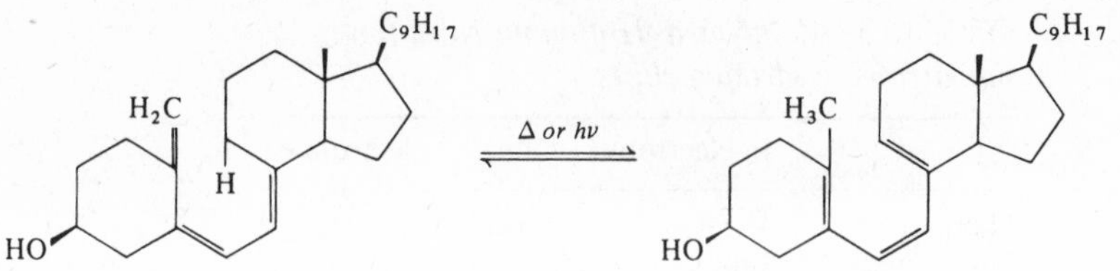

Vitamin D$_2$ (calciferol)          Precalciferol (*cf.* figure 16.9)

(Vitamin D$_3$, also called calciferol, has the 6-methylhept-2-yl side-chain, as in cholesterol, in place of C$_9$H$_{17}$. The D vitamins are produced photochemically in the skin from ergosterol via successive electrocyclic and sigmatropic reactions. Their deficiency, resulting in rickets, was most noticeable in slum children, who had little exposure to sunlight.)

The Cope and Claisen rearrangements are classed as [3,3] sigmatropic rearrangements, because the newly formed σ bond is three atoms removed from the original points of attachment of both π fragments:

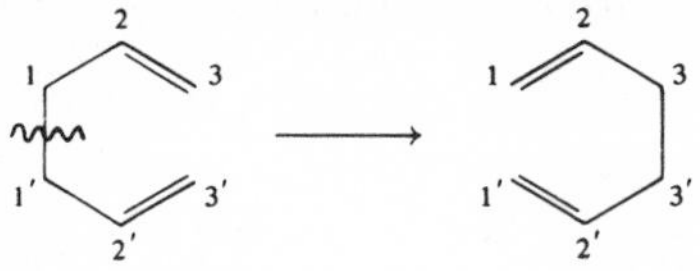

In principle each of the 3-carbon moieties could undergo either suprafacial or antarafacial addition, though supra-supra is sterically the most favourable. With this type of sigmatropic rearrangement the symmetry-allowed processes are the same as for cycloadditions (table 16.2).

## 16.3    Frontier orbital methods: cycloadditions
### 16.3.1    Woodward–Hoffmann rules

The early studies of Woodward and Hoffmann were presented in terms of the symmetry of the HOMO, and they continued to use orbital interactions to supplement their more comprehensive orbital correlations. Note in the correlation diagrams of figures 16.5 and 16.6, for example, that in an allowed reaction the energy of the HOMO falls whereas in a forbidden one it rises.

Not surprisingly, then, the frontier orbital approximation of PMO theory has been widely used for describing pericyclic reactions [71Acc57; 76MI]. Its application to intramolecular reactions is somewhat artificial because the molecular orbitals have to be dissected into HOMO and LUMO components. It is, however, particularly suitable for cycloadditions [75Acc361; 73SPC113], and it has been invaluable for probing secondary patterns of reactivity within the overall confines of the Woodward–Hoffmann rules, as discussed in the following sections.

Diene    Dienophile

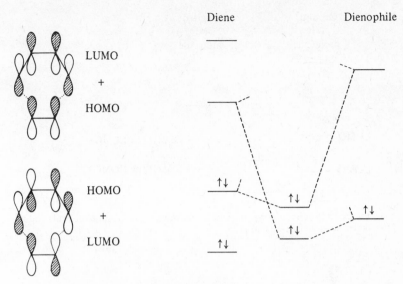

LUMO

+

HOMO

HOMO

+

LUMO

Figure 16.17. Frontier molecular orbital perturbations in Diels–Alder cycloaddition of butadiene to ethylene.

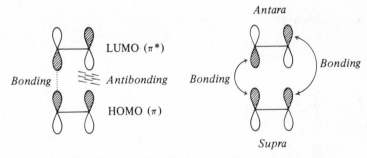

*Antara*

LUMO ($\pi^*$)

*Bonding*    *Antibonding*    *Bonding*    *Bonding*

HOMO ($\pi$)

*Supra*

Figure 16.18. Frontier molecular orbitals in cyclodimerisation of ethylene.

As the two reacting molecules approach each other it is easy to picture the molecular perturbation of the frontier orbitals. Figure 16.17 illustrates the Diels–Alder reaction. It does not matter whether we consider the diene–HOMO/dienophile–LUMO perturbation or diene–LUMO/dienophile–HOMO: in each case the mutual overlap at both points of contact is bonding as the two molecules approach.

By contrast the corresponding $[2s + 2s]$ addition engenders an antibonding interaction and the concerted reaction is forbidden, whereas $[2s + 2a]$ is allowed (figure 16.18).

For photochemical processes the pattern of frontier orbitals is different. Consider first the generalised interaction between two molecules, one in the ground state, one in the first excited state (figure 16.19). The

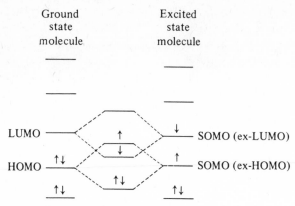

Figure 16.19. First order perturbations between a ground state molecule and a molecule in its first excited state.

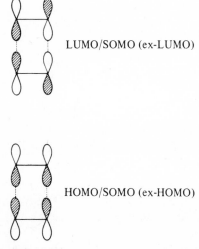

LUMO/SOMO (ex-LUMO)

HOMO/SOMO (ex-HOMO)

Figure 16.20. Bonding interactions in photochemical $[2s + 2s]$ cyclodimerisation.

important perturbations are now the first order interactions between the HOMO of the ground state molecule and the lower SOMO (singly occupied molecular orbital) of the excited molecule – that is, its erstwhile HOMO – and between the LUMO and the upper SOMO (ex-LUMO). For a $[2s + 2s]$ process these interactions are bonding (figure 16.20), and the photochemical reaction is allowed.

The extension of this approach to other cycloadditions is obvious, and the predictions accord with those of Woodward and Hoffmann as summarised in table 16.2.

### 16.3.2 Secondary stereochemistry

Although a pericyclic reaction is stereospecific there are many situations in which two alternative pericyclic reactions can take place, leading to stereoisomeric products. The Diels–Alder addition of maleic anhydride to cyclopentadiene to give *exo* (**16.12**) or *endo* (**16.13**) adducts is a classic example. The Woodward–Hoffmann rules say nothing about

(16.12)        (16.13)

this; both products are formed by normal [4s + 2s] cycloaddition. The *exo* adduct is thermodynamically the more stable, presumably for steric reasons. Nevertheless, it is the *endo* adduct which is formed the faster. The orbital interactions shown in figure 16.21 suggest a reason. The primary overlap is still that controlling the pericyclic process, but secondary bonding interactions, though they do not actually lead to the formation of new bonds, nevertheless lower the energy of the *endo* reaction path. Figure 16.22 shows an example of the opposite behaviour, with preferential formation of the *exo* adduct.

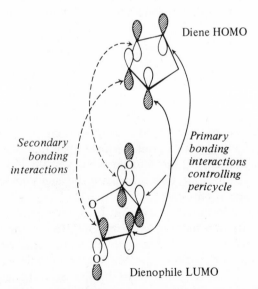

Figure 16.21. Secondary bonding interactions between frontier orbitals of cyclopentadiene and maleic anhydride favouring *endo* addition.

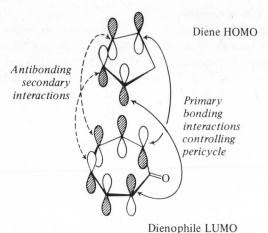

Diene HOMO

*Antibonding secondary interactions*

*Primary bonding interactions controlling pericycle*

Dienophile LUMO

Figure 16.22. In cycloaddition of tropone to cyclopentadiene the secondary interactions between frontier orbitals are antibonding in *endo* approach.

### 16.3.3   Reactivity in Diels–Alder reactions

Dienophiles bearing $\pi$ electron-withdrawing substituents, as in acrolein, are in general considerably more reactive than ethylene. The implication is that in these reactions the transition state is unsymmetrical, with a partial positive charge on the diene and partial negative charge on the dienophile (**16.14**). This accords with the picture given by FMO theory. The $\pi$ orbitals of acrolein resemble those of butadiene (figure 6.6;

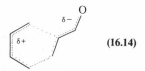

**(16.14)**

§6.1.3) modified by the presence of the oxygen atom: in particular, because the energy of an oxygen $2p$ orbital is lower than that of a carbon $2p$ orbital, the average energy of the acrolein orbitals is less than that of butadiene. Figure 16.23 illustrates this schematically, and compares the energy levels with those of ethylene.

There are two points to note in particular. First, the terminal carbon atom of acrolein contributes more to both the HOMO and the LUMO than does its inner neighbour, so there is indeed more bond making to this position in the initial butadiene–acrolein perturbation. Secondly, the energy difference between the butadiene–HOMO and the acrolein–LUMO is relatively small and this perturbation dominates the reaction (compare equation [6.8]; §6.4.1). This contrasts with the butadiene–

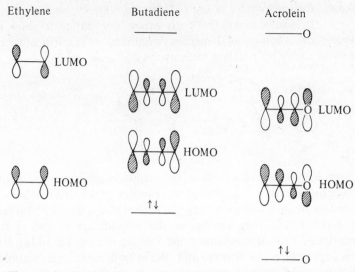

Figure 16.23. Frontier molecular orbitals of ethylene, butadiene and acrolein.

ethylene reaction, where the symmetrical array of orbitals demands that both of the HOMO/LUMO interactions are equivalent. It does not automatically follow that the combination of orbital coefficients and energy levels that happen in acrolein must generate the greater perturbation (and hence the faster reaction); the experimental facts suggest that they do, and detailed MO calculations concur.

This interpretation is further supported by the substituent effects in the cycloadditions of phencyclone with styrenes [80JO659]:

With unsubstituted styrene the FMO energies of diene and dienophile are similar. Electron-withdrawing substituents lower the energy of the styrene LUMO and the reaction is dominated by the diene–HOMO/dienophile–LUMO interaction (as in the butadiene–acrolein example). Electron-donating substituents, on the other hand, raise the energy of the styrene HOMO and increase the significance of the diene–LUMO/dienophile–HOMO interaction. Both effects accelerate the reaction: electron-deficient styrenes give a Hammett correlation with $\rho = 0.6$; electron-rich ones correlate with $\sigma^+$ with $\rho = -0.8$.

### 16.3.4    Regioselectivity in Diels–Alder and other cycloadditions

It is easy to use the 'curly arrow' convention to show that the Diels–Alder addition of 1-methoxybutadiene to acrolein must produce predominantly (in practice, exclusively) the '*ortho*' adduct:

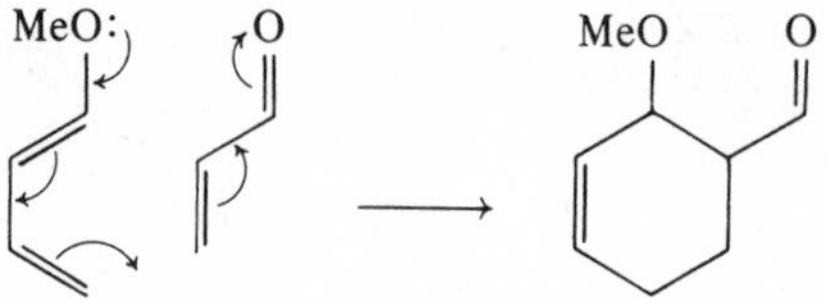

As represented above, of course, the arrows imply a two-step zwitterionic mechanism and not a pericyclic reaction at all. The arrows should therefore be taken as indicating the nature of the initial perturbation, or in other words they emphasise the importance of the electrostatic contribution to the dominant bond making process (*cf.* **16.14**). However, this argument is not always valid. When both diene and dienophile bear electron-withdrawing substituents one might think that the electrostatic interactions would swing the balance in favour of the '*meta*' adduct, but this is not found. Butadiene-1-carboxylic acid and acrylic acid react together to give predominantly the '*ortho*' adduct, even though this means that two electrophilic centres become bonded together:

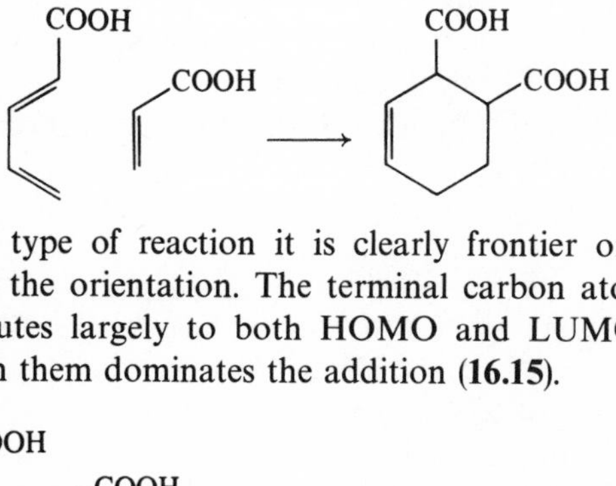

(90 per cent)

In this type of reaction it is clearly frontier orbital interactions that control the orientation. The terminal carbon atom in each component contributes largely to both HOMO and LUMO, and the interaction between them dominates the addition (**16.15**).

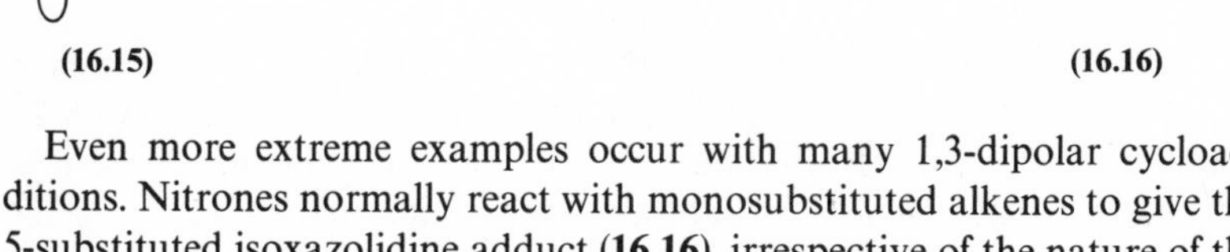

**(16.15)**                                          **(16.16)**

Even more extreme examples occur with many 1,3-dipolar cycloadditions. Nitrones normally react with monosubstituted alkenes to give the 5-substituted isoxazolidine adduct (**16.16**), irrespective of the nature of the substituent. For electron-withdrawing X groups the use of curly arrows would clearly point to the isomeric 4-substituted adduct, but FMO calculations reveal the importance of a very large carbon coefficient in the nitrone LUMO, and the interaction of this with the terminal carbon in the dipolarophile HOMO is enough to force the reaction into the observed orientation [*cf.* 73JA5798].

The difficulty with this approach, however, is that it requires a knowledge of the energies and coefficients of the frontier orbitals, and this removes it from the realm of 'pencil-and-paper' chemistry; it cannot be regarded as a simple substitute for use when curly arrows fail. For example, the dimersation of acrolein gives the '*ortho*' adduct (**16.17**) in conflict both with simple electrostatic predictions and with early FMO calculations, though later calculations gave the right result. There is something unsatisfactory about a method that gives alternative answers and permits the choice of the right one after the event.

(**16.17**)

## 16.4    The aromatic transition state

As far back as 1939 M. G. Evans drew an analogy between the electronic structures of benzene and the Diels–Alder transition state [39TF824]. In benzene each carbon atom is bonded to three other atoms and its remaining atomic orbital participates in a cyclical set of delocalised molecular orbitals. Exactly the same situation obtains in the Diels–Alder transition state. The only differences are that in the planar benzene molecule all the molecular orbitals are of the $\pi$ type whereas the Diels–Alder transition state is bent and there is some $\sigma$ character to the delocalised orbitals. The two systems are called *isoconjugate* (figure 16.24).

If the reaction were a two-step, non-concerted process, either via a biradical (**16.18**) or via a zwitterion, then the transition state would be acyclic. The relationship between the concerted cyclic transition state and

(**16.18**)

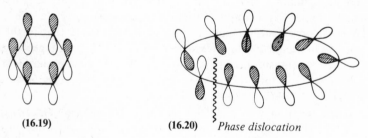

Benzene                    Diels–Alder TS

Figure 16.24. Isoconjugate delocalised systems of benzene and of the Diels–Alder transition state.

the non-concerted acyclic one is the same as that between benzene and the acyclic hexatriene (figure 6.14; §6.2.2). Just as benzene is more stable than hexatriene, so the cyclic aromatic transition state is more stable than the acyclic one. This concept has been summarised by Dewar [71AG761] in what he calls *Evans' principle*: 'Thermal pericyclic reactions take place preferentially via aromatic transition states'.

Before Evans' principle can be applied to pericyclic chemistry we need to broaden the definition of aromaticity by introducing the concept of *Möbius* aromatic systems as a counterpart to the usual *Hückel* type [71Acc272]. In a Hückel system one can arrange the constituent atomic orbitals so that they all overlap in phase (**16.19**; the shading is simply intended to show this in-phase overlap and not, as in the Woodward–Hoffmann and FMO discussions, to represent a particular molecular orbital). The rules for a Hückel system are: $4n + 2$ electrons – aromatic; $4n$ electrons – antiaromatic.

**(16.19)**              **(16.20)** *Phase dislocation*

A Möbius system is one in which the base set of atomic orbitals cannot be arranged without involving a phase dislocation at some point in the ring. Imagine a conjugated polyene which is bent round into a ring but is twisted through 180° in the process (**16.20**). Such a system has the same topology as a Möbius strip, such as can be made from a strip of paper by twisting one end through 180° and gluing the two ends together; the result is a piece of paper with only one side.

Table 16.4. *Rules for aromaticity*

|  | 4*n* electrons | 4*n* + 2 electrons |
| --- | --- | --- |
| Hückel | Antiaromatic | Aromatic |
| Möbius | Aromatic | Antiaromatic |

The rules for a Möbius system are the opposite of those for a Hückel system (table 16.4). If one inverts the phases of any one atomic orbital in either a Hückel or a Möbius system it can only change the phase dislocations two at a time, so one can generalise the definitions to say that a Hückel system has an even number of phase dislocations, a Möbius system an odd number.

Simple Möbius compounds are not known; they would presumably be highly strained. But Möbius transition states are possible and must be included in the application of Evans' principle.

### 16.4.1 Ground state pericyclic reactions

The thermal electrocyclic and cycloaddition reactions discussed in previous sections can equally well be described in terms of Evans' principle (figure 16.25).

Sigmatropic rearrangements, which are less easily handled by the other approaches, are also readily analysed by this method. A 1,5-hydrogen shift (six electrons) can proceed suprafacially via a Hückel aromatic transition state (**16.21**). A 1,3-hydrogen shift (four electrons) demands Möbius aromaticity and must be antarafacial (**16.22**).

(16.21)    (16.22)

The Cope rearrangement involves a chair-shaped transition state, similar to the chair of cyclohexane; it is a six-electron Hückel aromatic system (**16.23**). Each three-atom fragment reacts suprafacially. One could envisage an alternative mode of interaction, still symmetry-allowed, that resembled the boat form of cyclohexane. Electronically this would be isoconjugate with the antiaromatic bicyclohexatriene (**16.24**). Doering showed that the chair transition state is preferred by investigating the stereochemistry of the rearrangements of the diastereoisomeric 3,4-

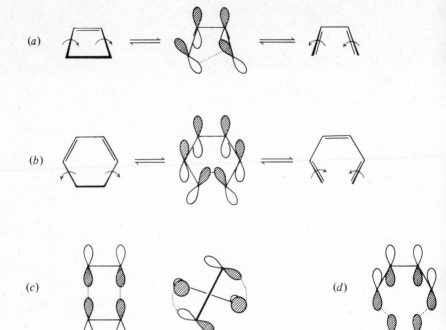

(a)

(b)

(c)

(d)

Figure 16.25. The application of the aromatic transition state concept (Evans' principle) to electrocyclic and cycloaddition reactions. (*a*) Four-electron electrocyclic reaction: Möbius aromaticity, with phase dislocation, requires conrotation. (*b*) Six-electron electrocyclic reaction: Hückel aromaticity, with all overlaps in phase, requires disrotation. (*c*) [2 + 2] cycloadditions, four electrons: supra-supra, with no phase dislocations, is Hückel antiaromatic, supra-antara, with one phase dislocation, is Möbius aromatic. (*d*) [4 + 2] cycloaddition, six electrons: supra-supra, with no phase dislocations, is Hückel aromatic.

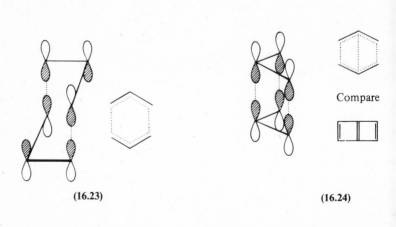

Compare

(16.23)

(16.24)

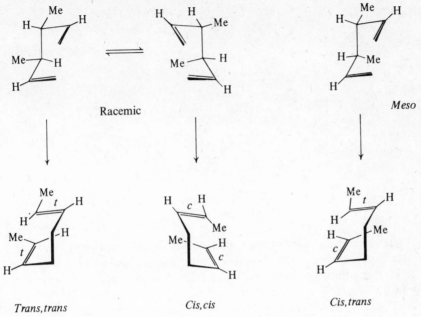

Figure 16.26. The stereochemistry of the Cope rearrangements of the
diastereoisomeric 3,4-dimethylhexa-1,5-dienes to the geometrical isomers of
octa-2,6-diene.

dimethylhexa-1,5-dienes, $CH_2=CH-CHMe-CHMe-CH=CH_2$ (figure
16.26). The *meso* compound gave specifically the *cis,trans*-octa-2,6-diene;
the racemate gave a mixture of the *cis,cis* and *trans,trans* isomers
[62T(18)67]. Reaction via the boat transition state would have given the
opposite results.

### 16.4.2  Evans' principle and photochemical reactions

The theory of conservation of orbital symmetry seems to account
very well for the usual contrast between thermal and photochemical
pericyclic processes. For example, the state diagram for the thermally
forbidden [2s + 2s] cyclodimerisation (figure 16.11) shows a direct correlation between the first excited states of the alkene pair and of the
cyclobutane, in apparent agreement with the observation that the reaction
is photochemically allowed.

In fact this is an oversimplification. Many photocatalysed processes
involve a direct conversion of excited reactants into the *ground state* of
the products. This is not possible unless the transition state for the
reaction has an electronic excited state that is only slightly more energetic
than its ground state; the small excess of energy can then be transferred
from electronic to vibrational and rotational excitation.

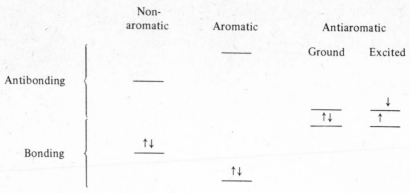

Figure 16.27. In antiaromatic systems the first excited state is only slightly more energetic than the ground state.

It is precisely this situation that occurs in an *antiaromatic* transition state. Compared with a non-aromatic system an aromatic one has bonding orbitals of lower energy and antibonding ones of higher energy. In an antiaromatic system the reverse is found, and the HOMO and LUMO lie close together (figure 16.27). Evans' principle can therefore be extended by adding: 'Photochemical pericyclic reactions take place preferentially via antiaromatic transition states'.

One should, however, retain a spirit of caution in applying any of these methods to photochemical processes. For one thing, it is even more difficult than with ground state reactions to prove that they are really concerted. And many photochemical reactions involve a triplet excited state (with the unpaired electrons having parallel spins) which can never be transformed directly into a normal electron-paired ground state.

## 16.5    Some other pericyclic reactions
### 16.5.1    Valence-bond isomers of benzene

How is it that the highly strained $(CH)_6$ compounds benzvalene (**16.25**) and prismane (**16.26**) are so reluctant to revert to their valence-bond isomer, benzene?

The benzvalene case can be better appreciated by considering first the simpler example of bicyclobutane. This readily undergoes thermal ring opening to butadiene:

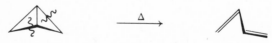

The process is an allowed $[2s + 2a]$ cycloelimination. (It is probably easier to visualise the reverse, which would involve the intramolecular cyclo-addition of the two double bonds as they approach each other ortho-gonally.) In benzvalene the extra two-carbon bridge precludes the antara-facial mode of opening and although it is some 250 kJ mol$^{-1}$ less stable than benzene it rearranges only slowly, with a half-life of ten days at room temperature (though the solid is liable to detonate) [67JA1031; 71JA3782].

The prismane to benzene isomerisation is not easy to visualise with only two-dimensional diagrams to help. In a schematic manner it can be shown as follows:

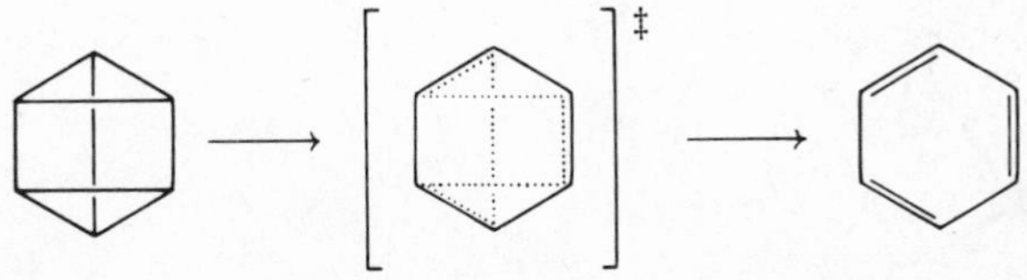

The transition state is comparable to that in the 'boat' form of the Cope rearrangement, involving two four-electron antiaromatic rings. Prismane, some 340 kJ mol$^{-1}$ less stable than benzene, is entirely stable at room temperature and even at 90 °C the half-life for its rearrangement to benzene is 11 hours [73JA2738]. This corresponds to a Gibbs energy of activation of about 120 kJ mol$^{-1}$, and emphasises the high energy associated with forbidden (or antiaromatic) transition states.

### 16.5.2   $[2 + 2]$ Cycloadditions of cumulenes

The simple $[2s + 2s]$ dimerisation of alkenes is thermally for-bidden. Nevertheless, a number of double-bonded compounds do readily undergo $[2 + 2]$ cycloadditions to generate four-membered rings. Some of these reactions are demonstrably two-stage unconcerted processes via biradicals or zwitterions (§16.1), but one important group does seem to be concerted, in apparent defiance of the normal rules. This is the addition of ketens (and some other hetero-allenes) to alkenes. Even with dienes the reaction gives a vinylcyclobutanone rather than a cyclohexanone:

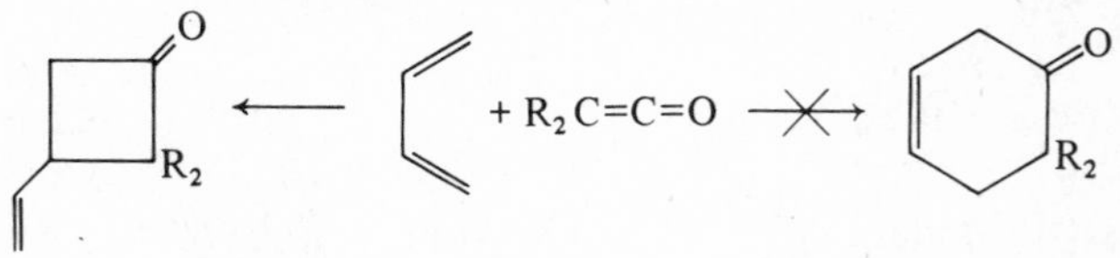

Woodward and Hoffmann rationalise this as a symmetry-allowed $[2s + 2a]$ cycloaddition controlled by and coalescent with a symmetry-allowed $[2s + 0s]$ cycloaddition. They suggest that a keten is best regarded as a zwitterion, possessing significant vinyl cation character:

$$R_2C{=}C{=}O \longleftrightarrow R_2C{=}\overset{+}{C}{-}\overset{-}{O}$$

Thus of the two orthogonal $p$ orbitals on the cumulative carbon atom, one (belonging to the C–C double bond) is occupied, and the other is largely vacant. This vacant orbital approaches the alkene double bond, initiating the $[2s + 0s]$ reaction (**16.27**). In so doing it aligns the carbon–carbon $\pi$ orbital of the keten in the correct orthogonal relationship for antarafacial reaction.

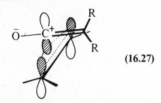

(16.27)

Dewar, on the other hand, explains the function of the carbonyl group in a different way. He supposes that the alkene approaches the carbon–carbon double bond of the keten diagonally, so that it attacks one $p$ orbital of the carbon–carbon $\pi$ system and one of the carbon–oxygen $\pi$ system (**16.28**). In the transition state the orthogonality of the keten $p$ orbitals is lost; the oxygen $p$ orbital is aligned between the two $p$ orbitals of the cumulative carbon atom, so coupling them together (**16.29**). This generates a continuous in-phase overlap over six orbitals, numbered in sequence in (**16.29**); all six electrons of the three original $\pi$ bonds are involved, and the system is Hückel aromatic.

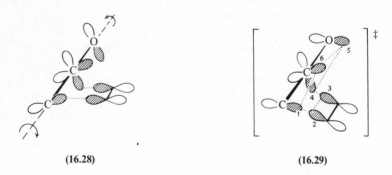

(16.28)          (16.29)

Allenes, too, undergo $[2 + 2]$ cycloadditions. Some of these are clearly biradical reactions [82JA3676]; others appear to follow a concerted mechanism [82JO2204] for which the aromatic transition state approach would seem to offer the simpler rationalisation.

### 16.5.3 Ene addition and retro-ene fragmentation

The ene reaction resembles both the Diels–Alder cycloaddition and the 1,5-sigmatropic shift of hydrogen. Addition to the enophile is suprafacial, and the transition state is a six-electron Hückel aromatic system:

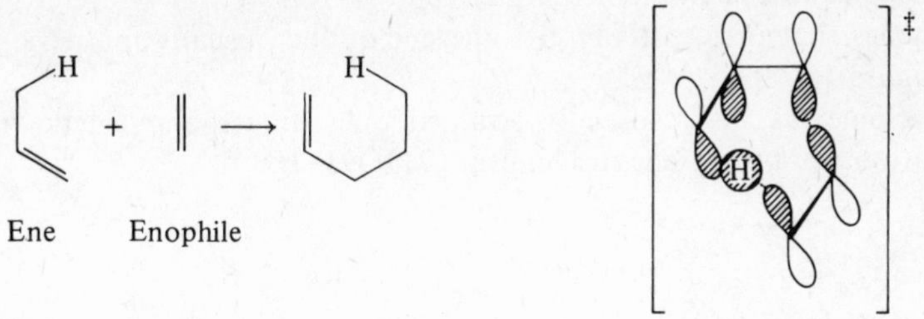

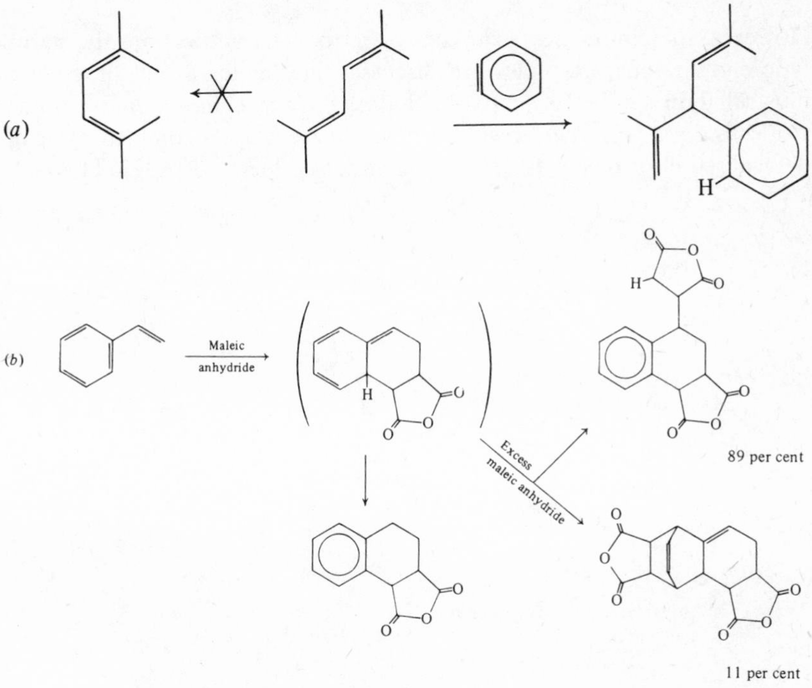

Ene    Enophile

(a)

(b)

Figure 16.28. The ene reaction of dienes: examples where ene addition is preferred to the Diels–Alder reaction.

(a) 2,5-Dimethylhexa-2,4-diene is sterically prevented from adopting the *s-cis* conformation necessary for Diels–Alder cycloaddition [60JO324].

(b) Styrene gives a 1:1 adduct with maleic anhydride; this has an isobenzene structure and can tautomerise to an aromatic tetralin. With excess of maleic anhydride 1:2 adducts are formed; the ene adduct is preferred to the Diels–Alder adduct in which aromaticity is irrevocably lost [55LA(595)1].

Reactivity in the ene reaction usually closely resembles that in the Diels–Alder; good enophiles are typically good dienophiles. A wide variety of alkenes can act as the ene component. Conjugated dienes can undergo either Diels–Alder or ene additions. In general the Diels–Alder reaction is considerably faster: the ene reaction with maleic anhydride may require 12 hours heating at 200 °C. However, there are examples of dienes that preferentially give the ene adduct, usually for steric reasons (figure 16.28).

Some alkenes, especially strained ones, undergo pyrolytic fragmentation via a retro-ene mechanism [71JO913]:

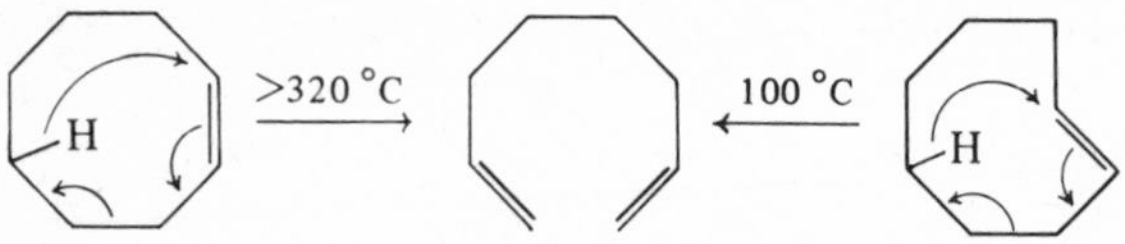

However, the commonest examples of retro-ene reactions are the pyrolytic eliminations; examples are discussed in chapter 8 and others are illustrated in figure 16.29. (Note that the Cope elimination of amine oxides is not a pericyclic reaction because there is not a complete cycle of delocalised electrons in the transition state [compare 78JA3927].)

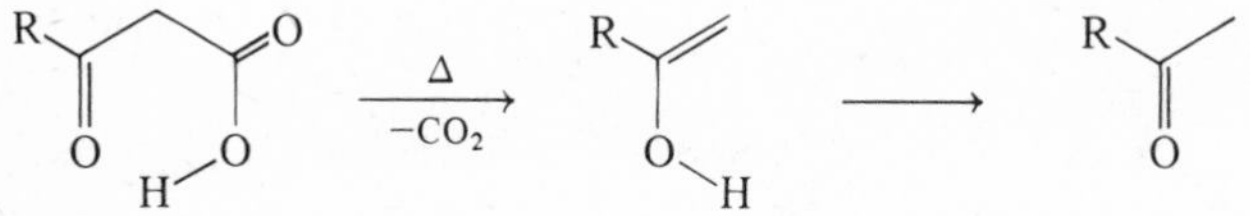

Decarboxylation of β-oxoacids

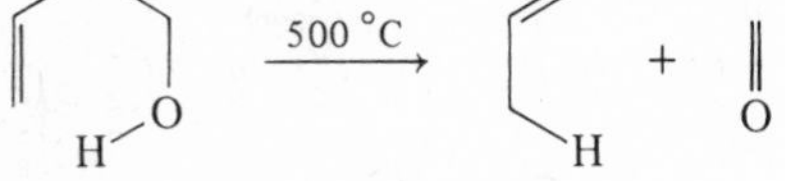

Pyrolysis of β-hydroxyalkenes

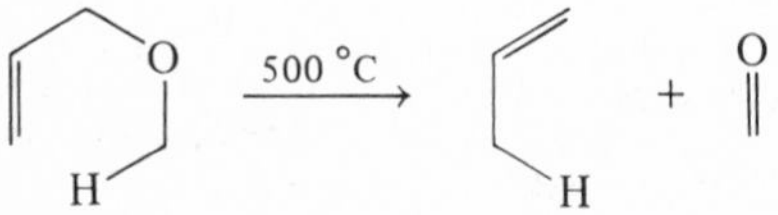

Pyrolysis of allyl ethers

Figure 16.29. Retro-ene fragmentations; compare also the pyrolytic eliminations discussed in chapter 8.

**16.6** **Further reading**

Concerted or non-concerted cycloadditions:

P. D. Bartlett, *Science*, 1968, **159**, 833.
R. Huisgen, *J. Org. Chem.*, 1976, **41**, 403; *Acc. Chem. Res.*, 1977, **10**, 117, 199; *Pure Appl. Chem.*, 1981, **53**, 171.
R. A. Firestone, *Tetrahedron*, 1977, **33**, 3009.
J. Sauer & R. Sustmann, *Angew. Chem. Internat. Ed.*, 1980, **19**, 779.

Conservation of orbital symmetry:

R. B. Woodward & R. Hoffmann, *Angew Chem. Internat. Ed.*, 1969, **8**, 781; also published as a monograph: *The Conservation of Orbital Symmetry*, Academic, New York, 1970.
T. L. Gilchrist & R. C. Storr, *Organic Reactions and Orbital Symmetry*, 2nd edn, Cambridge University Press, Cambridge, 1979.

Frontier molecular orbital approach:

K. N. Houk, *Acc. Chem. Res.*, 1975, **8**, 361.
K. Fukui, *Acc. Chem. Res.*, 1971, 4, 57; *Angew. Chem. Internat. Ed.*, 1982, **21**, 801.
I. Fleming, *Frontier Orbitals and Organic Chemical Reactions*, Wiley-Interscience, London, 1976.
R. Gleiter & L. A. Paquette, *Acc. Chem. Res.*, 1983, **16**, 328.

The aromatic transition state:

M. J. S. Dewar, *Angew. Chem. Internat. Ed.*, 1971, **10**, 761.
H. E. Zimmerman, *Acc. Chem. Res.*, 1971, **4**, 272.

General discussions and alternative theoretical approaches:

N. D. Epiotis, *Angew. Chem. Internat. Ed.*, 1974, **13**, 751.
W. C. Herndon, *Chem. Rev.*, 1972, **72**, 157.
S. I. Miller, *Adv. Phys. Org. Chem.*, 1968, **6**, 185.
K. N. Houk, *Survey Progr. Chem.*, 1973, **6**, 113.
A. P. Marchand & R. E. Lehr (eds), *Pericyclic Reactions*, 2 vols, Academic, New York, 1977.

Other topics:

W. C. Herndon, J. Feuer, W. B. Giles, D. Otteson & E. Silber (cycloadditions; PMO), and J. Michl (photochemical reactions), in *Chemical Reactivity and Reaction Paths*, ed. G. Klopman, Wiley-Interscience, New York, 1974.
E. N. Marvel, *Thermal Electrocyclic Reactions*, Academic, New York, 1980.
D. Ginsburg, *Tetrahedron*, 1983, **39**, 2095 (secondary orbital interactions).
C. W. Spangler, *Chem. Rev.*, 1976, **76**, 187 (sigmatropic rearrangements).
H. M. R. Hoffmann, *Angew. Chem. Internat. Ed.*, 1969, **8**, 556 (ene reaction).
G. Bianchi, C. De Micheli & R. Gandolfi, in *The Chemistry of Double-bonded Functional Groups*, ed. S. Patai, Wiley, London, 1977 (1,3,-dipolar cycloadditions).
E. Schaumann & R. Ketcham, *Angew. Chem. Internat. Ed.*, 1982, **21**, 225 ([2 + 2] cycloeliminations).

# Bibliography and references

Two books above all have played key roles in developing modern attitudes to physical and mechanistic organic chemistry; revised editions of both have appeared in recent years:

L. P. Hammett, *Physical Organic Chemistry*, 2nd edn, McGraw-Hill, New York, 1970.

C. K. Ingold, *Structure and Mechanism in Organic Chemistry*, 2nd edn, Bell, London, 1969.

Hammett, as his title implies, emphasises the physico-chemical aspects of the subject, Ingold the mechanistic.

Two other books, now sadly dated, are held in affectionate esteem by many thousands of organic chemists who first discovered the fascination of mechanistic studies within their pages. Despite their dissimilar titles they each encompass both sides of the subject, physical and mechanistic:

E. S. Gould, *Mechanism and Structure in Organic Chemistry*, Holt, Rinehart & Winston, New York, 1959.

J. Hine, *Physical Organic Chemistry*, 2nd edn, McGraw-Hill, New York, 1962.

Other relevant textbooks include:

K. B. Wiberg, *Physical Organic Chemistry*, Wiley, New York, 1964.

R. D. Gilliom, *Introduction to Physical Organic Chemistry*, Addison-Wesley, New York, 1970.

(Predominantly physico-chemical, with a relatively high mathematical content.)

R. A. Jackson, *Mechanism: an Introduction to the Study of Organic Reactions*, Clarendon, Oxford, 1972.

P. Sykes, *The Search for Organic Reaction Pathways*, Longman, London, 1972.

(More descriptive, less mathematical than Wiberg or Gilliom, but still with the emphasis on techniques whereby mechanisms are established and with less to say about their application to particular reaction types.)

R. W. Alder, R. Baker & J. M. Brown, *Mechanism in Organic Chemistry*, Wiley-Interscience, London, 1971.

T. M. Lowry & K. S. Richardson, *Mechanism and Theory in Organic Chemistry*, 2nd edn, Harper & Row, New York, 1981.

(Physical and mechanistic, but with the emphasis on the latter.)

J. March, *Advanced Organic Chemistry*, 2nd edn, McGraw-Hill, New York, 1977.

(An encyclopaedic work, combining descriptive and mechanistic organic chemistry, and providing a remarkably comprehensive collection of references.)

The series *Advances in Physical Organic Chemistry* (Academic, London), and *Progress in Physical Organic Chemistry* (Wiley-Interscience, New York) were both established in 1963 and have appeared at the rate of about one volume a year since then. They contain reviews on all aspects of physical organic chemistry, but the emphasis is more on the physico-chemical side. *Organic Reaction Mechanisms* (Interscience, London) has provided an annual survey of developments in mechanistic studies since 1965. Similar but much briefer summaries can be found in *Annual Reports on the Progress of Chemistry, Section B* (The Royal Society of Chemistry, London), which cover all aspects of organic chemistry.

## References

The references in this book are based on a system devised by Professor A. R. Katritzky, and I am grateful to him for permission to use it. Each reference is designated by a code: two digits (four for nineteenth-century references) denote the year; a letter code, largely self-explanatory, indicates the journal according to the table below; the volume number is given in parentheses, but only if more than one volume was published in the same year; the final digits are the page number. Thus [37JA96] refers to *J. Amer. Chem. Soc.*, 1937, **59**, 96. Italic numbers in parentheses after each reference give the page number on which the reference appears.

### *Journal Codes:*

| | |
|---|---|
| Acc | *Accounts of Chemical Research* |
| AG | *Angewandte Chemie (International Edition)* |
| AJ | *Australian Journal of Chemistry* |
| APO | *Advances in Physical Organic Chemistry* |
| AS | *Acta Chemica Scandanavica* |
| BB | *Bulletin des Sociétés Chimiques Belges* |
| BF | *Bulletin de la Société Chimique de France* |
| BJ | *Bulletin of the Chemical Society of Japan* |
| CB | *Chemische Berichte* |
| CC | *Chemical Communications* |
| CEd | *Journal of Chemical Education* |
| CI | *Chemistry and Industry (London)* |
| CJ | *Canadian Journal of Chemistry* |
| CR | *Chemical Reviews* |
| H | *Helvetica Chimica Acta* |
| J | *Journal of the Chemical Society* |
| J(B) | *Journal of the Chemical Society (Section B)* |
| J(P2) | *Journal of the Chemical Society (Perkin Transactions 2)* |
| JA | *Journal of the American Chemical Society* |
| JCP | *Journal of Chemical Physics* |
| JO | *Journal of Organic Chemistry* |
| JPC | *Journal of Physical Chemistry* |
| LA | *Liebigs Annalen der Chemie* |
| Med | *Journal of Medicinal Chemistry* |
| MI | *Miscellaneous (mainly references to books)* |
| NA | *Nature* |
| P | *Proceedings of the Chemical Society* |
| PAC | *Pure and Applied Chemistry* |
| PPO | *Progress in Physical Organic Chemistry* |
| PRS | *Proceedings of the Royal Society* |
| QR | *Quarterly Reviews of the Chemical Society* |
| R | *Recueil des Travaux Chimiques des Pays-Bas* |
| RCP | *Record of Chemical Progress* |
| SA | *Spectrochimica Acta* |
| SC | *Science* |
| SPC | *Survey of Progress in Chemistry* |
| T | *Tetrahedron* |
| TF | *Transactions of the Faraday Society* |
| TL | *Tetrahedron Letters* |
| ZP | *Zeitschrift für Physikalische Chemie* |

# Complete References

1870LA(153)256    V. Markownikov, *Liebigs Ann. Chem.*, 1870, **153**, 256 (*204*)

1889ZP(4)226    S. Arrhenius, *Zeit. Phys. Chem.*, 1889, **4**, 226 (see *J. Chem. Soc.*, 1889, **56**, 1003 for English abstract) (*24*)

1892LA(267)300    E. Schmidt, *Liebigs Ann. Chem.*, 1892, **267**, 300 (see *J. Chem. Soc.*, 1892, **62**, 808 for English abstract) (*204*)

00JA757    M. Gomberg, *J. Amer. Chem. Soc.*, 1900, **22**, 757 (*336*)

02LA(323)205    J. Meisenheimer, *Liebigs Ann. Chem.*, 1902, **323**, 205 (see *J. Chem. Soc.*, 1902, **82**, 795 for English abstract) (*288*)

03J(83)995    A. Lapworth, *J. Chem. Soc.*, 1903, **83**, 995 (*245*)

04J(85)1206    A. Lapworth, *J. Chem. Soc.*, 1904, **85**, 1206 (*245*)

12J(101)654    P. F. Frankland, *J. Chem. Soc.*, 1912, **101**, 654 (*197*)

23CI43    T. M. Lowry, *Chem. Ind. (London)*, 1923, 43 (*69*)

23J(123)44    H. Phillips, *J. Chem. Soc.*, 1923, **123**, 44 (*142*)

23J(123)1260    K. Ibbotson & J. Kenner, *J. Chem. Soc.*, 1923, **123**, 1260 (*294*)

23J(123)3006    R. G. W. Norrish, *J. Chem. Soc.*, 1923, **123**, 3006 (*196*)

23JA1014    T. D. Stewart & K. R. Edlund, *J. Amer. Chem. Soc.*, 1923, **45**, 1014 (*196*)

23R718    J. N. Brønsted, *Rec. Trav. Chim.*, 1923, **42**, 718 (*69*)

24ZP(108)185    J. N. Brønsted & K. J. Pederson, *Zeit. Phys. Chem.*, 1924, **108**, 185 (*77*)

25JA2340    A. W. Francis, *J. Amer. Chem. Soc.*, 1925, **47**, 2340 (*196*)

26LA(450)1    K. Kindler, *Liebigs Ann. Chem.*, 1926, **450**, 1 (*39*)

27J986    K. J. P. Orton & A. E. Bradfield, *J. Chem. Soc.*, 1927, 986 (*308*)

27J2761    F. G. Soper & D. R. Pryde, *J. Chem. Soc.*, 1927, 2761 (*310*)

27JPC1192    F. G. Soper, *J. Phys. Chem.*, 1927, **31**, 1192 (*310*)

28J998    K. J. P. Orton, F. G. Soper & G. Williams, *J. Chem. Soc.*, 1928, 998 (*309*)

28LA(461)132    P. Pfeiffer & R. Wizinger, *Liebigs Ann. Chem.*, 1928, **461**, 132 (*215*)

30J1032    C. K. Ingold, *J. Chem. Soc.*, 1930, 1032 (*58*)

32JA2721    L. P. Hammett & A. J. Deyrup, *J. Amer. Chem. Soc.*, 1932, **54**, 2721 (*83*)

33J984    C. K. Ingold & H. V. Kidd, *J. Chem. Soc.*, 1933, 984 (*311*)

33JA2468    M. S. Kharasch & F. R. Mayo, *J. Amer. Chem. Soc.*, 1933, **55**, 2468 (*206*)

33JA2531    M. S. Kharasch, M. C. McNab & F. R. Mayo, *J. Amer. Chem. Soc.*, 1933, **55**, 2531 (*204*)

34TF508    M. Polanyi & A. L. Szabo, *Trans. Faraday Soc.*, 1934, **30**, 508 (*267*)

35CR(17)125    L. P. Hammett, *Chem. Rev.*, 1935, **17**, 125 (*39*)

35J1525    E. D. Hughes, F. Juliusberger, S. Masterman, B. Topley & J. Weiss, *J. Chem. Soc.*, 1935, 1525 (*142*)

36JO170    F. W. Bergstrom, R. E. Wright, C. Chandler & W. A. Gilkey, *J. Org. Chem.*, 1936, **1**, 170 (*300*)

37J1252    W. A. Cowdrey, E. D. Hughes, C. K. Ingold, S. Masterman & A. D. Scott, *J. Chem. Soc.*, 1937, 1252 (*166*)

37JA96    L. P. Hammett, *J. Amer. Chem. Soc.*, 1937, **59**, 96 (*39, 45*)

37JA107    C. D. Hurd & I. Schmerling, *J. Amer. Chem. Soc.*, 1937, **59**, 107 (*315*)

37JA947    I. Roberts & G. E. Kimball, *J. Amer. Chem. Soc.*, 1937, **59**, 947 (*197*)

38JA1708    H. P. Treffers & L. P. Hammett, *J. Amer. Chem. Soc.*, 1938, **60**, 1708 (*272*)

38JA2687    A. R. Olsen & R. J. Miller, *J. Amer. Chem. Soc.*, 1938, **60**, 2687 (*271*)

38JPC513    J. G. Kirkwood & F. H. Westheimer, *J. Phys. Chem.*, 1938, **6**, 513 (*50*)

39J1188    T. P. Nevell, E. de Salas & C. L. Wilson, *J. Chem. Soc.*, 1939, 1188 (*173, 329*)

39JA1324    H. I. Bernstein & F. C. Whitmore, *J. Amer. Chem. Soc.*, 1939, **61**, 1324 (*321*)

39JO428    M. S. Kharasch, S. C. Kleiger & F. R. Mayo, *J. Org. Chem.*, 1939, **4**, 428 (*204*)

39TF824    M. G. Evans, *Trans. Faraday Soc.*, 1939, **35**, 824 (*375*)

40J960    L. C. Bateman, E. D. Hughes & C. K. Ingold, *J. Chem. Soc.*, 1940, 960 (*143*)

40J979    L. C. Bateman, M. G. Church, E. D. Hughes, C. K. Ingold & N. A. Taher, *J. Chem. Soc.*, 1940, 979 (*145*)

40J1011, 1017    L. C. Bateman, E. D. Hughes & C. K. Ingold, *J. Chem. Soc.*, 1940, 1011, 1017 (*143*)

40J1295    E. Coombs & D. P. Evans, *J. Chem. Soc.*, 1940, 1295 (*261*)

40JA1400    M. L. Crossley, R. H. Kienle & C. H. Benbrook, *J. Amer. Chem. Soc.*, 1940, **62**, 1400 (*297*)

41JA2459    A. R. Olsen & J. L. Hyde, *J. Amer. Chem. Soc.*, 1941, **63**, 2459 (*271*)

42J266    W. A. Waters, *J. Chem. Soc.*, 1942, 266 (*304*)

42JA900    G. W. Wheland, *J. Amer. Chem. Soc.*, 1942, **64**, 900 (*215*)

44JCP(12)469    G. B. Kistiakowsky & E. R. Van Artsdalen, *J. Chem. Phys.*, 1944, **12**, 469 (*339*)

45J129    I. D. Morton & P. W. Robertson, *J. Chem. Soc.*, 1945, 129 (*208*)

45JA704    M. S. Newman, H. G. Kuivila & A. B. Garrett, *J. Amer. Chem. Soc.*, 1945, **67**, 704 (*272*)

48J1514    P. Oxley & W. F. Short, *J. Chem. Soc.*, 1948, 1514 (*323*)

48J2055, 2058    M. L. Dhar, E. D. Hughes, C. K. Ingold (& S. Masterman), *J. Chem. Soc.*, 1948, 2055, 2058 (*190*)

48JA821    S. Winstein, E. Grunwald & L. L. Ingraham, *J. Amer. Chem. Soc.*, 1948, **70**, 821 (*169*)

48JA838    S. Winstein & R. Adams, *J. Amer. Chem. Soc.*, 1948, **70**, 838 (*171*)

48JA846    E. Grunwald & S. Winstein, *J. Amer. Chem. Soc.*, 1948, **70**, 846 (*104*)

49NA(163)599    L. Melander, *Nature*, 1949, **163**, 599 (*215*)

49PRS(A197)141    R. P. Bell & W. C. E. Higginson, *Proc. Royal Soc.*, 1949, **A197**, 141 (*78*)

50AS39    L. Smith & S. Skyle, *Acta Chem. Scand.*, 1950, **4**, 39 (*205*)

50J2400    A series of papers by E. D. Hughes, C. K. Ingold, R. J. Gillespie and others, *J. Chem. Soc.*, 1950, 2400–2684 (*213*)

50J2628    C. A. Bunton, E. D. Hughes, C. K. Ingold, D. I. H. Jacobs, G. J. Monkoff & R. I. Reed, *J. Chem. Soc.*, 1950, 2628 (*226*)

| | |
|---|---|
| 50JA1223 | H. C. Brown & R. S. Fletcher, *J. Amer. Chem. Soc.*, 1950, **72**, 1223 (*189*) |
| 50JCP(18)265 | H. C. Longuet-Higgins, *J. Chem. Phys.*, 1950, **18**, 265 (*124*) |
| 50NA(166)680 | C. A. Bunton, E. D. Hughes, C. K. Ingold & D. F. Meigh, *Nature*, 1950, **166**, 680 (*270*) |
| 51J3301 | N. B. Chapman & R. E. Parker, *J. Chem. Soc.*, 1951, 3301 (*292*) |
| 51JA1626 | M. L. Bender, *J. Amer. Chem. Soc.*, 1951, **73**, 1626 (*267*) |
| 51JA1958 | W. G. Young, S. Winstein & H. L. Goering, *J. Amer. Chem. Soc.*, 1951, **73**, 1958 (*167*) |
| 51JA2181 | J. D. Roberts, R. A. Clement & J. J. Drysdale, *J. Amer. Chem. Soc.*, 1951, **73**, 2181 (*65*) |
| 51JA4304 | E. R. Alexander & R. W. Kluiber, *J. Amer. Chem. Soc.*, 1951, **73**, 4304 (*315*) |
| 51JA5458 | E. Grunwald, *J. Amer. Chem. Soc.*, 1951, **73**, 5458 (*169*) |
| 52BB427 | P. J. C. Fierens & P. Verschelden, *Bull. Soc. Chim. Belges*, 1952, **61**, 427 (*160*) |
| 52J2488 | O. T. Benfey, E. D. Hughes & C. K. Ingold, *J. Chem. Soc.*, 1952, 2488 (*144*) |
| 52J4917 | C. A. Bunton & E. A. Halevi, *J. Chem. Soc.*, 1952, 4917 (*214*) |
| 52JA1147, 1154 | S. Winstein & D. Trifan, *J. Amer. Chem. Soc.*, 1952, **74**, 1147, 1154 (*328*) |
| 52JA1894 | H. C. Brown & M. Borkowski, *J. Amer. Chem. Soc.*, 1952, **74**, 1894 (*160*) |
| 52JA2193 | S. J. Cristol & N. L. Hause, *J. Amer. Chem. Soc.*, 1952, **74**, 2193 (*183*) |
| 52JA2282 | D. H. Smith, J. R. Schwartz & G. W. Wheland, *J. Amer. Chem. Soc.*, 1952, **74**, 2282 (*311*) |
| 52JA2729 | R. W. Taft, *J. Amer. Chem. Soc.*, 1952, **74**, 2729 (*58*) |
| 52JA3120 | R. W. Taft, *J. Amer. Chem. Soc.*, 1952, **74**, 3120 (*58*) |
| 52JA3341 | M. J. S. Dewar, *J. Amer. Chem. Soc.*, 1952, **74**, 3341 (and the following five papers) (*124*) |
| 52JA3353 | M. J. S. Dewar, *J. Amer. Chem. Soc.*, 1952, **74**, 3353 (*346*) |
| 52JA5866 | J. P. Ryan & P. R. O'Connor, *J. Amer. Chem. Soc.*, 1952, **74**, 5866 (*315*) |
| 52QR34 | E. D. Hughes & C. K. Ingold, *Quart. Rev.*, 1952, **6**, 34 (*309*) |
| 53CR(53)191 | H. H. Jaffé, *Chem. Rev.*, 1953, **53**, 191 (*55, 65*) |
| 53J3832 | E. D. Hughes, C. K. Ingold & R. Pasternak, *J. Chem. Soc.*, 1953, 3832 (*183*) |
| 53J3839 | E. D. Hughes, C. K. Ingold & J. B. Rose, *J. Chem. Soc.*, 1953, 3839 (*182, 185*) |
| 53JA136 | C. G. Swain, C. B. Scott & K. H. Lohmann, *J. Amer. Chem. Soc.*, 1953, **75**, 136 (*145*) |
| 53JA141 | C. G. Swain & C. B. Scott, *J. Amer. Chem. Soc.*, 1953, **75**, 141 (*163*) |
| 53JA2167 | J. D. Roberts & W. T. Moreland, *J. Amer. Chem. Soc.*, 1953, **75**, 2167 (*52*) |
| 53JA4231 | R. W. Taft, *J. Amer. Chem. Soc.*, 1953, **75**, 4231 (*58*) |
| 53JA6285 | H. C. Brown & M. Grayson, *J. Amer. Chem. Soc.*, 1953, **75**, 6285 (*234*) |
| 53SC(117)340 | J. E. Leffler, *Science*, 1953, **117**, 340 (*62*) |
| 54BJ423 | K. Fukui, T. Yonezawa & C. Nagata, *Bull. Chem. Soc. Japan*, 1954, **27**, 423 (*242*) |
| 54J2109 | N. B. Chapman, R. E. Parker & P. W. Soames, *J. Chem. Soc.*, 1954, 2109 (*292*) |
| 54J2918 | E. Gelles, E. D. Hughes & C. K. Ingold, *J. Chem. Soc.*, 1954, 2918 (*162*) |

| | |
|---|---|
| 54J2930 | P. B. D. de la Mare, E. D. Hughes, C. K. Ingold & Y. Pocker, *J. Chem. Soc.*, 1954, 2930 (*162*) |
| 54JA835 | A. T. Whatley & R. N. Pease, *J. Amer. Chem. Soc.*, 1954, **76**, 835 (*208*) |
| 54JA4385 | W. L. Petty & P. L. Nichols, *J. Amer. Chem. Soc.*, 1954, **76**, 4385 (*163*) |
| 54RCP111 | D. Y. Curtin, *Rec. Chem. Progr.*, 1954, **15**, 111 (*21*) |
| 55BB709 | J. Cortier, P. J. C. Fierens, M. Gilon & A. Halleux, *Bull. Soc. Chim. Belges*, 1955, **64**, 709 (*292*) |
| 55H1597, 1617, 1623 | H. Zollinger, *Helv. Chim. Acta*, 1955, **38**, 1597, 1617, 1623 (*237*) |
| 55J2926 | B. A. Bolto, J. Miller & V. A. Williams, *J. Chem. Soc.*, 1955, 2926 (*292*) |
| 55JA334 | G. S. Hammond, *J. Amer. Chem. Soc.*, 1955, **77**, 334 (*62*) |
| 55JA355 | D. Y. Curtin & M. C. Crew, *J. Amer. Chem. Soc.*, 1955, **77**, 355 (*321*) |
| 55JA1094 | R. F. Brown, N. M. v. Gulick & G. H. Schmid, *J. Amer. Chem. Soc.*, 1955, **77**, 1094 (*324*) |
| 55JA1397 | D. S. Noyce & W. A. Pryor, *J. Amer. Chem. Soc.*, 1955, **77**, 1397 (*262*) |
| 55JA3044 | N. C. Deno, J. J. Jaruzelski & A. Schriesheim, *J. Amer. Chem. Soc.*, 1955, **77**, 3044 (*85, 87*) |
| 55JA4540 | G. E. Hall, R. Piccolini & J. D. Roberts, *J. Amer. Chem. Soc.*, 1955, **77**, 4540 (*300*) |
| 55JA4638 | P. S. Skell & R. C. Woodworth, *J. Amer. Chem. Soc.*, 1955, **77**, 4638 (*207*) |
| 55JA5562 | S. Winstein & N. J. Holness, *J. Amer. Chem. Soc.*, 1955, **77**, 5562 (*21*) |
| 55JO726 | R. L. Burke & R. M. Herbst, *J. Org. Chem.*, 1955, **20**, 726 (*323*) |
| 55LA(595)1 | K. Alder & R. Schmitz-Josen, *Liebigs Ann. Chem.*, 1955, **595**, 1 (*383*) |
| 56J1572 | J. A. Leisten, *J. Chem. Soc.*, 1956, 1572 (*273*) |
| 56J3581 | M. J. S. Dewar, T. Mole & E. W. T. Warford, *J. Chem. Soc.*, 1956, 3581 (*241*) |
| 56J4633 | V. Gold, *J. Chem. Soc.*, 1956, 4633 (*158*) |
| 56JA87 | F. G. Bordwell & P. J. Boutan, *J. Amer. Chem. Soc.*, 1956, **78**, 87 (*65*) |
| 56JA592 | S. Winstein & M. Shatavsky, *J. Amer. Chem. Soc.*, 1956, **78**, 592 (*172*) |
| 56JA601 | J. D. Roberts, D. A. Semenow, H. E. Simmons & L. A. Carlsmith, *J. Amer. Chem. Soc.*, 1956, **78**, 601 (*299*) |
| 56JA611 | J. D. Roberts, C. W. Vaughan, L. A. Carlsmith & D. A. Semenow, *J. Amer. Chem. Soc.*, 1956, **78**, 611 (*299, 301*) |
| 56JA790 | D. J. Cram, F. D. Greene & C. H. Depuy, *J. Amer. Chem. Soc.*, 1956, **78**, 790 (*182*) |
| 56JA2190 | H. C. Brown, I. Moritani & M. Nakagawa, *J. Amer. Chem. Soc.*, 1956, **78**, 2190 (*186*) |
| 56JA2193 | H. C. Brown, I. Moritani & Y. Okamoto, *J. Amer. Chem. Soc.*, 1956, **78**, 2193 (*188*) |
| 56JA2199 | H. C. Brown & O. H. Wheeler, *J. Amer. Chem. Soc.*, 1956, **78**, 2199 (*187*) |
| 56JA2290 | H. Conroy & R. A. Firestone, *J. Amer. Chem. Soc.*, 1956, **78**, 2290 (*316*) |
| 56JA4347 | E. M. Kosower & S. Winstein, *J. Amer. Chem. Soc.*, 1956, **78**, 4347 (*171*) |
| 56JA4935 | A. Streitwieser, *J. Amer. Chem. Soc.*, 1956, **78**, 4935 (*169*) |
| 56JA6008 | J. G. Pritchard & F. A. Long, *J. Amer. Chem. Soc.*, 1956, **78**, 6008 (*110*) |

56JA6249     C. R. Smoot & H. C. Brown, *J. Amer. Chem. Soc.*, 1956, **78**, 6249 (*235*)

56JA6255     H. C. Brown & C. R. Smoot, *J. Amer. Chem. Soc.*, 1956, **78**, 6255 (*238*)

56JA6265     J. F. Bunnett & T. K. Brotherton, *J. Amer. Chem. Soc.*, 1956, **78**, 6265 (*301*)

56MI     R. W. Taft, in *Steric Effects in Organic Chemistry*, ed. M. S. Newman, Wiley, New York, 1956 (*58*)

57JA717     F. G. Bordwell & P. J. Boutan, *J. Amer. Chem. Soc.*, 1957, **79**, 717 (*65*)

57JA1652     M. L. Bender & B. W. Turnquest, *J. Amer. Chem. Soc.*, 1957, **79**, 1652 (*283*)

57JA2653     H. L. Goering & D. W. Larsen, *J. Amer. Chem. Soc.*, 1957, **79**, 2653 (*207*)

57JA2942     J. K. Kochi, *J. Amer. Chem. Soc.*, 1957, **79**, 2942 (*304*)

57JA3156     D. Y. Curtin & R. J. Crawford, *J. Amer. Chem. Soc.*, 1957, **79**, 3156 (*316*)

57JA3438     S. J. Cristol & E. F. Hoegger, *J. Amer. Chem. Soc.*, 1957, **79**, 3438 (*183*)

57JA3712     W. H. Saunders & R. A. Williams, *J. Amer. Chem. Soc.*, 1957, **79**, 3712 (*48*)

57JA4720     A. C. Cope, N. A. LeBel, H. H. Lee & W. R. Moore, *J. Amer. Chem. Soc.*, 1957, **79**, 4720 (*194*)

57JA6160     B. M. Benjamin, H. J. Schaeffer & C. J. Collins, *J. Amer. Chem. Soc.*, 1957, **79**, 6160 (*321*)

57JCP(27)1247     K. Fukui, T. Yonezawa & C. Nagata, *J. Chem. Phys.*, 1957, **27**, 1247 (*240, 242*)

57JO719     H. C. Brown & H. L. Young, *J. Org. Chem.*, 1957, **22**, 719 (*238*)

58H2274     R. Ernst, O. A. Stamm & H. Zollinger, *Helv. Chim. Acta*, 1958, **41**, 2274 (*237*)

58J36     P. B. D. de la Mare & S. Galandauer, *J. Chem. Soc.*, 1958, 36 (*205*)

58J403     C. A. Bunton, T. Hadwick, D. R. Llewellyn & Y. Pocker, *J. Chem. Soc.*, 1958, 403 (*318*)

58J1691     R. P. Bell & M. J. Smith, *J. Chem. Soc.*, 1958, 1691 (*261*)

58J2420     C. A. Bunton & G. Stedman, *J. Chem. Soc.*, 1958, 2420 (*214*)

58J2982     C. Beard & W. J. Hickinbottom, *J. Chem. Soc.*, 1958, 2982 (*308*)

58J3398     A. Maccoll, *J. Chem. Soc.*, 1958, 3398 (*194*)

58JA169     S. Winstein & G. C. Robinson, *J. Amer. Chem. Soc.*, 1958, **80**, 169 (*147*)

58JA4979     H. C. Brown & Y. Okamoto, *J. Amer. Chem. Soc.*, 1958, **80**, 4979 (*47, 65*)

58JA5388     M. L. Bender & M. C. Neveu, *J. Amer. Chem. Soc.*, 1958, **80**, 5388 (*283*)

58JA5539     D. S. Noyce & W. L. Reed, *J. Amer. Chem. Soc.*, 1958, **80**, 5539 (*262*)

58JA6020     J. F. Bunnett & J. J. Randall, *J. Amer. Chem. Soc.*, 1958, **80**, 6020 (*292*)

59BJ971     Y. Yukawa & Y. Tsuno, *Bull. Chem. Soc. Japan*, 1959, **32**, 971 (*48*)

59CB192     R. Huisgen & J. Sauer, *Chem. Ber.*, 1959, **92**, 192 (*300*)

59CI332     Y. Pocker, *Chem. Ind. (London)*, 1959, 332 (*318*)

59JA475     W. P. Jencks, *J. Amer. Chem. Soc.*, 1959, **81**, 475 (*255*)

59JA628     M. Stiles, D. Wolf & G. V. Hudson, *J. Amer. Chem. Soc.*, 1959, **81**, 628 (*262–3*)

59JA3308     H. C. Brown & G. Marino, *J. Amer. Chem. Soc.*, 1959, **81**, 3308 (*238*)

| | |
|---|---|
| 59JA3315 | S. U. Choi & H. C. Brown, *J. Amer. Chem. Soc.*, 1959, **81**, 3315 (*235*) |
| 59JA5352 | R. W. Taft, S. Ehrenson, I. C. Lewis & R. E. Glick, *J. Amer. Chem. Soc.*, 1959, **81**, 5352 (*65*) |
| 59R815 | H. van Bekkum, P. E. Verkade & B. M. Wepster, *Rec. Trav. Chim.*, 1959, **78**, 815 (*65*) |
| 59T(5)127 | R. B. Turner, *Tetrahedron*, 1959, **5**, 127 (*186*) |
| 59T(5)194 | W. M. Schubert, J. M. Craven, R. G. Minton & R. B. Murphy, *Tetrahedron*, 1959, **5**, 194 (*56*) |
| 60AS577 | P. Salomaa, *Acta Chem. Scand.*, 1960, **14**, 577 (*271*) |
| 60J2983 | R. P. Bell & P. T. McTigue, *J. Chem. Soc.*, 1960, 2983 (*261*) |
| 60J4054 | D. V. Banthorpe, E. D. Hughes & C. K. Ingold, *J. Chem. Soc.*, 1960, 4054 (*188*) |
| 60J4885 | J. R. Knowles, R. O. C. Norman & G. K. Radda, *J. Chem. Soc.*, 1960, 4885 (*238*) |
| 60JA1773 | B. M. Anderson & W. P. Jencks, *J. Amer. Chem. Soc.*, 1960, **82**, 1773 (*44*) |
| 60JA1778 | W. P. Jencks & J. Carriuolo, *J. Amer. Chem. Soc.*, 1960, **82**, 1778 (*282*) |
| 60JA2357 | D. Y. Curtin & R. J. Harder, *J. Amer. Chem. Soc.*, 1960, **82**, 2357 (*11*) |
| 60JA2515 | H. L. Goering & M. M. Pombo, *J. Amer. Chem. Soc.*, 1960, **82**, 2515 (*167*) |
| 60JA4708 | H. C. Brown & G. Zweifel, *J. Amer. Chem. Soc.*, 1960, **82**, 4708 (*209*) |
| 60JA5858 | T. C. Bruice & U. K. Pandit, *J. Amer. Chem. Soc.*, 1960, **82**, 5858 (*285*) |
| 60JO324 | E. M. Arnett, *J. Org. Chem.*, 1960, **25**, 324 (*383*) |
| 60JPC1805 | R. W. Taft, *J. Phys. Chem.*, 1960, **64**, 1805 (*49, 65*) |
| 60P84 | D. N. Kershaw & J. A. Leisten, *Proc. Chem. Soc.*, 1960, 84 (*273*) |
| 61JA1743 | W. P. Jencks & J. Carriuolo, *J. Amer. Chem. Soc.*, 1961, **83**, 1743 (*281*) |
| 61JA3647 | G. G. Smith, F. D. Bagley & R. Taylor, *J. Amer. Chem. Soc.*, 1961, **83**, 3647 (*194*) |
| 61JA3846 | W. N. White & W. K. Fife, *J. Amer. Chem. Soc.*, 1961, **83**, 3846 (*316*) |
| 61JA4571 | G. A. Olah, S. J. Kuhn & S. H. Flood, *J. Amer. Chem. Soc.*, 1961, **83**, 4571 (*220*) |
| 61MI | A. Streitwieser, *Molecular Orbital Theory for Organic Chemists*, Wiley, New York, 1961 (*240*) |
| 61T(16)146 | E. Havinga & J. L. M. A. Schaltmann, *Tetrahedron*, 1961, **16**, 146 (*358*) |
| 62AG225 | J. F. Bunnett, *Angew. Chem. Internat. Ed.*, 1962, **1**, 225 (*178, 180, 187*) |
| 62AJ467 | K. R. Adam, I. Lauder & V. R. Stimson, *Austral. J. Chem.*, 1962, **15**, 467 (*270*) |
| 62CI1287 | G. Kohnstam, A. Queen & T. Ribar, *Chem. Ind. (London)*, 1962, 1287 (*157*) |
| 62CJ1981 | L. R. C. Barclay, N. D. Hall & G. A. Cooke, *Canad. J. Chem.*, 1962, **40**, 1981 (*271*) |
| 62JA826 | E. H. Cordes & W. P. Jencks, *J. Amer. Chem. Soc.*, 1962, **84**, 826 (*260*) |
| 62JA1026 | W. Griezerstein, R. A. Bonelli & J. A. Brieux, *J. Amer. Chem. Soc.*, 1962, **84**, 1026 (*289*) |
| 62JA3539 | M. J. S. Dewar & P. J. Grisdale, *J. Amer. Chem. Soc.*, 1962, **84**, 3539 (*57*) |

62JA3548    M. J. S. Dewar & P. J. Grisdale, *J. Amer. Chem. Soc.*, 1962, **84**, 3548 *(57)*

62JA4817    R. Taylor, G. G. Smith & W. H. Wetzel, *J. Amer. Chem. Soc.*, 1962, **84**, 4817 *(194)*

62P265    C. K. Ingold, *Proc. Chem. Soc.*, 1962, 265 *(175)*

62T(18)67    W. v. E. Doering & W. R. Roth, *Tetrahedron*, 1962, **18**, 67 *(379)*

63APO35    L. M. Stock & H. C. Brown, *Adv. Phys. Org. Chem.*, 1963, **1**, 35 *(49, 220)*

63J5854    C. A. Bunton & M. D. Carr, *J. Chem. Soc.*, 1963, 5854 *(319)*

63JA37    M. L. Bender & M. C. Chen, *J. Amer. Chem. Soc.*, 1963, **85**, 37 *(272)*

63JA1148    N. Kornblum, R. Seltzer & P. Haberfield, *J. Amer. Chem. Soc.*, 1963, **85**, 1148 *(165)*

63JA1702    M. Cocivera & S. Winstein, *J. Amer. Chem. Soc.*, 1963, **85**, 1702 *(151)*

63JA1949    A. C. Cope & A. S. Mehta, *J. Amer. Chem. Soc.*, 1963, **85**, 1949 *(184)*

63JA2245    M. J. S. Dewar & R. C. Fahey, *J. Amer. Chem. Soc.*, 1963, **85**, 2245 *(203)*

63JA2843    E. H. Cordes & W. P. Jencks, *J. Amer. Chem. Soc.*, 1963, **85**, 2843 *(259)*

63JA2976    G. A. Russell, A. Ito & D. G. Hendry, *J. Amer. Chem. Soc.*, 1963, **85**, 2976 *(346)*

63JA3142    R. E. Pearson & J. C. Martin, *J. Amer. Chem. Soc.*, 1963, **85**, 3142 *(344)*

63JA3754    R. F. Bridges & G. A. Russell, *J. Amer. Chem. Soc.*, 1963, **85**, 3754 *(344)*

63JO278    R. T. Conley, *J. Org. Chem.*, 1963, **28**, 278 *(323)*

63JPC737    R. H. Boyd, *J. Phys. Chem.*, 1963, **67**, 737 *(92)*

63LA(661)1    K. Dimroth, C. Reichardt, T. Siepmann & F. Bohlmann, *Liebigs Ann. Chem.*, 1963, **661**, 1 *(105)*

63MI    M. J. S. Dewar, in *Molecular Rearrangements*, vol. 1, ed. P. de Mayo, Interscience, New York, 1963 *(314)*

63T1939    V. A. Mironov, E. V. Sobolev & A. N. Elizarova, *Tetrahedron*, 1963, **19**, 1939 *(367)*

63TL1017    R. Huisgen & R. Knorr, *Tetrahedron Lett.*, 1963, 1017 *(8)*

64AG1    M. Eigen, *Angew. Chem. Internat. Ed.*, 1964, **3**, 1 *(257)*

64AJ953    L. C. Gruen & P. T. McTigue, *Austral. J. Chem.*, 1964, **17**, 953 *(260–1)*

64CJ1681    R. Stewart & J. P. O'Donnell, *Canad. J. Chem.*, 1964, **42**, 1681 *(92)*

64CJ1957    K. Yates, J. B. Stevens & A. R. Katritzky, *Canad. J. Chem.*, 1964, **42**, 1957 *(85, 86)*

64J2864    D. V. Banthorpe, E. D. Hughes & C. K. Ingold, *J. Chem. Soc.*, 1964, 2864 *(312)*

64JA120    H. L. Goering & J. F. Levy, *J. Amer. Chem. Soc.*, 1964, **86**, 120 *(149)*

64JA622    L. K. Montgomery, K. Schueller & P. D. Bartlett, *J. Amer. Chem. Soc.*, 1964, **86**, 622 *(348)*

64JA837    J. F. Krisch & W. P. Jencks, *J. Amer. Chem. Soc.*, 1964, **86**, 837 *(282)*

64JA2357    G. A. Russell & R. C. Williamson, *J. Amer. Chem. Soc.*, 1964, **86**, 2357 *(344)*

64JA3368    C. Walling & P. Wagner, *J. Amer. Chem. Soc.*, 1964, **86**, 3368 *(343)*

64JA5188    H. D. Holtz & L. M. Stock, *J. Amer. Chem. Soc.*, 1964, **86**, 5188 *(52)*

| 64PPO63 | W. P. Jencks, *Progr. Phys. Org. Chem.*, 1964, **2**, 63 (*255*) |
|---|---|
| 64PPO323 | C. D. Ritchie & W. F. Sager, *Progr. Phys. Org. Chem.*, 1964, **2**, 323 (*57*) |
| 65AJ117 | J. Miller & K. W. Wong, *Austral. J. Chem.*, 1965, **18**, 117 (*295*) |
| 65APO123 | D. Samuel & B. L. Silver, *Adv. Phys. Org. Chem.*, 1965, **3**, 123 (*268*) |
| 65JA1534 | S. Seltzer, *J. Amer. Chem. Soc.*, 1965, **87**, 1534 (*350*) |
| 65JA3173 | A. M. Wenthe & E. H. Cordes, *J. Amer. Chem. Soc.*, 1965, **87**, 3173 (*253*) |
| 65JA3401 | W. H. Saunders, S. R. Fahrenholtz, E. A. Caress, J. P. Loew & M. Schreiber, *J. Amer. Chem. Soc.*, 1965, **87**, 3401 (*188*) |
| 65JA5172 | R. C. Fahey & C. Schubert, *J. Amer. Chem. Soc.*, 1965, **87**, 5172 (*202*) |
| 65JA5209 | J. F. Bunnett & C. Bernasconi, *J. Amer. Chem. Soc.*, 1965, **87**, 5209 (*290*) |
| 65T1993 | P. Brown & R. C. Cookson, *Tetrahedron*, 1965, **21**, 1993 (*35*) |
| 65TL963 | G. B. R. de Graff, H. J. den Hertog & W. C. Melger, *Tetrahedron Lett.*, 1965, 963 (*302*) |
| 65TL1207 | R. E. K. Winter, *Tetrahedron. Lett.*, 1965, 1207 (*357*) |
| 66APO1 | R. P. Bell, *Adv. Phys. Org. Chem.*, 1966, **4**, 1 (*246*) |
| 66APO73 | H. H. Greenwood & R. McWeeny, *Adv. Phys. Org. Chem.*, 1966, **4**, 73 (*240*) |
| 66BJ2274 | Y. Yukawa, Y. Tsuno & M. Sawada, *Bull. Chem. Soc. Japan*, 1966, **39**, 2274 (*48*, *65*) |
| 66CJ1899 | J. F. Bunnett & F. P. Olsen, *Canad. J. Chem.*, 1966, **44**, 1899 (*88*) |
| 66CJ1917 | J. F. Bunnett & F. P. Olsen, *Canad. J. Chem.*, 1966, **44**, 1917 (*91*) |
| 66J(B)310 | K. C. Ho, J. Miller & K. W. Wong, *J. Chem. Soc. B*, 1966, 310 (*295*) |
| 66J(B)705 | D. J. McLennan, *J. Chem. Soc. B*, 1966, 705 (*181*) |
| 66J(B)842 | A. A. Humffray & J. J. Ryan, *J. Chem. Soc. B*, 1966, 842 (*49*) |
| 66JA747 | J. W. Thanassi & T. C. Bruice, *J. Amer. Chem. Soc.*, 1966, **88**, 747 (*285*) |
| 66JA5254 | S. Proskow, H. E. Simmons & T. L. Cairns, *J. Amer. Chem. Soc.*, 1966, **88**, 5254 (*349*) |
| 66JA5525 | J. Hine, *J. Amer. Chem. Soc.*, 1966, **88**, 5525 (*182*) |
| 66JA5851 | H. C. Brown & R. L. Sharp, *J. Amer. Chem. Soc.*, 1966, **88**, 5851 (*209*) |
| 66JO4234 | K. B. Gash & G. U. Yuen, *J. Org. Chem.*, 1966, **31**, 4234 (*169*) |
| 67APO237 | S. L. Johnson, *Adv. Phys. Org. Chem.*, 1967, **5**, 237 (*268*, *279*) |
| 67CEd89 | K. M. Ibne-Rasa, *J. Chem. Educ.*, 1967, **44**, 89 (*163*) |
| 67CJ911 | D. Dolman & R. Stewart, *Canad. J. Chem.*, 1967, **45**, 911 (*92*) |
| 67J(B)1235 | C. D. Johnson, A. R. Katritzky & N. Shakir, *J. Chem. Soc. B*, 1967, 1235 (*85*, *86*) |
| 67JA213 | J. J. Christiansen, R. M. Izatt & L. D. Hansen, *J. Amer. Chem. Soc.*, 1967, **89**, 213 (*98*) |
| 67JA582 | G. J. Gleicher & P. v. R. Schleyer, *J. Amer. Chem. Soc.*, 1967, **89**, 582 (*161*) |
| 67JA1031 | K. E. Wilzbach, J. S. Ritscher & L. Kaplan, *J. Amer. Chem. Soc.*, 1967, **89**, 1031 (and see also 1968, **90**, 2732) (*381*) |
| 67JA2686 | K. Yates & R. A. McClelland, *J. Amer. Chem. Soc.*, 1967, **89**, 2686 (*274*) |
| 67JA3228 | T. H. Fife, *J. Amer. Chem. Soc.*, 1967, **89**, 3228 (*249*) |
| 67JA4744 | G. A. Olah & G. M. Bollinger, *J. Amer. Chem. Soc.*, 1967, **89**, 4744 (*197*) |
| 67JA4857 | A. R. Fersht & A. J. Kirby, *J. Amer. Chem. Soc.*, 1967, **89**, 4857 (*284*) |

| | |
|---|---|
| 67JA5677 | F. W. Baker, R. C. Parish & L. M. Stock, *J. Amer. Chem. Soc.*, 1967, **89**, 5677 (*52*) |
| 67JA6308 | S. Linke, G. T. Tisue & W. Lwowski, *J. Amer. Chem. Soc.*, 1967, **89**, 6308 (*325*) |
| 67JA6701 | J. L. Coke & M. P. Cooke, *J. Amer. Chem. Soc.*, 1967, **89**, 6701 (*184*) |
| 67R865 | C. W. F. Kort & H. Cerfontain, *Rec. Trav. Chim.*, 1967, **86**, 865 (*233*) |
| 67SA(23A)2279 | J. E. Dubois & F. Garnier, *Spectrochim. Acta*, 1967, **23A**, 2279 (*201*) |
| 68CC449 | D. H. Froemsdorf, W. Dowd, W. A. Gifford & S. Meyerson, *Chem. Commun.*, 1968, 449 (*182*) |
| 68J(B)142 | D. Cook & A. J. Parker, *J. Chem. Soc. B*, 1968, 142 (*159*) |
| 68J(B)515 | V. Gold, D. G. Oakenfull & T. Riley, *J. Chem. Soc. B*, 1968, 515 (*283*) |
| 68J(B)800 | R. G. Coombes, R. B. Moodie & K. Schofield, *J. Chem. Soc. B*, 1968, 800 (*223*) |
| 68JA223 | G. Klopman, *J. Amer. Chem. Soc.*, 1968, **90**, 223 (*136*) |
| 68JA319 | R. G. Pearson, H. Sobel & J. Songstad, *J. Amer. Chem. Soc.*, 1968, **90**, 319 (*163*) |
| 68JA336 | C. F. Wilcox & C. Leung, *J. Amer. Chem. Soc.*, 1968, **90**, 336 (*50, 52*) |
| 68JA947 | G. A. Olah & J. M. Bollinger, *J. Amer. Chem. Soc.*, 1968, **90**, 947 (*197*) |
| 68JA2105 | P. C. Myhre, M. Beug & L. L. James, *J. Amer. Chem. Soc.*, 1968, **90**, 2105 (*215*) |
| 68JA2587 | G. A. Olah, J. M. Bollinger & J. Brinich, *J. Amer. Chem. Soc.*, 1968, **90**, 2587 (*197*) |
| 68JA2622 | W. P. Jencks & M. Gilchrist, *J. Amer. Chem. Soc.*, 1968, **90**, 2622 (*282*) |
| 68JA4081 | T. H. Fife & L. K. Jao, *J. Amer. Chem. Soc.*, 1968, **90**, 4081 (*250*) |
| 68JA4498 | R. Srinivasan, *J. Amer. Chem. Soc.*, 1968, **90**, 4498 (*357*) |
| 68JA5049 | R. Alexander, E. C. F. Ko, A. J. Parker & T. J. Broxton, *J. Amer. Chem. Soc.*, 1968, **90**, 5049 (*102*) |
| 68JA5848 | S. A. Shain & J. F. Kirsch, *J. Amer. Chem. Soc.*, 1968, **90**, 5848 (*268*) |
| 68JA6213 | H. Goering & K. Humski, *J. Amer. Chem. Soc.*, 1968, **90**, 6213 (*332*) |
| 68JA6546 | A. Diaz, I. Lazdins & S. Winstein, *J. Amer. Chem. Soc.*, 1968, **90**, 6546 (*171*) |
| 68JPC1866 | H. E. O'Neal & S. W. Benson, *J. Phys. Chem.*, 1968, **72**, 1866 (*350*) |
| 68MIa, b | J. Miller, *Aromatic Nucleophilic Substitution*, Elsevier, Amsterdam, 1968: (*a*) p. 77; (*b*) p. 144 (*294–5*) |
| 68MIc | P. Y. Sollenberger & R. B. Martin, in *The Chemistry of the Amino Group*, ed. S. Patai, Interscience, London, 1968 (*255*) |
| 68R24 | C. W. F. Kort & H. Cerfontain, *Rec. Trav. Chim.*, 1968, **87**, 24 (*233*) |
| 68SC(159)833 | P. D. Bartlett, *Science*, 1968, **159**, 833 (*348*) |
| 69AG781 | R. B. Woodward & R. Hoffmann, *Angew. Chem. Internat. Ed.*, 1969, **8**, 781; also reprinted as *The Conservation of Orbital Symmetry*, Academic, New York, 1970 (*351*) |
| 69CJ1555 | S. McLean, C. J. Webster & R. J. D. Rutherford, *Canad. J. Chem.*, 1969, **47**, 1555 (*367*) |
| 69CR1 | A. J. Parker, *Chem. Rev.*, 1969, **69**, 1 (*101, 162, 164*) |
| 69CR33 | A. Maccoll, *Chem. Rev.*, 1969, **69**, 33 (*194*) |
| 69CR543 | C. J. Collins, *Chem. Rev.*, 1969, **69**, 543 (*334*) |

| | |
|---|---|
| 68J(B)1 | J. G. Hoggett, R. B. Moodie & K. Schofield, *J. Chem. Soc. B*, 1969, 1 (*226*) |
| 69JA615 | M. Charton, *J. Amer. Chem. Soc.*, 1969, **91**, 615 (*60*) |
| 69JA1469 | J. H. Rolston & K. Yates, *J. Amer. Chem. Soc.*, 1969, **91**, 1469 (*196-7, 199*) |
| 69JA2160 | P. G. Gassman & D. S. Patton, *J. Amer. Chem. Soc.*, 1969, **91**, 2160 (*173*) |
| 69JA2993, 3003 | B. Holmquist & T. C. Bruice, *J. Amer. Chem. Soc.*, 1969, **91**, 2993, 3003 (*273*) |
| 69JA6057 | J. L. Kurz & J. M. Farrar, *J. Amer. Chem. Soc.*, 1969, **91**, 6057 (*98*) |
| 69JA6090 | P. R. Rony, *J. Amer. Chem. Soc.*, 1969, **91**, 6090 (*247*) |
| 69JA6654 | C. D. Johnson, A. R. Katritzky & S. A. Shapiro, *J. Amer. Chem. Soc.*, 1969, **91**, 6654 (*85*) |
| 69JA7144 | O. S. Tee, *J. Amer. Chem. Soc.*, 1969, **91**, 7144 (*182*) |
| 69JA7224 | R. A. Marcus, *J. Amer. Chem. Soc.*, 1969, **91**, 7224 (*81*) |
| 69JA7508 | J. M. Harris, F. L. Schadt, P. v. R. Schleyer & C. J. Lancelot, *J. Amer. Chem. Soc.*, 1969, **91**, 7508 (*171*) |
| 69JO2018 | W. A. Pryor, U. Tonellato, D. L. Fuller & S. Jumonville, *J. Org. Chem.*, 1969, **34**, 2018 (*344*) |
| 69MI | C. K. Ingold, *Structure and Mechanism in Organic Chemistry*, 2nd edn, Bell, London, 1969 (*104, 141, 175, 265*) |
| 69SPC53 | J. F. Bunnett, *Survey Progr. Chem.*, 1969, **5**, 53 (*178, 180, 187*) |
| 70AG830 | C. Rüchardt, *Angew. Chem. Internat. Ed.*, 1970, **9**, 830 (*341*) |
| 70CR295 | D. V. Banthorpe, *Chem. Rev.*, 1970, **70**, 295 (*221-2*) |
| 70J(B)274 | R. A. More O'Ferrall, *J. Chem. Soc. B*, 1970, 274 (*179*) |
| 70J(B)338 | B. J. Gregory, R. B. Moodie & K. Schofield, *J. Chem. Soc. B.*, 1970, 338 (*323*) |
| 70J(B)347 | R. G. Coombes, D. H. G. Crout, J. G. Hoggett, R. B. Moodie & K. Schofield, *J. Chem. Soc. B.*, 1970, 347 (*239*) |
| 70J(B)797 | P. F. Christy, J. H. Ridd & N. D. Stears, *J. Chem. Soc. B*, 1970, 797 (*223*) |
| 70JA1681 | T. H. Fife & L. H. Brod, *J. Amer. Chem. Soc.*, 1970, **92**, 1681 (*250*) |
| 70JA2417 | J. A. Orvik & J. F. Bunnett, *J. Amer. Chem. Soc.*, 1970, **92**, 2417 (*291*) |
| 70JA2540 | J. L. Fry, J. M. Harris, R. C. Bingham & P. v. R. Schleyer, *J. Amer. Chem. Soc.*, 1970, **92**, 2540 (*152*) |
| 70JA2549 | P. G. Gassmann & A. F. Fentiman, *J. Amer. Chem. Soc.*, 1970, **92**, 2549 (*173*) |
| 70JA2816 | R. C. Fahey & M. W. Monahan, *J. Amer. Chem. Soc.*, 1970, **92**, 2816 (*203*) |
| 70JA3210 | A. J. Kresge, *J. Amer. Chem. Soc.*, 1970, **92**, 3210 (*81*) |
| 70JA5926 | F. G. Bordwell, W. J. Boyle & K. C. Yee, *J. Amer. Chem. Soc.*, 1970, **92**, 5926 (*81*) |
| 70JA5956 | R. F. Pratt & T. C. Bruice, *J. Amer. Chem. Soc.*, 1970, **92**, 5956 (*273*) |
| 70JA7401 | H. L. Goering, R. G. Briody & G. Sandrock, *J. Amer. Chem. Soc.*, 1970, **92**, 7401 (*150*) |
| 70JA7463, 7464 | J. K. Kim & J. F. Bunnett, *J. Amer. Chem. Soc.*, 1970, **92**, 7463, 7464 (*305*) |
| 70JA7596 | P. C. Myhre & G. D. Andrews, *J. Amer. Chem. Soc.*, 1970, **92**, 7596 (*207*) |
| 70PPO129 | C. C. Lee, *Progr. Phys. Org. Chem.*, 1970, **7**, 129 (*334*) |
| 70QR473 | P. D. Bartlett, *Quart. Rev.*, 1970, **24**, 473 (*349*) |
| 71Acc57 | K. Fukui, *Acc. Chem. Res.*, 1971, **4**, 57 (*368*) |
| 71Acc136 | K. Yates, *Acc. Chem. Res.*, 1971, **4**, 136 (*274*) |

| | |
|---|---|
| 71Acc240 | G. A. Olah, *Acc. Chem. Res.*, 1971, **4**, 240 (*219, 221*) |
| 71Acc248 | J. H. Ridd, *Acc. Chem. Res.*, 1971, **4**, 248 (*222*) |
| 71Acc272 | H. E. Zimmerman, *Acc. Chem. Res.*, 1971, **4**, 272 (*376*) |
| 71AG761 | M. J. S. Dewar, *Angew. Chem. Internat. Ed.*, 1971, **10**, 761 (*376*) |
| 71BJ1632 | T. Imamoto, Y. Tsuno & Y. Yukawa, *Bull. Chem. Soc. Japan*, 1971, **44**, 1632 (*327*) |
| 71BJ2776 | T. Imamoto, S. G. Kim, Y. Tsuno & Y. Yukawa, *Bull. Chem. Soc. Japan*, 1971, **44**, 2776 (*327*) |
| 71CC1601 | Y. Yukawa & T. Ando, *Chem. Commun.*, 1971, 1601 (*323*) |
| 71CEd427 | A. R. Katritzky & R. D. Topsom, *J. Chem. Educ.*, 1971, **48**, 427 (*50*) |
| 71CR247 | P. Gray, A. A. Herod & A. Jones, *Chem. Rev.*, 1971, **71**, 247 (*341*) |
| 71J(B)460 | F. W. Fowler, A. R. Katritzky & R. J. D. Rutherford, *J. Chem. Soc. B*, 1971, 460 (*105*) |
| 71JA2445 | R. C. Fahey & C. A. McPherson, *J. Amer. Chem. Soc.*, 1971, **93**, 2445 (*204*) |
| 71JA3633 | A. Padwa, W. Koehn, J. Masaracchia, C. L. Osborn & D. J. Trecker, *J. Amer. Chem. Soc.*, 1971, **93**, 3633 (*361*) |
| 71JA3782 | T. J. Katz, E. J. Wang & N. Acton, *J. Amer. Chem. Soc.*, 1971, **93**, 3782 (*381*) |
| 71JA5765 | H. C. Brown & C. J. Kim, *J. Amer. Chem. Soc.*, 1971, **93**, 5765 (*171*) |
| 71JO913 | J. K. Crandall & R. J. Watkins, *J. Org. Chem.*, 1971, **36**, 913 (*384*) |
| 71MI*a* | J. G. Hoggett, R. B. Moodie, J. R. Penton & K. Schofield, *Nitration and Aromatic Reactivity*, Cambridge University Press, Cambridge, 1971 (*242*) |
| 71MI*b* | A. J. Kirby & A. R. Fersht, *Progr. Bioorg. Chem.*, 1971, **1**, 1 (*285*) |
| 71PPO235 | M. Charton, *Progr. Phys. Org. Chem.*, 1971, **8**, 235 (*60*) |
| 71TF1995 | R. P. Bell, W. H. Sachs & R. L. Tranter, *Trans. Faraday Soc.*, 1971, **67**, 1995 (*34*) |
| 72AG874 | P. Rys, P. Skrabal & H. Zollinger, *Angew. Chem. Internat. Ed.*, 1972, **11**, 874 (*222–3*) |
| 72CEd400 | L. M. Stock, *J. Chem. Educ.*, 1972, **49**, 400 (*50*) |
| 72CR705 | W. P. Jencks, *Chem. Rev.*, 1972, **72**, 705 (*252*) |
| 72J(P2)127 | S. R. Hartshorn, R. B. Moodie & K. Stead, *J. Chem. Soc. Perkin 2*, 1972, 127 (*223*) |
| 72JA1010 | H. L. Goering & J. V. Clevenger, *J. Amer. Chem. Soc.*, 1972, **94**, 1010 (*332*) |
| 72JA1148 | R. Alexander, A. J. Parker, J. H. Sharp & W. E. Waghorne, *J. Amer. Chem. Soc.*, 1972, **94**, 1148 (*101*) |
| 72JA3080 | R. Golden & L. M. Stock, *J. Amer. Chem. Soc.*, 1972, **94**, 3080 (*50*) |
| 72JA3907 | F. G. Bordwell & W. J. Boyle, *J. Amer. Chem. Soc.*, 1972, **94**, 3907 (*82*) |
| 72JA6083 | D. J. Pasto, B. Lepeska & T. C. Cheng, *J. Amer. Chem. Soc.*, 1972, **94**, 6083 (*209*) |
| 72JO1770 | M. F. Ruasse & J. E. Dubois, *J. Org. Chem.*, 1972, **37**, 1770 (*199–200*) |
| 72MI*a* | A. J. Kirby, in *Comprehensive Chemical Kinetics*, vol. 10, eds C. H. Bamford & C. F. H. Tipper, Elsevier, Amsterdam, 1972 (*279*) |
| 72MI*b* | P. R. Story & B. C. Clark, in *Carbonium Ions*, vol. 3, eds G. A. Olah & P. v. R. Schleyer, Wiley-Interscience, New York, 1972 (*171*) |
| 72MI*c* | N. B. Chapman & J. Shorter (eds), *Advances in Linear Free Energy Relationships*, Plenum, London, 1972 (*60*) |

| | |
|---|---|
| 72TL1241 | J. J. Dannenberg, D. R. Weinwurzel, K. Dill & B. J. Goldberg, *Tetrahedron Lett.*, 1972, 1241 (*334*) |
| 73Acc41 | R. E. Barnett, *Acc. Chem. Res.*, 1973, **6**, 41 (*247*) |
| 73Acc46 | R. A. Sneen, *Acc. Chem. Res.*, 1973, **6**, 46 (*150*) |
| 73Acc53 | M. Saunders, P. Vogel, E. L. Hagen & J. Rosenfeld, *Acc. Chem. Res.*, 1973, **6**, 53 (*333*) |
| 73Acc377 | H. C. Brown, *Acc. Chem. Res.*, 1973, **6**, 377 (*330*) |
| 73AG173 | G. A. Olah, *Angew. Chem. Internat. Ed.*, 1973, **12**, 173 (*172, 332*) |
| 73J(P2)160 | F. T. Boyle & R. A. Y. Jones, *J. Chem. Soc. Perkin 2*, 1973, 160 (*88*) |
| 73J(P2)1321 | H. M. Gilow & J. H. Ridd, *J. Chem. Soc. Perkin 2*, 1973, 1321 (*231*) |
| 73J(P2)1893 | M. H. Abraham, *J. Chem. Soc. Perkin 2*, 1973, 1893 (*150*) |
| 73J(P2)1915 | N. C. Marziano, G. M. Cimino & R. C. Passerini, *J. Chem. Soc. Perkin 2*, 1973, 1915 (*89*) |
| 73JA408 | B. G. Cox & A. J. Parker, *J. Amer. Chem. Soc.*, 1973, **95**, 408 (*103*) |
| 73JA2738 | T. J. Katz & N. Acton, *J. Amer. Chem. Soc.*, 1973, **95**, 2738 (*381*) |
| 73JA4754 | W. A. Pryor, W. H. Davis & J. P. Stanley, *J. Amer. Chem. Soc.*, 1973, **95**, 4754 (*344*) |
| 73JA5054, 5055 | R. Huisgen & G. Steiner, *J. Amer. Chem. Soc.*, 1973, **95**, 5054, 5055 (*349*) |
| 73JA5056 | G. Steiner & R. Huisgen, *J. Amer. Chem. Soc.*, 1973, **95**, 5056 (*349*) |
| 73JA5798 | J. Sims & K. N. Houk, *J. Amer. Chem. Soc.*, 1973, **95**, 5798 (*375*) |
| 73JO493 | J. E. Dubois & M. F. Ruasse, *J. Org. Chem.*, 1973, **38**, 493 (*199*) |
| 73Med1207 | C. Hansch, A. Leo, S. H. Unger, K. H. Kim, D. Nikaitani & E. J. Lien, *J. Med. Chem.*, 1973, **16**, 1207 (*65*) |
| 73MIa | R. P. Bell, *The Proton in Chemistry*, 2nd edn, Chapman & Hall, London, 1973, chapter 5 (*78*) |
| 73MIb | F. Franks (ed), *Water: a Comprehensive Treatise*, vol. 3, Plenum, New York, 1973 (*98–9*) |
| 73MIc | E. S. Huyser, in *The Chemistry of the Carbon–Halogen Bond*, ed. S. Patai, Wiley-Interscience, London, 1973 (*346*) |
| 73PPO1 | S. Ehrenson, R. T. C. Brownlee & R. W. Taft, *Progr. Phys. Org. Chem.*, 1973, **10**, 1 (*54*) |
| 73SPC113 | K. N. Houk, *Survey Progr. Chem.*, 1973, **6**, 113 (*368*) |
| 74Acc361 | P. B. D. de la Mare, *Acc. Chem. Res.*, 1974, **7**, 361 (*230*) |
| 74AG1 | B. Chevrier & R. Weiss, *Angew. Chem. Internat. Ed.*, 1974, **13**, 1 (*235*) |
| 74AG47 | J. B. Hendrickson, *Angew. Chem. Internat. Ed.*, 1974, **13**, 47 (*347*) |
| 74CC293 | J. W. Chapman & A. N. Strachan, *Chem. Commun.*, 1974, 293 (*214*) |
| 74CR581 | E. H. Cordes & H. G. Bull, *Chem. Rev.*, 1974, **74**, 581 (*247–8*) |
| 74J(P2)600 | N. C. Marziano, J. H. Rees & J. H. Ridd, *J. Chem. Soc. Perkin 2*, 1974, 600 (*224*) |
| 74J(P2)1373 | D. J. McLennan & R. J. Wong, *J. Chem. Soc. Perkin 2*, 1974, 1373 (*177*) |
| 74JA430 | R. A. Bartsch & K. E. Wiegers, *J. Amer. Chem. Soc.*, 1974, **96**, 430 (*188*) |
| 74JA480 | R. A. Abramovitch & E. P. Kyba, *J. Amer. Chem. Soc.*, 1974, **96**, 480 (*324*) |
| 74JA2862 | D. G. Lee & M. H. Sador, *J. Amer. Chem. Soc.*, 1974, **96**, 2862 (*276*) |
| 74JA3565 | G. A. Olah, P. W. Westerman, E. G. Melby & Y. K. Mo, *J. Amer. Chem. Soc.*, 1974, **96**, 3565 (*197*) |

| | |
|---|---|
| 74JA4335 | R. C. Hahn & D. L. Strack, *J. Amer. Chem. Soc.*, 1974, **96**, 4335 (*227–8*) |
| 74JA4484 | J. M. Harris, A. Becker, J. F. Fagan & F. A. Walden, *J. Amer. Chem. Soc.*, 1974, **96**, 4484 (*146*) |
| 74JA4534 | R. C. Fahey, C. A. McPherson & R. A. Smith, *J. Amer. Chem. Soc.*, 1974, **96**, 4534 (*204*) |
| 74JA7222 | S. B. Hanna, C. Jermini, H. Loewenschuss & H. Zollinger, *J. Amer. Chem. Soc.*, 1974, **96**, 7222 (*82, 237*) |
| 74JA7638 | F. K. Fong, *J. Amer. Chem. Soc.*, 1974, **96**, 7638 (*332*) |
| 74JA7986 | S. Rosenberg, S. M. Silver, J. M. Sayer & W. P. Jencks, *J. Amer. Chem. Soc.*, 1974, **96**, 7986 (*256–7*) |
| 74JA7998 | J. M. Sayer, B. Pinsky, A. Schonbrunn & W. Washtien, *J. Amer. Chem. Soc.*, 1974, **96**, 7998 (*256*) |
| 74JO2441 | M. F. Ruasse & J. E. Dubois, *J. Org. Chem.*, 1974, **39**, 2441 (*199*) |
| 74T1733 | H. S. Mosher, *Tetrahedron*, 1975, **30**, 1733 (*320*) |
| 75Acc239 | R. A. Bartsch, *Acc. Chem. Res.*, 1975, **8**, 239 (*188*) |
| 75Acc361 | K. N. Houk, *Acc. Chem. Res.*, 1975, **8**, 361 (*368*) |
| 75APO1 | T. H. Fife, *Adv. Phys. Org. Chem.*, 1975, **11**, 1 (*286*) |
| 75APO177 | G. M. Kramer, *Adv. Phys. Org. Chem.*, 1975, **11**, 177 (*332*) |
| 75J(P2)648 | J. W. Barnett, R. B. Moodie, K. Schofield & J. B. Weston, *J. Chem. Soc. Perkin 2*, 1975, 648 (*240*) |
| 75J(P2)1113 | B. Capon & K. Nimmo, *J. Chem. Soc. Perkin 2*, 1975, 1113 (*250*) |
| 75J(P2)1365 | M. H. Abraham, P. L. Grellier & M. J. Hogarth, *J. Chem. Soc. Perkin 2*, 1975, 1365 (*159*) |
| 75J(P2)1802 | H. B. Amin & R. Taylor, *J. Chem. Soc. Perkin 2*, 1975, 1802 (*194*) |
| 75JA783, 791, 796, 799 | C. G. Swain and others, *J. Amer. Chem. Soc.*, 1975, **97**, 783, 791, 796, 799 (*296*) |
| 75JA2431 | A. C. Day, *J. Amer. Chem. Soc.*, 1975, **97**, 2431 (*351*) |
| 75JA2477 | I. N. Feit, I. K. Breger, A. M. Capobianco, T. W. Cooke & L. F. Gitlin, *J. Amer. Chem. Soc.*, 1975, **97**, 2477 (*192*) |
| 75JA3102 | D. A. Winey & E. R. Thornton, *J. Amer. Chem. Soc.*, 1975, **97**, 3102 (*179*) |
| 75JA5975 | P. C. Hiberty, *J. Amer. Chem. Soc.*, 1975, **97**, 5975 (*195*) |
| 75JA6615 | R. L. Yates, N. D. Epiotis & F. Bernandi, *J. Amer. Chem. Soc.*, 1975, **97**, 6615 (*167*) |
| 75JO1835 | R. Levine & R. E. Biehl, *J. Org. Chem.*, 1975, **40**, 1835 (*302*) |
| 75T2463 | P. Deslongchamps, *Tetrahedron*, 1975, **31**, 2463 (*138*) |
| 75T2999 | D. J. McLennan, *Tetrahedron*, 1975, **31**, 2999 (*191*) |
| 75TL3371 | K. L. Shepard, *Tetrahedron Lett.*, 1975, 3371 (*300*) |
| 76Acc41 | G. A. Olah, *Acc. Chem. Res.*, 1976, **9**, 41 (*332*) |
| 76Acc281 | D. J. McLennan, *Acc. Chem. Res.*, 1976, **9**, 281 (*150, 167*) |
| 76AJ787 | D. J. McLennan & R. J. Wong, *Austral. J. Chem.*, 1976, **29**, 787 (*177*) |
| 76CC734 | J. E. Baldwin, *Chem. Commun.*, 1976, 734 (*138*) |
| 76CC736 | J. E. Baldwin, J. Cutting, W. Dupont, L. Kruse, L. Silberman & R. C. Thomas, *Chem. Commun.*, 1976, 736 (*137–8*) |
| 76CC871 | B. Capon, K. Nimmo & G. L. Reid, *Chem. Commun.*, 1976, 871 (*254*) |
| 76J(P2)280 | S. de B. Norfolk & R. Taylor, *J. Chem. Soc. Perkin 2*, 1976, 280 (*194*) |
| 76J(P2)1089 | R. B. Moodie, K. Schofield & J. B. Weston, *J. Chem. Soc. Perkin 2*, 1976, 1089 (*227*) |
| 76JA318 | D. H. Aue, H. M. Webb & M. T. Bowers, *J. Amer. Chem. Soc.*, 1976, **98**, 318 (*98–9*) |
| 76JA488 | R. A. Cox & R. Stewart, *J. Amer. Chem. Soc.*, 1976, **98**, 488 (*92*) |
| 76JA1573 | M. Cocivera, C. A. Fyfe, A. Effio, S. P. Vaish & H. E. Chen, *J. Amer. Chem. Soc.*, 1976, **98**, 1573 (*256*) |

76JA2190    R. E. Townshend, G. Ramunni, G. A. Segal, W. J. Hehre & L. Saleem, *J. Amer. Chem. Soc.*, 1976, **98**, 2190 *(351)*

76JA2865    K. Humski, V. Sendijarević & V. J. Shiner, *J. Amer. Chem. Soc.*, 1976, **98**, 2865 *(151)*

76JA3301    R. G. Bergstrom, R. G. M. Landells, G. H. Wahl & H. Zollinger, *J. Amer. Chem. Soc.*, 1976, **98**, 3301 *(297)*

76JA7371    M. Cocivera & A. Effio, *J. Amer. Chem. Soc.*, 1976, **98**, 7371 *(256)*

76JA7658    T. W. Bentley & P. v. R. Schleyer, *J. Amer. Chem. Soc.*, 1976, **98**, 7658 *(152)*

76JA7667    F. L. Schadt, T. W. Bentley & P. v. R. Schleyer, *J. Amer. Chem. Soc.*, 1976, **98**, 7667 *(152)*

76JO403    R. Huisgen, *J. Org. Chem.*, 1976, **41**, 403 *(351)*

76JO3201    D. M. Muir & A. J. Parker, *J. Org. Chem.*, 1976, **41**, 3201 *(192)*

76JO3364    F. P. Ballistreri, E. Maccarone & A. Mamo, *J. Org. Chem.*, 1976, **41**, 3364 *(156)*

76MI    I. Fleming, *Frontier Orbitals and Organic Chemical Reactions*, Wiley-Interscience, London, 1976 *(368)*

76PPO49    T. Fujita & T. Nishioka, *Progr. Phys. Org. Chem.*, 1976, **12**, 49 *(60)*

76PPO91    S. H. Unger & C. Hansch, *Progr. Phys. Org. Chem.*, 1976, **12**, 91 *(67)*

76T179    H. C. Brown, *Tetrahedron*, 1976, **32**, 179 *(330)*

77Acc125    G. Bartoli & P. E. Todesco, *Acc. Chem. Res.*, 1977, **10**, 125 *(161)*

77APO(14)69    A. Pross, *Adv. Phys. Org. Chem.*, 1977, **14**, 69 *(62)*

77APO(15)1    J. Hine, *Adv. Phys. Org. Chem.*, 1977, **15**, 1 *(182)*

77CC608    S. Acevedo & K. Bowden, *Chem. Commun.*, 1977, 608 *(51)*

77CJ1468    C. A. Fyfe, A. Koll, S. W. H. Damji, C. D. Malkiewich & P. A. Forte, *Canad. J. Chem.*, 1977, **55**, 1468 *(291)*

77CR639    A. R. Katritzky & R. D. Topsom, *Chem. Rev.*, 1977, **77**, 639 *(54, 66)*

77J(P2)306, 309, 845    N. C. Marziano and others, *J. Chem. Soc. Perkin 2*, 1977, 306, 309, 845 *(89)*

77J(P2)873    M. H. Abraham & D. J. McLennan, *J. Chem. Soc. Perkin 2*, 1977, 873 *(150)*

77J(P2)1693    R. B. Moodie, P. N. Thomas & K. Schofield, *J. Chem. Soc. Perkin 2*, 1977, 1693 *(238)*

77J(P2)1753    D. J. McLennan, *J. Chem. Soc. Perkin 2*, 1977, 1753 *(178)*

77J(P2)1758    A. Grout, D. J. McLennan & I. H. Spackman, *J. Chem. Soc. Perkin 2*, 1977, 1758 *(178)*

77JA377    M. J. S. Dewar, R. C. Haddon, A. Kormonicki & H. Rzepa, *J. Amer. Chem. Soc.*, 1977, **99**, 377 *(332)*

77JA3387    V. Lucchini, G. Modena, G. Scorrano & U. Tonellato, *J. Amer. Chem. Soc.*, 1977, **99**, 3387 *(91)*

77JA4090    C. F. Bernasconi, R. H. de Rossi & P. Schmid, *J. Amer. Chem. Soc.*, 1977, **99**, 4090 *(291)*

77JA4219    W. N. Olmstead & J. I. Brauman, *J. Amer. Chem. Soc.*, 1977, **99**, 4219 *(164)*

77JA4417    M. J. S. Dewar & L. E. Wade, *J. Amer. Chem. Soc.*, 1977, **99**, 4417 *(316, 351)*

77JA5516    C. L. Perrin, *J. Amer. Chem. Soc.*, 1977, **99**, 5516 *(224)*

77JA5687    M. Charton, *J. Amer. Chem. Soc.*, 1977, **99**, 5687 *(57)*

77JA6699    W. B. Chiao & W. H. Saunders, *J. Amer. Chem. Soc.*, 1977, **99**, 6699 *(184)*

77JA7653    D. J. Hupe & D. Wu, *J. Amer. Chem. Soc.*, 1977, **99**, 7653 *(82)*

77JA8119    D. W. Goetz, H. B. Schlegel & L. C. Allen, *J. Amer. Chem. Soc.*, **99**, 8119 *(332)*

| | |
|---|---|
| 77JO205 | S. Alunni, E. Baciocchi & P. Perucci, *J. Org. Chem.*, 1977, **42**, 205 (*180*) |
| 77JO534 | C. T. Wang & E. J. Grubbs, *J. Org. Chem.*, 1977, **42**, 534 (*50*) |
| 77JO698 | D. H. Wertz & N. L. Allinger, *J. Org. Chem.*, 1977, **42**, 698 (*194*) |
| 77JO916 | R. W. Taft & L. S. Levitt, *J. Org. Chem.*, 1977, **42**, 916 (*57*) |
| 77T3009 | R. A. Firestone, *Tetrahedron*, 1977, **33**, 3009 (*351*) |
| 78Acc1 | J. Hine, *Acc. Chem. Res.*, 1978, **11**, 1 (*247*) |
| 78Acc413 | J. F. Bunnett, *Acc. Chem. Res.*, 1978, **11**, 413 (*305–6*) |
| 78APO1 | J. H. Ridd, *Adv. Phys. Org. Chem.*, 1978, **16**, 1 (*224*) |
| 78APO51 | J. M. Tedder & J. C. Walton, *Adv. Phys. Org. Chem.*, 1978, **16**, 51 (*206*) |
| 78CC180 | R. B. Moodie, K. Schofield & G. D. Tobin, *Chem. Commun.*, 1978, 180 (*230*) |
| 78JA973 | C. E. Barnes & P. C. Myhre, *J. Amer. Chem. Soc.*, 1978, **100**, 973 (*230*) |
| 78JA1873 | J. F. Bunnett & S. J. Shafer, *J. Amer. Chem. Soc.*, 1978, **100**, 1873 (*306*) |
| 78JA3139 | J. M. Harris, D. L. Mount & D. J. Raber, *J. Amer. Chem. Soc.*, 1978, **100**, 3139 (*330*) |
| 78JA3143 | H. C. Brown, M. Ravindranathan, F. J. Chloupek & I. Rothberg, *J. Amer. Chem. Soc.*, 1978, **100**, 3143 (*330*) |
| 78JA3861 | R. A. Cox & K. Yates, *J. Amer. Chem. Soc.*, 1978, **100**, 3861 (*89*) |
| 78JA3927 | H. Kwart, T. J. George, R. Louw & W. Ultee, *J. Amer. Chem. Soc.*, 1978, **100**, 3927 (*384*) |
| 78JA5650 | M. J. S. Dewar, S. Olivella & H. S. Rzepa, *J. Amer. Chem. Soc.*, 1978, **100**, 5650 (*351*) |
| 78JA5954 | N. Å. Bergman, Y. Chiang & A. Kresge, *J. Amer. Chem. Soc.*, 1978, **100**, 5954 (*33, 257–8*) |
| 78JA5956 | M. M. Cox & W. P. Jencks, *J. Amer. Chem. Soc.*, 1978, **100**, 5956 (*33*) |
| 78JA6119 | W. M. Ching & R. G. Kallen, *J. Amer. Chem. Soc.*, 1978, **100**, 6119 (*246*) |
| 78JA7027 | R. A. McClelland & M. Ahmad, *J. Amer. Chem. Soc.*, 1978, **100**, 7027 (*250*) |
| 78JA7037 | R. Eliason & M. M. Kreevoy, *J. Amer. Chem. Soc.*, 1978, **100**, 7037 (*250*) |
| 78JA7645 | M. F. Ruasse, A. Argile & J. E. Dubois, *J. Amer. Chem. Soc.*, 1978, **100**, 7645 (*201*) |
| 78JA7765 | R. W. Taft, M. Taagepera, J. L. M. Abboud, J. F. Wolf, D. J. DeFrees, W. J. Hehre, J. E. Bartmess & R. T. McIver, *J. Amer. Chem. Soc.*, 1978, **100**, 7765 (*57, 66*) |
| 78JA8133 | R. C. Seib, V. J. Shiner, V. Sendijarević & K. Humski, *J. Amer. Chem. Soc.*, 1978, **100**, 8133 (*151*) |
| 78JO1843 | A. J. Parker, U. Mayer, R. Schmid & V. Gutmann, *J. Org. Chem.*, 1978, **43**, 1843 (*101*) |
| 78JO3667 | H. C. Brown, M. Ravindranathan, C. G. Rao, F. J. Chloupek & M. H. Rei, *J. Org. Chem.*, 1978, **43**, 3667 (*332*) |
| 78T1619 | E. L. Motell, A. W. Boone & W. H. Fink, *Tetrahedron*, 1978, **34**, 1619 (*33*) |
| 78T2057 | J. R. Beck, *Tetrahedron*, 1978, **34**, 2057 (*293*) |
| 78T3553 | J. A. MacPhee, A. Panaye & J. E. Dubois, *Tetrahedron*, 1978, **34**, 3553 (*67*) |
| 79CC1131 | J. F. King & G. T. Y. Tsang, *Chem. Commun.*, 1979, 1131 (*161*) |
| 79CJ2944 | R. A. Cox & K. Yates, *Canad. J. Chem.*, 1979, **57**, 2944 (*91, 278*) |
| 79CJ2960 | R. A. Cox, M. F. Goodman & K. Yates, *Canad. J. Chem.*, 1979, **57**, 2960 (*278*) |

| 79J(P2)133 | R. B. Moodie, K. Schofield & P. G. Taylor, *J. Chem. Soc. Perkin 2*, 1979, 133 (*214*) |
|---|---|
| 79J(P2)537 | G. H. E. Nieuwdorp. C. L. de Ligny & H. C. van Houwelingen, *J. Chem. Soc. Perkin 2*, 1979, 537 (*55*) |
| 79J(P2)618 | J. C. Giffney & J. H. Ridd, *J. Chem. Soc. Perkin 2*, 1979, 618 (*226*) |
| 79J(P2)1317 | T. O. Bamkole, J. Hirst & I. Onyido, *J. Chem. Soc. Perkin 2*, 1979, 1317 (*295*) |
| 79J(P2)1451 | R. G. Coombes, J. G. Goldring & P. Hadjigeorgiou, *J. Chem. Soc. Perkin 2*, 1979, 1451 (*230*) |
| 79JA522 | E. M. Arnett, C. Petro & P. v. R. Schleyer, *J. Amer. Chem. Soc.*, 1979, **101**, 522 (*152, 332*) |
| 79JA783 | M. J. S. Dewar & G. P. Ford, *J. Amer. Chem. Soc.*, 1979, **101**, 783 (*222*) |
| 79JA1176 | R. A. Bartsch, R. A. Read, D. T. Larsen, D. K. Roberts, K. J. Scott & B. R. Cho, *J. Amer. Chem. Soc.*, 1979, **101**, 1176 (*188*) |
| 79JA1295 | J. R. Keefe, J. Morey, C. A. Palmer & J. C. Lee, *J. Amer. Chem. Soc.*, 1979, **101**, 1295 (*82*) |
| 79JA2107 | R. M. Magid & O. S. Fruchey, *J. Amer. Chem. Soc.*, 1979, **101**, 2107 (*167*) |
| 79JA2669 | M. Ahmad, R. G. Bergstrom, M. J. Cashen, Y. Chiang, A. J. Kresge, R. A. McClelland & M. F. Powell, *J. Amer. Chem. Soc.*, 1979, **101**, 2669 (see also correction in *J. Amer. Chem. Soc.*, 1982, **104**, 1156) (*250*) |
| 79JA2845 | R. D. Bach, R. C. Badger & T. J. Lang, *J. Amer. Chem. Soc.*, 1979, **101**, 2845 (*185*) |
| 79JA3288 | P. R. Young & W. P. Jencks, *J. Amer. Chem. Soc.*, 1979, **101**, 3288 (*155*) |
| 79JA3300 | M. H. O'Leary & J. F. Marlier, *J. Amer. Chem. Soc.*, 1979, **101**, 3300 (*269*) |
| 79JA4672 | J. L. Jensen, L. R. Herold, P. A. Lenz, S. Trusty, V. Sergi, K. Bell & P. Rogers, *J. Amer. Chem. Soc.*, 1979, **101**, 4672 (*250*) |
| 79JA4678 | P. R. Young & P. E. McMahon, *J. Amer. Chem. Soc.*, 1979, **101**, 4678 (*246*) |
| 79JA5532 | D. J. DeFrees, M. Taagepera, B. A. Levi, S. K. Pollack, K. D. Summerhays, R. W. Taft, M. Wolfsberg & W. J. Hehre, *J. Amer. Chem. Soc.*, 1979, **101**, 5532 (*36, 152*) |
| 79JA6693 | J. J. Gajewski & N. D. Conrad, *J. Amer. Chem. Soc.*, 1979, **101**, 6693 (*351*) |
| 79JA7594 | S. B. Kaldor & W. H. Saunders, *J. Amer. Chem. Soc.*, 1979, **101**, 7594 (*34*) |
| 79JO745 | C. C. Greig, C. D. Johnson, S. Rose & P. G. Taylor, *J. Org. Chem.*, 1979, **44**, 745 (*91*) |
| 79JO753 | C. D. Johnson, S. Rose & P. G. Taylor, *J. Org. Chem.*, 1979, **44**, 753 (*91*) |
| 79JO903 | M. Charton, *J. Org. Chem.*, 1979, **44**, 903 (*57*) |
| 79JO1173 | M. F. Ruasse, J. E. Dubois & A. Argile, *J. Org. Chem.*, 1979, **44**, 1173 (*197*) |
| 79JO1463 | J. F. Bunnett, S. Sridharan & W. P. Cavin, *J. Org. Chem.*, 1979, **44**, 1463 (*191*) |
| 79JO2108 | V. J. Shiner, D. A. Nollen & K. Humski, *J. Org. Chem.*, 1979, **44**, 2108 (*151*) |
| 79JO2642 | P. Cogolli, F. Maiolo, L. Testaferri, M. Tingoli & M. Tiecco, *J. Org. Chem.*, 1979, **44**, 2642 (*288*) |
| 79JO2758 | M. F. Ruasse, A. Argile, E. Bienvenue-Gotz & J. E. Dubois, *J. Org. Chem.*, 1979, **44**, 2758 (*197*) |

| 79JO3344 | M. R. MacLaury & A. Saracino, *J. Org. Chem.*, 1979, **44**, 3344 (*181*) |
|---|---|
| 79JO4766 | J. Bromilow, R. T. C. Brownlee, V. O. Lopez & R. W. Taft, *J. Org. Chem.*, 1979, **44**, 4766 (*54*) |
| 79JO4770 | M. J. Tremelling. S. P. Hopper & P. C. Mendelowitz, *J. Org. Chem.*, 1979, **44**, 4770 (*188*) |
| 80Acc76 | J. H. Bowie, *Acc. Chem. Res.*, 1980, **13**, 76 (*99*, *164*) |
| 80APO1 | M. J. Perkins, *Adv. Phys. Org. Chem.*, 1980, **17**, 1 (*9*) |
| 80CC926 | P. Helsby & J. H. Ridd, *Chem. Commun.*, 1980, 926 (*229*) |
| 80J(P2)250 | Y. Karton & A. Pross, *J. Chem. Soc. Perkin 2*, 1980, 250 (*158*) |
| 80J(P2)260 | A. C. Scott, J. M. Tedder, J. C. Walton & S. Mhatre, *J. Chem. Soc. Perkin 2*, 1980, 260 (*337*) |
| 80J(P2)985 | W. F. Reynolds, *J. Chem. Soc. Perkin 2*, 1980, 985 (*50*) |
| 80J(P2)1244 | T. W. Bentley, C. T. Bowen, W. Parker & C. I. F. Watt, *J. Chem. Soc. Perkin 2*, 1980, 1244 (*152*) |
| 80J(P2)1350 | U. Berg, R. Gallo, G. Klatte & J. Metzger, *J. Chem. Soc. Perkin 2*, 1980, 1350 (*60*) |
| 80JA1763 | A. Komornicki, J. D. Goddard & H. F. Schaefer, *J. Amer. Chem. Soc.*, 1980, **102**, 1763 (*351*) |
| 80JA6463 | V. P. Vitullo, J. Grabowski & S. Sridharan, *J. Amer. Chem. Soc.*, 1980, **102**, 6463 (*155*) |
| 80JA6466 | J. L. Palmer & W. P. Jencks, *J. Amer. Chem. Soc.*, 1980, **102**, 6466 (*259*) |
| 80JA6472 | J. L. Palmer & W. P. Jencks, *J. Amer. Chem. Soc.*, 1980, **102**, 6472 (*254*) |
| 80JA6867 | M. Saunders & M. R. Kates, *J. Amer. Chem. Soc.*, 1980, **102**, 6867 (*322*) |
| 80JA7988 | D. F. DeTar, *J. Amer. Chem. Soc.*, 1980, **102**, 7988 (*57*) |
| 80JO659 | M. Yasuda, K. Harano & K. Kanematsu, *J. Org. Chem.*, 1980, **45**, 659 (*373*) |
| 80JO818 | A. Pross, L. Radom & R. W. Taft, *J. Org. Chem.*, 1980, **45**, 818 (*58*) |
| 80JO1056 | G. Kemister, A. Pross, L. Radom & R. W. Taft, *J. Org. Chem.*, 1980, **45**, 1056 (*58*) |
| 80JO1401 | K. Yates & H. W. Leung, *J. Org. Chem.*, 1980, **45**, 1401 (*202*) |
| 80JO3539 | A. R. Stein, M. Tencer, E. A. Moffatt, R. Dawe & J. Sweet, *J. Org. Chem.*, 1980, **45**, 3539 (*157*) |
| 80JO5166 | D. F. DeTar, *J. Org. Chem.*, 1980, **45**, 5166 (*60*) |
| 80JO5306 | L. C. Vishwakarma & A. Fry, *J. Org. Chem.*, 1980, **45**, 5306 (*210*) |
| 80MI | R. P. Bell, *The Tunnel Effect in Chemistry*, Chapman & Hall, London, 1980 (*33*) |
| 80T701 | J. M. Tedder, *Tetrahedron*, 1980, **36**, 701 (*206*) |
| 80T1173 | J. I. Seeman, E. B. Sanders & W. A. Farone, *Tetrahedron*, 1980, **36**, 1173 (*21*) |
| 80T1901 | R. M. Magid, *Tetrahedron*, 1980, **36**, 1901 (*167*) |
| 80T1993 | A. Maquestiau, R. Flammang, M. Flammang-Barbieux, H. Mispreuve, I. Howe & J. H. Beynon, *Tetrahedron*, 1980, **36**, 1993 (*320*) |
| 81Acc306 | B. Capon, A. K. Ghosh & D. M. A. Grieve, *Acc. Chem. Res.*, 1981, **14**, 306 (*269*) |
| 81CC737 | V. P. Vitullo, J. Grabowski & S. Sridharan, *Chem. Commun.*, 1981, 737 (*157*) |
| 81CJ2116 | R. A. Cox & K. Yates, *Canad. J. Chem.*, 1981, **59**, 2116 (*89*) |
| 81J(P2)94 | M. R. Draper & J. H. Ridd, *J. Chem. Soc. Perkin 2*, 1981, 94 (*224*) |
| 81J(P2)409 | C. D. Johnson, I. Roberts & P. G. Taylor, *J. Chem. Soc. Perkin 2*, 1981, 409 (*56*) |

81J(P2)518    F. Al-Omran, K. Fujiwara, J. C. Giffney & J. H. Ridd, *J. Chem. Soc. Perkin 2*, 1981, 518 (*226, 229*)
81J(P2)848    H. W. Gibbs, L. Main, R. B. Moodie & K. Schofield, *J. Chem. Soc. Perkin 2*, 1981, 848 (*224*)
81J(P2)1070   N. C. Marziano, A. Tomasin & P. G. Traverso, *J. Chem. Soc. Perkin 2*, 1981, 1070 (*89*)
81J(P2)1084   E. S. Lewis, C. C. Shen & R. A. More O'Ferrall, *J. Chem. Soc. Perkin 2*, 1981, 1084 (*63*)
81J(P2)1127   S. Dintürk & R. A. Jackson, *J. Chem. Soc. Perkin 2*, 1981, 1127 (*345*)
81J(P2)1201   T. A. Bamkole, J. Hirst & I. Onyido, *J. Chem. Soc. Perkin 2*, 1981, 1201 (*295*)
81J(P2)1316   D. J. McLennan, *J. Chem. Soc. Perkin 2*, 1981, 1316 (*152*)
81J(P2)1358   A. K. Manglik, R. B. Moodie, K. Schofield, E. Dedeoglu, A. Dutly & P. Rys, *J. Chem. Soc. Perkin 2*, 1981, 1358 (*228*)
81J(P2)1459   C. Galli, *J. Chem. Soc. Perkin 2*, 1981, 1459 (*304*)
81JA39        R. D. Topsom, *J. Amer. Chem. Soc.*, 1981, **103**, 39 (*50*)
81JA946       C. Paradisi & J. F. Bunnett, *J. Amer. Chem. Soc.*, 1981, **103**, 946 (*155*)
81JA978       D. K. Bohme & G. I. Mackay, *J. Amer. Chem. Soc.*, 1981, **103**, 978 (*100*)
81JA2457      J. R. Keefe & W. P. Jencks, *J. Amer. Chem. Soc.*, 1981, **103**, 2457 (*177*)
81JA2783      S. Fukuzumi & J. K. Kochi, *J. Amer. Chem. Soc.*, 1981, **103**, 2783 (*202*)
81JA5466      T. W. Bentley, C. T. Bowen, D. H. Morten & P. v. R. Schleyer, *J. Amer. Chem. Soc.*, 1981, **103**, 5466 (*152–4*)
81JA6912      R. A. McClelland & G. Patel, *J. Amer. Chem. Soc.*, 1981, **103**, 6912 (*269*)
81JA6924      B. Chawla, S. K. Pollack, C. B. Lebrilla, M. J. Kamlet & R. W. Taft, *J. Amer. Chem. Soc.*, 1981, **103**, 6924 (*107*)
81JA7140      C. Galli & J. F. Bunnett, *J. Amer. Chem. Soc.*, 1981, **103**, 7140 (*306*)
81JA7240      S. Fukuzumi & J. K. Kochi, *J. Amer. Chem. Soc.*, 1981, **103**, 7240 (*232*)
81JO635       S. P. McManus, *J. Org. Chem.*, 1981, **46**, 635 (*163*)
81JO661       R. W. Taft, N. J. Pienta, M. J. Kamlet & E. M. Arnett, *J. Org. Chem.*, 1981, **46**, 661 (*107*)
81JO3053      M. H. Abraham, R. W. Taft & M. J. Kamlet, *J. Org. Chem.*, 1981, **46**, 3053 (*108*)
81JO3533      G. A. Olah, A. P. Fung, S. C. Narong & J. A. Olah, *J. Org. Chem.*, 1981, **46**, 3533 (*224*)
81JO4242      K. C. Brown, F. J. Romano & W. H. Saunders, *J. Org. Chem.*, 1981, **46**, 4242 (*176*)
81JO4247      D. J. Miller & W. H. Saunders, *J. Org. Chem.*, 1981, **46**, 4247 (*34, 176*)
81PAC171      R. Huisgen, *Pure Appl. Chem.*, 1981, **53**, 171 (*351*)
81PAC189      A. J. Kresge, *Pure Appl. Chem.*, 1981, **53**, 189 (*257*)
81PPO1        A. Pross & L. Radom, *Progr. Phys. Org. Chem.*, 1981, **13**, 1 (*45, 58*)
81PPO119      M. Charton, *Progr. Phys. Org. Chem.*, 1981, **13**, 119 (*52, 66*)
81PPO485      M. J. Kamlet, J. L. M. Abboud & R. W. Taft, *Progr. Phys. Org. Chem.*, 1981, **13**, 485 (*106–7*)
82AG401       J. M. Tedder, *Angew. Chem. Internat. Ed.*, 1982, **21**, 401 (*206*)
82CC273       K. Bowden & M. Hojatti, *Chem. Commun.*, 1982, 273 (*50*)
82J(P2)923    M. H. Abraham, M. J. Kamlet & R. W. Taft, *J. Chem. Soc. Perkin 2*, 1982, 923 (*107*)

82J(P2)1025 G. W. L. Ellis & C. D. Johnson, *J. Chem. Soc. Perkin 2*, 1982, 1025 (*91*)

82J(P2)1139 C. Galli, *J. Chem. Soc. Perkin 2*, 1982, 1139 (*304*)

82JA187 A. Pross & S. S. Shaik, *J. Amer. Chem. Soc.*, 1982, **104**, 187 (*192*)

82JA907 C. S. Yannoni, V. Macho & P. C. Myhre, *J. Amer. Chem. Soc.*, 1982, **104**, 907 (*332*)

82JA1958 V. Lucchini, G. Modena, G. Scorrano, R. A. Cox & K. Yates, *J. Amer. Chem. Soc.*, 1982, **104**, 1958 (*89*)

82JA2501 H. J. Shine, H. Zmuda, K. H. Park, H. Kwart, A. G. Horgan & M. Brechbiel, *J. Amer. Chem. Soc.*, 1982, **104**, 2501 (*313*)

82JA3676 D. J. Pasto, P. F. Heid & S. E. Warren, *J. Amer. Chem. Soc.*, 1982, **104**, 3676 (*382*)

82JA4793 S. Ehrenson, *J. Amer. Chem. Soc.*, 1982, **104**, 4793 (*51*)

82JA4907 D. J. Nelson & H. C. Brown, *J. Amer. Chem. Soc.*, 1982, **104**, 4907 (*210*)

82JA5181 H. J. Shine, H. Zmuda, H. Kwart, A. G. Horgan & M. Brechbiel, *J. Amer. Chem. Soc.*, 1982, **104**, 5181 (*314–5*)

82JA6836 H. Olsen, *J. Amer. Chem. Soc.*, 1982, **104**, 6836 (*350*)

82JA7105 G. A. Olah, G. K. S. Prakash, M. Arvanaghi & F. A. L. Anet, *J. Amer. Chem. Soc.*, 1982, **104**, 7105 (*332*)

82JA7599 S. Fukuzumi & J. K. Kochi, *J. Amer. Chem. Soc.*, 1982, **104**, 7599 (*202, 232*)

82JO736 M. H. Whangbo & K. R. Stewart, *J. Org. Chem.*, 1982, **47**, 736 (*186*)

82JO1734 M. J. Kamlet & R. W. Taft, *J. Org. Chem.*, 1982, **47**, 1734 (*108*)

82JO2204 D. J. Pasto & P. F. Heid, *J. Org. Chem.*, 1982, **47**, 2204 (*382*)

82JO2318 A. J. Hoefnagel & B. M. Wepster, *J. Org. Chem.*, 1982, **47**, 2318 (*51*)

82JO2517 H. Meislich & S. J. Jasne, *J. Org. Chem.*, 1982, **47**, 2517 (*167*)

82JO3224 F. G. Bordwell & D. L. Hughes, *J. Org. Chem.*, 1982, **47**, 3224 (*81*)

82JO3237 E. Baciocchi, R. Ruzziconi & G. V. Sebastiani, *J. Org. Chem.*, 1982, **47**, 3237 (*178, 185*)

82T313 J. M. Tedder, *Tetrahedron*, 1982, **38**, 313 (*341, 344*)

82T1615 M. Chastrette & J. Caretto, *Tetrahedron*, 1982, **38**, 1615 (*105*)

82T1847 J. M. Lluch & J. Bertran, *Tetrahedron*, 1982, **38**, 1847 (*351*)

83Acc207 R. M. Pitzer, *Acc. Chem. Res.*, 1983, **16**, 207 (*10*)

83Acc394 R. A. McClelland & L. J. Santry, *Acc. Chem. Res.*, 1983, **16**, 394 (*268*)

83CR83 J. I. Seeman, *Chem. Rev.*, 1983, **83**, 83 (*21*)

83J(P2)75 C. Bloomfield, A. K. Manglik, R. B. Moodie, K. Schofield & G. D. Tobin, *J. Chem. Soc. Perkin 2*, 1983, 75 (*230*)

83JA265 J. R. Keefe & W. P. Jencks, *J. Amer. Chem. Soc.*, 1983, **105**, 265 (see also correction on p. 6195) (*177*)

83JA719 P. C. Hibberty, G. Ohanessian & H. B. Schlegel, *J. Amer. Chem. Soc.*, 1983, **105**, 719 (*351*)

83JA2481 C. H. DePuy, E. W. Della, J. Filley, J. J. Grabowski & V. M. Bierbaum, *J. Amer. Chem. Soc.*, 1983, **105**, 2481 (*165*)

83JA4767 K. A. Engdahl, H. Bivehed, P. Ahlberg & W. H. Saunders, *J. Amer. Chem. Soc.*, 1983, **105**, 4767 (*247*)

83JA5509 M. Henchman, J. F. Paulson & P. M. Hierl, *J. Amer. Chem. Soc.*, 1983, **105**, 5509 (*100*)

83JA5915 K. Raghavachari, R. C. Haddon, P. v. R. Schleyer & H. F. Schaefer, *J. Amer. Chem. Soc.*, 1983, **105**, 5915 (*332*)

83JA6185 M. Yoshimine, A. D. McLean, B. Liu, D. J. DeFrees & J. S. Binkley, *J. Amer. Chem. Soc.*, 1983, **105**, 6185 (*332*)

83JO191   T. J. Broxton & M. J. McLeish, *J. Org. Chem.*, 1983, **48**, 191 (*298*)

83JO302   L. K. Tan & S. Brownstein, *J. Org. Chem.*, 1983, **48**, 302 (*236*)

83JO2877   M. J. Kamlet, J. L. M. Abboud, M. H. Abraham & R. W. Taft, *J. Org. Chem.* 1983, **48**, 2877 (*107*)

83PAC1281  V. Gold, *Pure Appl. Chem.*, 1983, **55**, 1281 (*207, 343*)

83PPO205  D. E. Sunkyo & W. J. Hehre, *Progr. Phys. Org. Chem.*, 1983, **14**, 205 (*152*)

83T193   L. Testaferri, M. Tiecco, M. Tingoli, D. Chianelli & M. Montanucci, *Tetrahedron*, 1983, **39**, 193 (*288*)

83TL2851  J. P. Bradley, T. C. Jarvis, C. D. Johnson, P. D. McDonnell & T. A. P. Weatherstone, *Tetrahedron Lett.*, 1983, 2851 (*138*)

84JA203   M. J. S. Dewar & A. B. Pierini, *J. Amer. Chem. Soc.*, 1984, **106**, 203 (*351*)

84JA209   M. J. S. Dewar, *J. Amer. Chem. Soc.*, 1984, **106**, 209 (*351*)

# Subject index